NEW TREATMENT STRATEGIES FOR DENGUE AND OTHER FLAVIVIRAL DISEASES

The Novartis Foundation is an international scientific and educational charity (UK Registered Charity No. 313574). Known until September 1997 as the Ciba Foundation, it was established in 1947 by the CIBA company of Basle, which merged with Sandoz in 1996, to form Novartis. The Foundation operates independently in London under English trust law. It was formally opened on 22 June 1949.

The Foundation promotes the study and general knowledge of science and in particular encourages international co-operation in scientific research. To this end, it organizes internationally acclaimed meetings (typically eight symposia and allied open meetings and 15–20 discussion meetings each year) and publishes eight books per year featuring the presented papers and discussions from the symposia. Although primarily an operational rather than a grant-making foundation, it awards bursaries to young scientists to attend the symposia and afterwards work with one of the other participants.

The Foundation's headquarters at 41 Portland Place, London W1B 1BN, provide library facilities, open to graduates in science and allied disciplines. Media relations are fostered by regular press conferences and by articles prepared by the Foundation's Science Writer in Residence. The Foundation offers accommodation and meeting facilities to visiting scientists and their societies. Information on all Foundation activities can be found at http://www.novartisfound.org.uk

The Novartis Institute for Tropical Diseases (NITD) is a small-molecule drug discovery research institute dedicated to finding new drugs for the treatment of tropical diseases. The NITD is designed as part of Novartis' Corporate Citizenship efforts to improve access of medicines to the developing world. Novartis believes in the importance of a long-term commitment to helping reduce this global disease burden and thereby improving the prosperity and health of populations of developing nations. NITD was set up as a Public-Private Partnership between Novartis and the Singapore Economic Development Board (EDB). Dengue fever and tuberculosis were collectively selected as the diseases on which to focus, with the possibility to expand to other disease areas in later years.

For more information, please visit www.nitd.novartis.com

Novartis Foundation Symposium 277

NEW TREATMENT STRATEGIES FOR DENGUE AND OTHER FLAVIVIRAL DISEASES

2006

John Wiley & Sons, Ltd

Published in 2006 by John Wiley & Sons Ltd,
The Atrium, Southern Gate, Chichester PO19 8SQ, UK

National 01243 779777
International (+44) 1243 779777
e-mail (for orders and customer service enquiries): cs-books@wiley.co.uk
Visit our Home Page on http://eu.wiley.com

Other Wiley Editorial Offices

John Wiley & Sons Inc., 111 River Street, Hoboken, NJ 07030, USA

Jossey-Bass, 989 Market Street, San Francisco, CA 94103-1741, USA

Wiley-VCH Verlag GmbH, Boschstr. 12, D-69469 Weinheim, Germany

John Wiley & Sons Australia Ltd, 33 Park Road, Milton, Queensland 4064, Australia

John Wiley & Sons (Asia) Pte Ltd, 2 Clementi Loop #02-01, Jin Xing Distripark, Singapore 129809

John Wiley & Sons Canada Ltd, 6045 Freemont Blvd, Mississauga, Ontario, Canada L5R 4J3

Wiley also publishes its books in a variety of electronic formats. Some content that appears in print may not be available in electronic books.

Novartis Foundation Symposium 277
x +266 pages, 57 figures, 19 tables

British Library Cataloguing in Publication Data

A catalogue record for this book is available from the British Library

ISBN-13 978-0-470-01643-5
ISBN-10 0-470-01643-4

Typeset in 10½ on 12½ pt Garamond by SNP Best-set Typesetter Ltd., Hong Kong.
Printed and bound in Great Britain by T. J. International Ltd, Padstow, Cornwall.
This book is printed on acid-free paper responsibly manufactured from sustainable forestry, in which at least two trees are planted for each one used for paper production.

Contents

Novartis Foundation Symposium on New treatment strategies for dengue and other flaviviral diseases, held at the Novartis Institute for Tropical Diseases in Singapore, 26–27 September 2005

Editors: Gregory Bock (Organizer) and Jamie Goode

This symposium is based on a proposal made by Shubash Vasudevan, Thomas Keller, David Jans and Pad Padmanabhan

Participants

Bruno Canard AFMB UMR 6098 CNRS, 31 Chemin Joseph Aiguier, F-13402 Marseille cedex 20, France

Nathalie Charlier University of Leuven, Rega Institute of Medical Research, Laboratory of Virology, Minderbroedersstraat 10, B-3000 Leuven, Belgium

Thomas G. Evans Infectious Diseases Translational Medicine, Novartis Institute for Biomedical Research, 100 Technology Square 4153, Cambridge, MA 02139, USA

David Fairlie Centre for Drug Design and Development, Institute for Molecular Bioscience, The University of Queensland, Brisbane, Queensland 4072, Australia

Jeremy Farrar Oxford University Clinical Research Unit, The Hospital for Tropical Diseases, 190 Ben Ham Tu, Qan 5, Ho Chi Minh City, Vietnam

Marie Flamand Laboratory of Virology & Infectious Disease, The Rockefeller University, 1230 York Avenue Box 64, New York, NY 10021, USA

Andrea Gamarnik Fundacion Instituto Le Loir, Buenos Aires 1405, Argentina

Feng Gu Novartis Institute for Tropical Diseases, 10 Biopolis Road, Chromos #05-01, Singapore 138670

Duane J. Gubler Asia-Pacific Institute for Tropical Medicine and Infectious Diseases, Department of Tropical Medicine and Medical Microbiology, John A. Burns School of Medicine, 651 Ilalo Street, BSB 3rd Floor, Honolulu, HI 96813, USA

Scott Halstead Pediatric Dengue Vaccine Initiative, 5824 Edson Lane, N. Bethesda, MD 20852, USA

Eva Harris Division of Infectious Diseases, School of Public Health, University of California, Berkeley, 140 Warren Hall, Berkeley, California 94720-7360, USA

Franz X. Heinz Institute of Virology, Medical University of Vienna, A1095 Vienna, Austria

Paul Herrling Novartis International AG, WSJ-200.204, CH-4002 Basel, Switzerland

Martin Hibberd Population Genetics (Infectious Disease), Genome Institute of Singapore, Genome #02-01, 60 Biopolis Street, Singapore 138672

Edward Holmes Department of Biology, The Pennsylvania State University, Mueller Laboratory, University Park, PA 16802, USA

Joachim Hombach Initiative for Vaccine Research IVR, World Health Organization, 20 Avenue Appia, CH-1211 Geneva 27, Switzerland

David A. Jans Nuclear Signalling Laboratory, Department of Biochemistry. & Molecular Biology, PO Box 13D, Monash University, Victoria 3800, Australia

Thomas Keller Novartis Institute for Tropical Diseases, 10 Biopolis Road, Chromos #05-01, Singapore 138670

Richard Kuhn Department of Biological Sciences, Purdue University, 915 West State St, West Lafayette, IN 47907-2054, USA

Myoung-Ok Kwon Novartis International AG, WSJ-200.296, CH-4002 Basel, Switzerland

Julien Lescar School of Biological Sciences, Nanyang Technological University, 60 Nanyang Drive, Singapore 637551

Siew Pheng Lim Novartis Institute for Tropical Diseases, 10 Biopolis Road, Chromos #05-01, Singapore 138670

Prida Malasit Medical Molecular Biology Unit, Office of Research and Development, Faculty of Medicine, Siriraj Hospital, Mahidol University, 2 Prannok Road, Bangkoknoi, Bangkok 10700, Thailand

Johan Neyts Rega Institute for Medical Research, Department of Chemotherapy, K. U. Leuven, Minderbroedersstraat 10, B-3000 Leuven, Belgium

Mary Ng Mah Lee Department of Microbiology, National University of Singapore, 5 Science Drive 2, Singapore 117597

R. Pad Padmanabhan Department of Microbiology and Immunology, Georgetown University Medical Centre, SW309 Medical-Dental Building, 3900 Reservoir Rd, Washington DC 20057, USA

Charles Rice *(Chair)* Laboratory of Virology and Infectious Disease, The Rockefeller University, 1230 York Avenue, Box 64, New York, NY 10021, USA

Vijaya Satchidanandam Department of Microbiology and Cell Biology, Room 254A, Sir C.V. Raman Avenue, Indian Institute of Science, Bangalore 560012, India

Wouter Schul Novartis Institute for Tropical Diseases, 10 Biopolis Road, Chromos #05-01, Singapore 138670

Gavin Screaton Dean's Office, 2nd Floor CWB, Imperial College Hammersmith Campus, Du Cane Road, London W12 0NN, UK

John Stephenson Department for Infectious and Tropical Diseases, London School of Hygiene and Tropical Medicine, Keppel St, London WC1E 7HT, UK

Subhash Vasudevan Novartis Institute for Tropical Diseases, 10 Biopolis Road, #05-01 Chromos, Singapore 138670

Paul Young Department of Microbiology and Parasitology, School of Molecular and Microbial Sciences, University of Queensland, St. Lucia, QLD 4072, Australia

Rolf Zinkernagel University of Zurich, University Hospital, Department of Pathology, Institute of Experimental Immunology, Schmelzbergstrasse 12, CH-8091 Zürich, Switzerland

Chair's introduction

Charles Rice

Laboratory of Virology and Infectious Disease, The Rockefeller University, 1230 York Avenue, Box 64, New York, NY 10021, USA

There are three strategies that we can consider if we are thinking about controlling flavivirus infections. The one that has worked well in the past has been mosquito control. Then vaccination was used, with vaccines such as those against the yellow fever 17D strain. This is still an effective means of prevention when it is actually implemented. The third strategy, which is the main focus of this meeting, is therapy, although we shouldn't neglect these other aspects of flavivirus control.

The following lists some of the issues that I think we should consider during our discussions. What is the scientific feasibility of a particular kind of approach for dealing with dengue fever, and is it likely to be effective? How much is it going to cost? Ideally, we don't want therapies to be so expensive that they will be restricted to developed countries. Related to this, what is the target population going to be for a particular kind of therapy or prevention?

With respect to therapies, what about antibodies? We are not going to be talking too much about these, but they have been proven to work. What about small molecules? There are state-of-the-art technologies that can be employed to elucidate target structures, develop biochemical assays, screen for small molecules and find those with good pharmacological properties. If we can develop these kinds of therapies, how are they going to be used? Are they more applicable in a prophylactic setting, or in a post-exposure context? One of the key issues for therapeutics is the diagnosis of flavivirus infections. In order to implement post-exposure therapeutics, we will need rapid diagnosis. Diagnostic methods will probably have to be low-tech and low cost.

Another important issue in therapeutics is the nature of the best small molecule targets: we will be touching on a number of the flavivirus proteins involved in host cell entry. Are viral targets best, or should we also go after host targets? The latter might be better in providing a broad protection against different flaviviruses. Should we be considering targets that are highly specific at all, where resistance might be an issue?

In terms of vaccines, what are the various strategies that can be considered? In the case of dengue, what is the feasibility of the 'holy grail'—a tetravalent vaccine that provides long-lasting protective immunity? We would like to highlight some of the vaccine progress on 'simpler' flavivirus diseases such as Japanese encephalitis. Again, what is the cost of vaccination? This will be important for implementation in developing countries. One of the things we run into in the USA and other countries is the complexity of liability issues associated with vaccines. We are now also seeing this also in small molecule therapeutics.

Another area we will discuss is that of hi-tech approaches such as microarrays and proteomics. What can these kinds of techniques tell us about flavivirus biology? They can be applied to state-of-the-art epidemiology and evolution studies, and they can be used to study the effects of virus infection at the organismal and cellular levels. One thing that we would like to see come from such studies is a better understanding of the pathogenic mechanisms of the more severe sequelae of dengue infection. Might this provide us with new therapeutic insights in a post-exposure setting? Much of the pathology associated with these diseases is after the peak of viraemia.

I have been working on hepatitis C in recent years, and in terms of preclinical models it is clear that the flavivirus field has a tremendous advantage. We have biochemical assays for a number of the virus-specific enzymes, and viruses have never been difficult to coax to replicate in cell culture, so you can actually do cell-based evaluation of potential antivirals.

Finally, another issue which I would like to invoke some discussion on is where we stand in terms of the animal models. We have models that can be used to support flavivirus infection, but these don't necessarily recapitulate the pathology in human disease. How important is it to try to develop models that do that? These are some of the issues that I hope we can address in our time together here. Let's begin!

Dengue/dengue haemorrhagic fever: history and current status

Duane J. Gubler

Director, Asia-Pacific Institute of Tropical Medicine and Infectious Diseases, John A. Burns School of Medicine, University of Hawaii, USA

Abstract. Dengue fever (DF) is an old disease; the first record of a clinically compatible disease being recorded in a Chinese medical encyclopaedia in 992. As the global shipping industry expanded in the 18th and 19th centuries, port cities grew and became more urbanized, creating ideal conditions for the principal mosquito vector, *Aedes aegypti.* Both the mosquitoes and the viruses were thus spread to new geographic areas causing major epidemics. Because dispersal was by sailing ship, however, there were long intervals (10–40 years) between epidemics. In the aftermath of World War II, rapid urbanization in Southeast Asia led to increased transmission and hyperendemicity. The first major epidemics of the severe and fatal form of disease, dengue haemorrhagic fever (DHF), occurred in Southeast Asia as a direct result of this changing ecology. In the last 25 years of the 20th century, a dramatic global geographic expansion of epidemic DF/DHF occurred, facilitated by unplanned urbanization in tropical developing countries, modern transportation, lack of effective mosquito control and globalization. As we go into the 21st century, epidemic DF/DHF is one of the most important infectious diseases affecting tropical urban areas. Each year there are an estimated 50–100 million dengue infections, 500 000 cases of DHF that must be hospitalized and 20 000–25 000 deaths, mainly in children. Epidemic DF/DHF has an economic impact on the community of the same order of magnitude as malaria and other important infectious diseases. There are currently no vaccines nor antiviral drugs available for dengue viruses; the only effective way to prevent epidemic DF/DHF is to control the mosquito vector, *Aedes aegypti.*

2006 New treatment strategies for dengue and other flaviviral diseases. Wiley, Chichester (Novartis Foundation Symposium 277) p 3–22

History

The first reports of major epidemics of an illness thought to possibly be dengue occurred on three continents (Asia, Africa and North America) in 1779 and 1780 (Rush 1789, Hirsch 1883, Pepper 1941, Howe 1977). However, reports of illnesses compatible with dengue fever occurred even earlier. The earliest record found to date was in a Chinese ‘encyclopaedia of disease symptoms and remedies,’ first published during the Chin Dynasty (AD 265 to 420) and formally edited in AD 610

(Tang Dynasty) and again in 992 during the Northern Sung Dynasty (Nobuchi 1979). Outbreaks of illness in the West Indies in 1635 and in Panama in 1699 could also have been dengue (Howe 1977, McSherry 1982). Thus, a dengue-like illness had a wide geographic distribution before the 18th century, when major epidemics of dengue-like illness occurred widely. It is uncertain that the epidemics in Batavia (Jakarta), Indonesia, and Cairo, Egypt, in 1779 were actually dengue (Carey 1971).

At some point in the past, probably with the clearing of the forests and development of human settlements, dengue viruses moved out of the jungle and into a rural environment, where they were, and still are, transmitted to humans by peridomestic mosquitoes such as *Aedes albopictus*. Migration of people and commerce ultimately moved the viruses into the villages, towns, and cities of tropical Asia, where the viruses were most likely transmitted sporadically by *Aedes albopictus* and other closely related peridomestic *Stegomyia* mosquito species.

The slave trade between West Africa and the Americas, and the resulting commerce, were responsible for the introduction and the widespread geographic distribution of an African mosquito, *Aedes aegypti*, in the New World during the 17th, 18th and 19th centuries. This species became highly adapted to humans and urban environments and was spread throughout the tropics of the world by sailing ships. The species first infested port cities and then moved inland as urbanization expanded. Because *Ae. aegypti* had evolved to become intimately associated with humans, preferring to feed on them and to share their dwellings, this species became a very efficient epidemic vector of dengue and yellow fever viruses (Gubler 1997). Therefore, when these viruses were introduced into port cities infested with *Ae. aegypti*, epidemics occurred. It was in this setting that major epidemics of dengue fever occurred during the 18th, 19th and early 20th centuries, as the global shipping industry developed and port cities were urbanized in response to increased commerce and ocean traffic. The last major dengue pandemic began during World War II and continues through the present (Gubler 1997, Halstead 1992).

The earliest known use of the word *dengue* to describe an illness was in Spain in 1801 (Soler 1949). However, the most likely origin of the word is from Swahili (Christie et al 1872, Christie 1881). In both the 1823 and 1870 epidemics of dengue-like illness in Zanzibar and the East African coast, the disease was called *Ki-Dinga pepo*. From this came the name *dinga* or *denga*, which was used to describe the illness in both epidemics. Christie (Christie et al 1872, Christie 1881) speculates that the name denga was taken via the slave trade to the New World, where it was called 'Dandy fever' or 'The Dandy' in the St. Thomas epidemic of 1827. The illness was first called *dunga* in Cuba during the 1828 epidemic, but later changed to dengue, the name by which it has been known ever since (Munoz 1828). Most likely, the Spanish recognized the disease in Cuba as the same one that was called dengue in Spain in 1801. If the word *dengue* did originate in East Africa from *dinga* or *denga*, this suggests the disease was occurring before the 1823 epidemics described by Christie. This is not unlikely since epidemics were reported in Africa, the Middle East and Spain in the late 1700s.

With documentation that yellow fever was transmitted by mosquitoes, many early workers suspected that dengue fever was also mosquito-borne. In the previrology era, work was slow and relied on use of human volunteers. Work done by Graham (1903), Bancroft (1906) and Cleland et al (1918) documented dengue transmission by mosquitoes.

Although it had been shown that dengue fever was caused by a filterable agent, (Ashburn et al 2004, Siler et al 1926) the first dengue viruses were not isolated until the 1940s, during World War II (Kimura & Hotta 1944, Hotta 1952, Sabin & Schlesinger 1945, Sabin 1952). Dengue fever was a major cause of morbidity among Allied and Japanese soldiers in the Pacific and Asian theatres. Sabin and his group were able to show that some virus strains from three geographic locations (Hawaii, New Guinea and India) were antigenically similar (Sabin & Schlesinger, Sabin 1952). This virus was called dengue 1 (DENV-1), and the Hawaii virus was designated as the prototype strain (Haw-DENV-1). Another antigenically distinct virus strain isolated from New Guinea was called dengue 2 (DENV-2), and the New Guinea C strain (NGC-DENV-2) was designated the prototype. The Japanese virus isolated by Kimura and Hotta (Kimura & Hotta 1944, Hotta 1952) was subsequently shown to be DENV-1 as well. Two more serotypes, dengue 3 (DENV-3) and dengue 4 (DENV-4), were later isolated from patients with a haemorrhagic disease during an epidemic in Manila, in 1956 (Hammon et al 1960). Since these original isolates were made, thousands of dengue viruses have been isolated from all parts of the tropics; all have fit into the four-serotype classification.

The occurrence of severe and fatal haemorrhagic disease associated with dengue infections is not unique to the twentieth century. Patients with disease clinically compatible with dengue haemorrhagic fever (DHF) have been reported sporadically since 1780, when such cases were observed in the Philadelphia epidemic (Rush 1789). Significant numbers of cases of haemorrhagic disease were associated with several subsequent epidemics, including Charters Towers, Australia, in 1897, Beirut in 1910, Taiwan in 1916, Greece in 1928 and Taiwan in 1931 (Copanaris 1928, Akashi 1932, Halstead & Papaevangelou 1980, Rosen 1986, Hare 1898, Koizumi et al 1916). However, epidemic occurrences such as these were relatively rare, and the long intervals between them made each a unique event that was not considered important in terms of a long-term, continuous public health problem. Understanding the emergence of dengue and DHF as a global public health problem in the last half of the 20th century requires a review of the ecological and demographic changes that occurred in the Asian and American tropics during this period. The detailed history of dengue has been recently reviewed (Gubler 1997).

Natural history

There are four dengue virus serotypes: DENV-1, DENV-2, DENV-3 and DENV-4. They belong to the genus *Flavivirus*, family *Flaviviridae* (of which yellow fever is the type species), which contains approximately 56 viruses (ICTV 2005).

Humans are infected with dengue viruses by the bite of an infective *Ae. aegypti* mosquito (Gubler 1988). *Ae. aegypti* is a small, black-and-white, highly domesticated urban mosquito that prefers to lay its eggs in artificial containers commonly found in and around homes in the tropics, for example, flower vases, old automobile tires, buckets that collect rainwater, and trash in general. Containers used for water storage, especially 55-gallon drums and cement cisterns, are especially important in producing large numbers of adult mosquitoes in close proximity to dwellings where people live and work. The adult mosquitoes prefer to rest indoors, are unobtrusive, and prefer to feed on humans during daylight hours. The female mosquitoes are very nervous feeders, disrupting the feeding process at the slightest movement, only to return to the same or a different person to continue feeding moments later. Because of this behaviour, *Ae. aegypti* females will often feed on several persons during a single blood meal and, if infective, may transmit dengue virus to multiple persons in a short period of time even if they only probe without taking blood (Gubler & Rosen 1976). It is not uncommon to see several members of the same household become ill with dengue fever within a 24 to 36 h time frame, suggesting transmission by a single infective mosquito (D. J. Gubler, unpublished data). It is this behaviour that makes *Ae. aegypti* such an efficient epidemic vector. Inhabitants of dwellings in the tropics are rarely aware of the presence of this mosquito, making its control difficult.

After a person is bitten by an infective mosquito, the virus undergoes an incubation period of 3–14 days (average, 4–7 days), after which the person may experience acute onset of fever accompanied by a variety of non-specific signs and symptoms. During this acute febrile period, which may be as short as 2 days and as long as 10, dengue viruses may circulate in the peripheral blood. If other *Ae. aegypti* mosquitoes bite the ill person during this febrile viraemic stage, those mosquitoes may become infected and subsequently transmit the virus to other uninfected persons, after an extrinsic incubation period of 8 to 12 days (Gubler 1988).

Changing disease patterns

The disease pattern associated with dengue, which was characterized by relatively infrequent epidemics until the 1940s, changed with the ecological disruption in Southeast Asia during and after World War II. The economic development and urbanization in the post-war years created ideal conditions for increased transmission of mosquito-borne diseases, and in this setting a global pandemic of dengue began. With increased epidemic transmission and movement of people within and between countries, hyperendemicity (the cocirculation of multiple dengue virus serotypes) developed in Southeast Asian cities, and epidemic DHF, a newly described disease, emerged (World Health Organization 1997, Halstead 1980). The first known epidemic of DHF occurred in Manila in 1953–1954, but within 20 years the disease had spread throughout Southeast Asia. By the mid-1970s, DHF had become a leading cause of hospitalization and death among children in the region (World Health Organization 1997). In the 1980s and 1990s, dengue transmission in Asia further intensified; epidemic DHF increased in incidence and expanded geographically west into India, Pakistan, Sri Lanka, and the Maldives, and east into China (Gubler 1997, Halstead 1980, 1992, World Health Organization 1997). At the same time, the geographic distribution of epidemic DHF was expanding into new regions—the Pacific islands in the 1970s and 1980s and the American tropics in the 1980s and 1990s (Gubler 1993, 1997, Halstead 1992, 1980, Gubler & Trent 1994, Gubler & Clark 1995, Rosen 1982, Barnes & Rosen 1974, Guzman et al 1984, Pinheiro 1989).

Epidemiological changes in the Americas have been the most dramatic. In the 1960s and most of the 1970s, epidemic dengue was rare in the American region because the principal mosquito vector, *Ae. aegypti*, had been eradicated from most of Central and South America (Gubler 1987, 1989, 1997, 1993, Pinheiro 1989). The eradication program was discontinued in the early 1970s, and this species then began to reinvade those countries from which it had been eradicated. By the 1990s, *Ae. aegypti* had regained the geographic distribution it had before eradication was initiated (Fig. 1).

Epidemic dengue invariably followed after reinfestation of a country by *Ae. aegypti*. By the 1980s, the American region was experiencing major epidemics of dengue fever in countries that had been free of the disease for 35–130 years (Gubler & Trent 1994, Gubler 1993, Pinheiro 1989, Gubler 1987, 1989). With increased epidemic activity came the development of hyperendemicity in American countries and the emergence of epidemic DHF, much as had occurred in Southeast Asia 25 years earlier (Pinheiro 1989, Gubler 1987, 1989, 1993). From 1981 to 2005, 28 American countries reported laboratory-confirmed DHF (Gubler & Trent 1994, Gubler & Clark 1995, Rosen 1982, Barnes & Rosen 1974, Guzman et al 1984, Gubler 1989, 1993, 2002, Pinheiro 1989) (Fig. 2).

While Africa has not yet had a major epidemic of DHF, sporadic cases of severe disease have occurred as epidemic DF has increased markedly in the past 25 years. Before the 1980s, little was known of the distribution of dengue viruses in Africa. Since then, however, major epidemics caused by all four serotypes have occurred in both East and West Africa (Gubler 1997, 2002). In 2006, dengue viruses and

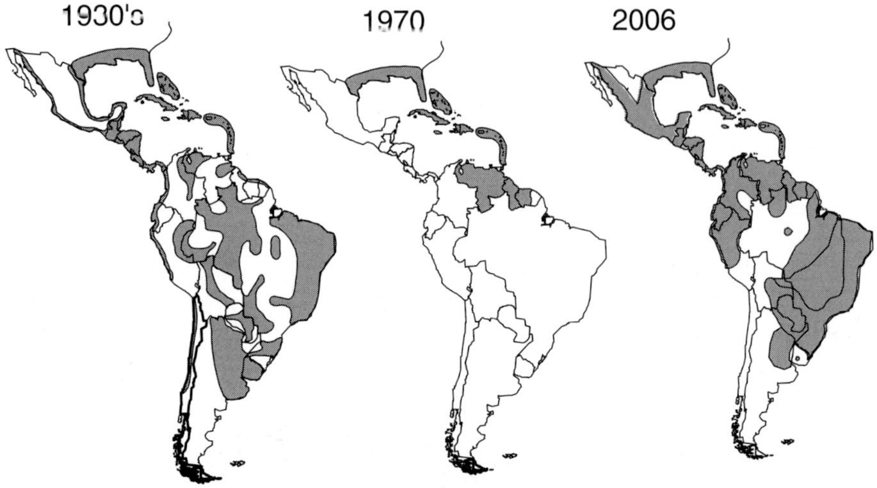

FIG. 1. Distribution of *Aedes aegypti* mosquitoes in the Americas in 1930, 1970 and 2006.

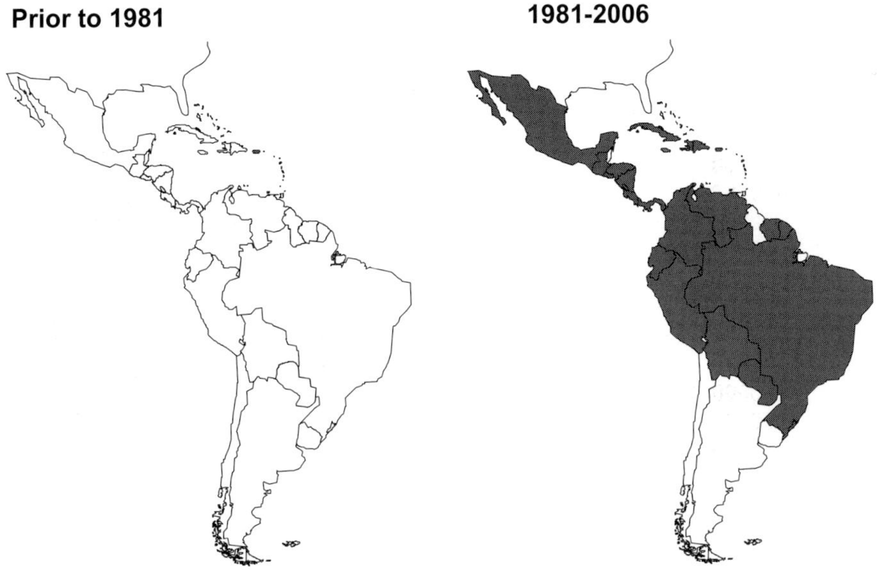

FIG. 2. Expanding geographic distribution in the Americas from 1981 to 2006.

Ae. aegypti mosquitoes have a worldwide distribution in the tropics with over 2.5 billion people living in dengue-endemic areas (Fig. 3) (Gubler 1997, 2002).

Currently, DF causes more illness and death than any other arboviral disease of humans. The number of cases of DEN/DHF reported to WHO has increased dramatically in the past two decades (Fig. 4).

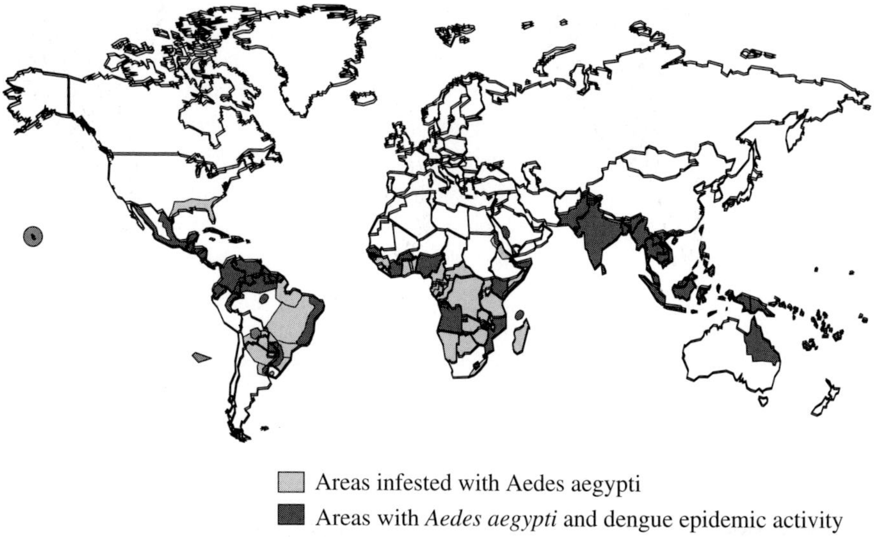

FIG. 3. Global distribution of *Aedes aegypti* mosquitoes and recent epidemic dengue.

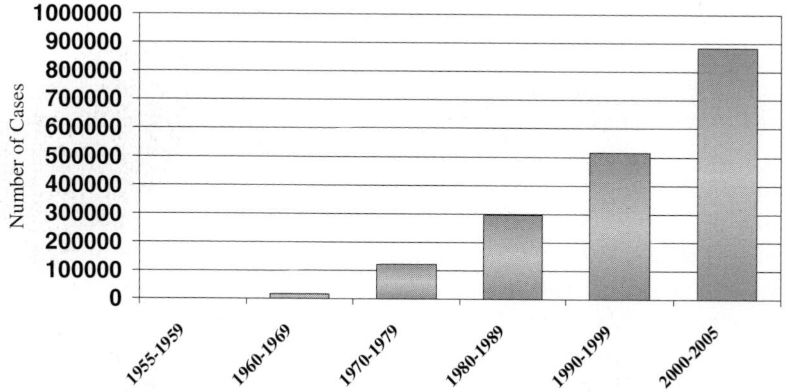

FIG. 4. Global average dengue fever/dengue haemorrhagic fever cases reported to WHO annually, by decade.

Each year, an estimated 50–100 million dengue infections and several hundred thousand cases of DHF occur, depending on epidemic activity (Gubler & Clark 1995, Gubler 2002, Monath 1994, World Health Organization 2000). DHF is a leading cause of hospitalization and death among children in many Southeast Asian countries (World Health Organization 1997).

Factors responsible for increased incidence

The emergence of epidemic dengue and DHF as a global public health problem in the past 25 years is closely associated with demographic and societal changes that have occurred over the past 50 years (Gubler & Trent 1994, Gubler & Clark 1995, Gubler 2002). A major factor has been the unprecedented population growth which has been the primary driving force for unplanned and uncontrolled urbanization, especially in tropical developing countries. The substandard housing and the deterioration in water, sewer and waste management systems associated with unplanned urbanization have created ideal conditions for increased transmission of mosquito-borne diseases in tropical urban centres.

A second major factor has been the lack of effective mosquito control in dengue-endemic areas (Gubler & Trent 1994, Gubler & Clark 1995, Gubler 1989, 2002). Emphasis during the past 25 years has been on space spraying with insecticides to kill adult mosquitoes; this has not been effective (Gubler 1989, Newton & Rieter 1992) and, in fact, has been detrimental to prevention and control efforts by giving citizens of the community and government officials a false sense of security (Gubler 1989). Additionally, the geographic distribution and population densities of *Ae. aegypti* have increased, especially in urban areas of the tropics, because of increased numbers of mosquito larval habitats in the domestic environment. The latter include non-biodegradable plastics and used automobile tires, both of which have increased dramatically during this same period of time.

Another major factor in the global emergence of dengue and DHF is globalization and increased movement of humans, animals and commodities via aeroplane, which provides the ideal mechanism for the transport of dengue and other urban pathogens between population centres of the world (Gubler & Trent 1994, Gubler & Clark 1995, Gubler 1989, 2002). For instance in 2004, an estimated 1 billion persons travelled somewhere via aeroplane. Many travellers become infected while visiting tropical dengue endemic areas, but become ill after returning home, resulting in a constant movement of dengue viruses in infected humans to all areas of the world, and ensuring repeated introductions of new dengue virus strains and serotypes into areas where the mosquito vectors occur. The result is increased epidemic activity, the development of hyperendemicity, and the emergence of epidemic DHF (Fig. 5).

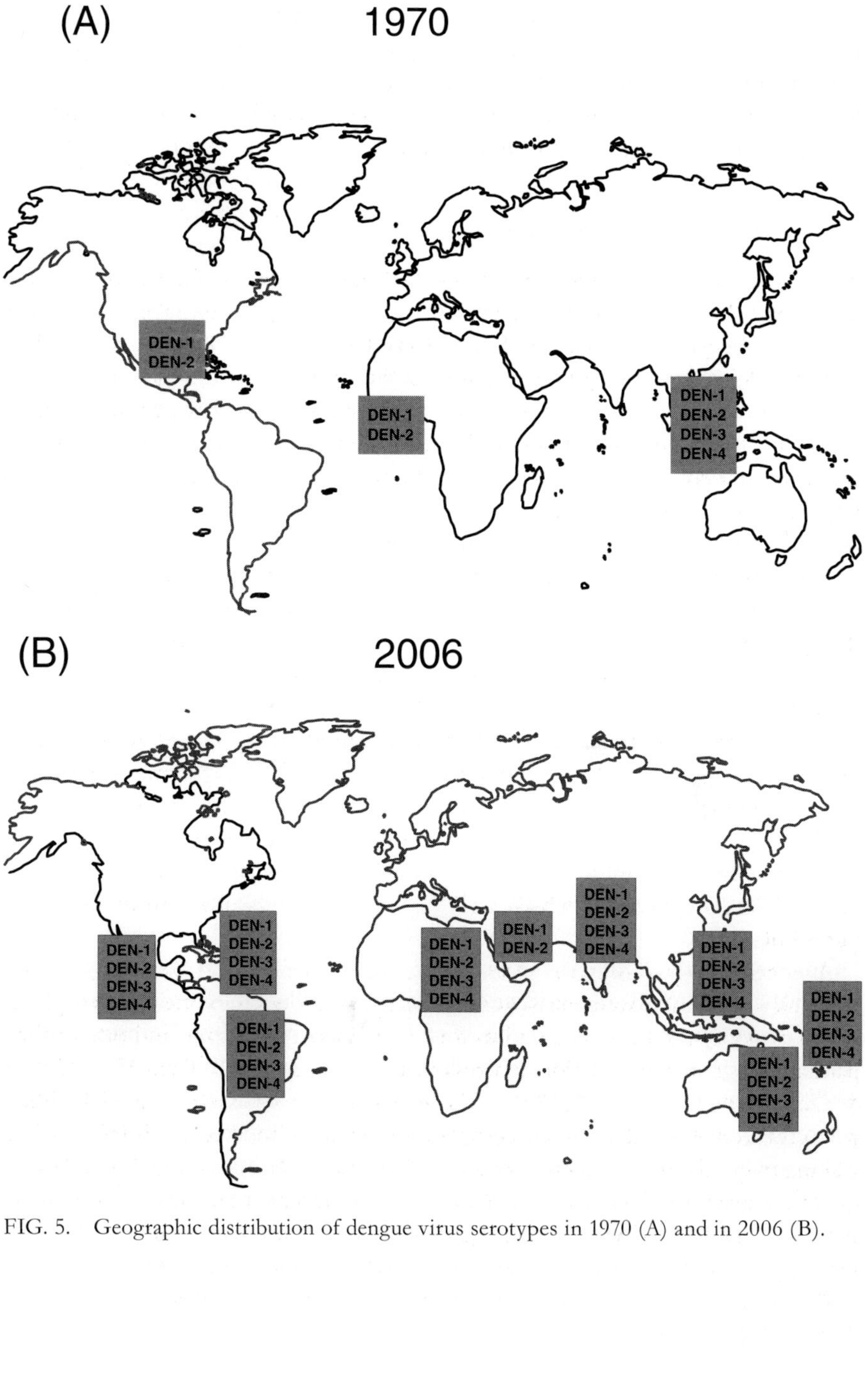

FIG. 5. Geographic distribution of dengue virus serotypes in 1970 (A) and in 2006 (B).

The USA and Europe are not immune to the introduction of dengue viruses. Each year for the past 25 years, imported dengue cases to the USA have been documented by the Centers for Disease Control and Prevention (CDC) (Gubler 1996, Rigau-Pérez et al 1994, CDC 2004, unpublished data). These cases represent introductions of all four virus serotypes from all tropical regions of the world. Most dengue introductions into the USA come from the American and Asian tropics, and reflect the increased number of Americans travelling to those areas. Overall, from 1976 to 2003, 3697 suspected cases of imported dengue were reported to the CDC (Gubler 1996, Rigau-Pérez et al 1994, CDC 2004, unpublished data). Although adequate blood samples were received from only a portion of these patients, 875 (24%) were confirmed as dengue. A similar increase in reported dengue has been seen in Europe in recent years (Wichmann et al 2003).

These cases represent only the tip of the iceberg because most physicians in the United States and Europe have a low index of suspicion for dengue, which is often not included in the differential diagnosis, even if the patient recently travelled to a tropical country (Gubler 1996, Wichmann et al 2003). As a result, many imported dengue cases are never reported. It is important to increase awareness of dengue and DHF among physicians in temperate areas, however, because the disease can be life-threatening. For example, two cases of the severe form of DHF, dengue shock syndrome, were described in Swedish tourists returning from holiday in Asia (Wittesjo et al 1993). In the USA, severe disease also occurs among imported cases of dengue (CDC 1995). It is important, therefore, that physicians in the USA and Europe consider dengue in the differential diagnosis of viral syndrome in all patients with a travel history to any tropical area.

The potential for epidemic dengue transmission in the USA and Europe still exists, since both have infestations of at least one of the principal mosquito vectors. On eight occasions in the past 25 years (in 1980 after an absence of 35 years, and in 1986, 1995, 1997, 1998, 1999, 2001 and 2005), autochthonous transmission occurred in the USA, secondary to importation of the virus in humans. Of interest was the 2001 Hawaii outbreak, which was the first dengue transmission in that state in 56 years (Effler et al 2005) caused by DEN-1 virus introduced from Tahiti where a major epidemic of DHF was occurring. Transmission in Hawaii was sporadic and illness mild; 122 cases were confirmed (Effler et al 2005). Although the outbreaks in the USA have been small, they underscore the potential for dengue transmission in areas where two competent mosquito vectors occur (Gubler & Trent 1994). *Ae. aegypti*, the most important and efficient epidemic vector of dengue viruses, has been in the USA for over 200 years and has been responsible for transmitting major epidemics in the past (Ehrankramz et al 1971). Currently, this species is found only in the Gulf Coast states from Texas to Florida, although small foci have recently been reported in Arizona. *Ae. albopictus*, another, but less efficient epidemic vector of dengue viruses, was introduced to the continental USA in the early 1980s and has since become widespread in the eastern half of the country. Although CDC has ceased surveillance, at last count it occurred in 1044 counties in 36 of the continental states (C. G. Moore, Colorado State University, 2004, personal communication); this species has also been found in Hawaii for over 90 years. *Ae. albopictus* has recently been introduced and has become established in several European countries. Both *Ae. aegypti* and *Ae. albopictus* can transmit dengue viruses to humans and their presence increases the risk of autochthonous dengue transmission, secondary to imported cases (Gubler 1988, 1996).

Prevention

Prevention and control of dengue fever/DHF currently depends on controlling the mosquito vector, *Ae. aegypti*, in and around the home where most transmission occurs. Space sprays with insecticides to kill adult mosquitoes are usually ineffective, unless they are sprayed indoors where the mosquitoes are resting. The most effective way to control the mosquitoes that transmit dengue is larval control, including eliminating, cleaning or chemically treating water-holding containers that serve as the larval habitats for *Ae. aegypti* in the domestic environment (Gubler 1989, World Health Organization 2000, Newton & Rieter 1992, Reiter & Gubler 1997). At present, there is no vaccine for dengue viruses, although several candidates are at various stages of development (Kinney & Huang 2001, Chang et al 2004). To be effective, a dengue vaccine must protect against all four virus serotypes, i.e. be a tetravalent formulation. For effective use in dengue endemic countries, a dengue vaccine should be safe for use in children 9–12 months of age, must be economical and should provide long-lasting protective immunity (ideally >10 years). Currently, there are at least six tetravalent candidate dengue vaccines that are in or near clinical trials in humans. The Pediatric dengue Vaccine Initiative funded by the Bill and Melinda Gates Foundation, was founded to facilitate bringing one or more of these promising candidate vaccines to fruition (Accelerating the Development and Introduction of a Dengue Vaccine for Poor Children. Hosted by: Children's Hospital No. 1 and Pasteur Institute of Ho Chi Minh City, December 5–8, 2001, Ho Chi Minh City, Vietnam).

There is no completely effective method of preventing dengue infection in travellers to tropical areas. The risk of infection can be significantly reduced, however, by understanding the basic behaviour and habits of the mosquito vectors and by taking a few simple precautions, such as using aerosol bomb insecticides to kill adult mosquitoes indoors, using a repellent containing diethylmetatoluamide (DEET) on exposed skin, and wearing protective clothing treated with a similar repellent. The risk of exposure may be lower in modern, air-conditioned hotels with well-kept grounds, and in rural areas.

References

Akashi K 1932 A dengue epidemic in the Tainan district of Taiwan in 1931 (in Japanese). Taiwan No Ikai, J Med Assoc Taiwan 31:767–777

Ashburn PM, Craig CF, US Army Board for the Study of Tropical Diseases 2004 Experimental investigations regarding the etiology of dengue fever. 1907. J Infect Dis 189:1747–1783

Bancroft TL 1906 On the aetiology of dengue fever. Aust Med Gazette 25:17–18

Barnes WJ, Rosen L 1974 Fatal hemorrhagic disease and shock associated with primary dengue infection on a Pacific island. Am J Trop Med Hyg 23:495–506

Carey DE 1971 Chikungunya and dengue: a case of mistaken identity? J Hist Med Allied Sci 26:243–262

Centers for Disease Control and Prevention (CDC) 1995 Imported dengue—United States, 1993–1994. Morb Mortal Wkly Rep 44:353–356

Chang GJ, Kuno G, Purdy ED, Davis BS 2004 Recent advancement in flavivirus vaccine development. Expert Rev Vaccines 3:199–220

Christie J 1872 Remarks on 'Kidinga Pepo'. A peculiar form of exanthematous disease. BMJ 577–579

Christie J 1881 On epidemics of dengue fever: Their diffusion and etiology. Glasgow Med J 16:161–176

Cleland JB, Bradley B, McDonald W 1918 Dengue fever in Australia. Its history and clinical course, its experimental transmission by *Stegomyia fasciata*, and the results of inoculation and other experiments. J Hyg 16:317–418

Copanaris P 1928 L'epidemie de dengue in Grèce au cours de l'été 1928. Office Int Hyg Public Bull 20:1590–1601

Effler P, Pang L, Kitsutani P et al 2005 Dengue fever, 2001–2002. Hawaii, Emerg Infect Dis 11:742–749

Ehrankramz NJ, Ventura AK, Cuadrado RR et al 1971 Pandemic dengue in Caribbean countries and the southern United States: past, present and potential problems. N Engl J Med 285:1460–1469

Graham H 1903 The dengue: a study of its pathology and mode of propagation. J Trop Med 6:209–214

Gubler DJ 1987 Dengue and dengue hemorrhagic fever in the Americas. PR Health Sci J 6:107–111

Gubler DJ 1988 Dengue. In Monath TP (ed): Epidemiology of arthropod-borne viral diseases. CRC Press, Boca Raton, Fla p 223–260

Gubler DJ 1989 *Aedes aegypti* and *Aedes aegypti*–borne disease control in the 1990s: top down or bottom up. Charles Franklin Craig Lecture. Am J Trop Med Hyg 40:571–578

Gubler DJ 1993 Dengue and dengue hemorrhagic fever in the Americas. P Thoncharoen (ed) In: Dengue Hemorrhagic Fever. World Health Organization Monograph, Regional Publication, SEARO, No. 22. New Delhi, India

Gubler DJ 1996 Arboviruses as imported disease agents: the need for increased awareness. Arch Virol 11:21–32

Gubler DJ 1997 Dengue and dengue haemorrhagic fever: Its history and resurgence as a global public health problem. In: Gubler DJ, Kuno G (eds) Dengue and dengue hemorrhagic fever. CAB International, London

Gubler DJ 2002 Epidemic dengue/dengue hemorrhagic fever as a public health, social and economic problem in the 21st century. Trends Micriobiol 10:100–103

Gubler DJ, Rosen L 1976 A simple technique for demonstrating transmission of dengue viruses by mosquitoes without the use of vertebrate hosts. Am J Trop Med Hyg 25:146–150

Gubler DJ, Trent DW 1994 Emergence of epidemic dengue/dengue hemorrhagic fever as a public health problem in the Americas. Infect Agents Dis 2:383–393

Gubler DJ, Clark GG 1995 Dengue/dengue hemorrhagic fever: the emergence of a global health problem. Emerg Infect Dis 1:55–57

Guzman MG, Kouri GP, Bravo J et al 1984 Dengue hemorrhagic fever in Cuba. II. Clinical investigations. Trans Soc Trop Med Hyg 78:239–241

Halstead SB 1980 Dengue hemorrhagic fever—public health problem and a field for research. Bull World Health Organ 58:1–21

Halstead SB 1992 The XXth century dengue pandemic: need for surveillance and research. Rapp Trimest Stat Sanit Mond 45:292–298

Halstead SB, Papaevangelou G 1980 Transmission of dengue 1 and 2 viruses in Greece in 1928. Am J Trop Med Hyg 29:635–637

Hammon W McD, Rudnick A, Sather G, Rogers KD, Morse LJ 1960 New hemorrhagic fevers of children in the Philippines and Thailand. Trans Assoc Am Physicians 73:140–155

Hare FE 1898 The 1897 epidemic of dengue in North Queensland. Aust Med Gazette 17:98–107

Hirsch A 1883 Dengue, a comparatively new disease: Its symptoms. In: Handbook of geographical and historical pathology, vol 1. Sydenham Society, London p 55–81

Hotta S 1952 Experimental studies on dengue 1. Isolation, identification and modification of the virus. J Infect Dis 90:1–9

Howe GM 1977 A world geography of human diseases. Academic Press, New York, p 302–317

ICTV 2005 Virus taxonomy: classification and nomenclature of viruses. In: C.M. Fauquet, M.A. Mayo, J. Maniloff, J. Desselberger and L.A. Ball (eds) Elsevier Academic Press, San Diego

Kimura R, Hotta S 1944 Studies on dengue fever (VI). On the inoculation of dengue virus into mice (in Japanese). Nippon Igaku No. 3379:629–633

Kinney RM, Huang CYH 2001 Development of new vaccines against dengue fever and Japanese encephalitis. Intervirology 44:176–197

Koizumi M, Yamaguchi K, Tonomura K 1916 Dengue fever. Nisshin Igaku 6:955–1004

McSherry JA 1982 Some medical aspects of the Darien scheme: was it dengue? Scott Med J 27:183–184

Monath TP 1994 Dengue: the risk to developed and developing countries. Proc Natl Acad Sci USA 91:2395–2400

Munoz JA 1828 Memoria sobre la epidemia que ha sufrido esta ciudad, nombrada vulgarmente al Dengue. Oficina del Gobierno y Capitania General, Havana

Newton EAC, Rieter P 1992 A model of the transmission of dengue fever with an evalution of the impact of ultra-low volume (ULV) insecticide application on dengue epidemics. Am J Trop Med Hyg 47:709–720

Nobuchi H 1979 The symptoms of a dengue-like illness recorded in a Chinese medical encyclopedia (in Japanese). Kanpo Rinsho 26:422–425

Pepper OHP 1941 A note on David Bylon and dengue. Ann Med Hist 3rd Ser 3:363–368

Pinheiro FP 1989 Dengue in the Americas, 1980–1987. Epidemiol Bull 10:1–8

Reiter P, Gubler DJ 1997 Surveillance and control of urban dengue vectors. In Gubler DJ, Kuno G (eds) Dengue and dengue hemorrhagic fever. London, CAB International

Rigau-Pérez JG, Gubler DJ, Vorndam AV, Clark 1994 Dengue surveillance—United States, 1986–1992. MMWR 43:7–19

Rosen L 1982 Dengue—An overview. In Mackenzie JS (ed) Viral diseases in Southeast Asia and the Western Pacific. Academic Press, Sydney, Australia p 484–493

Rosen L 1986 Dengue in Greece in 1927 and 1928 and the pathogenesis of dengue hemorrhagic fever: New data and a different conclusion. Am J Trop Med Hyg 35:642–653

Rush AB 1789 An account of the bilious remitting fever, as it appeared in Philadelphia in the Summer and Autumn of the year. In: Medical inquiries and observation. Prichard & Hall, Philadelphia p 104–117
Sabin AB 1952 Research on dengue during World War II. Am J Trop Med Hyg 1:30–50
Sabin AB, Schlesinger RW 1945 Production of immunity to dengue with virus modified by propagation in mice. Science 101:640–642
Siler JF, Hall MW, Hitchens AP 1926 Dengue, its history, epidemiology, mechanism of transmission, etiology, clinical manifestations, immunity and prevention. Philippine J Sci 29:1–302
Soler B, Pascual P, Petinto P 1949 La duquesa de alba y su tiempo. Madrid, Ediciones y Publicaciones Españoles
The dengue epidemic in Greece 1928 League Nations monthly. Epidemiol Rep 7:335–338
Wichmann O, Muehlberger N, Jelinek T 2003 Dengue–the underestimated risk in travelers. Dengue Bulletin 27:126–137
Wittesjo B, Eitrem R, Niklasson B 1993 Dengue fever among Swedish tourists. Scand J Infect Dis 25:699–704
World Health Organization 1997 Dengue hemorrhagic fever: diagnosis, treatment and control, 2nd edn, World Health Organization, Geneva
World Health Organization 2000 Strengthening implementation of the global strategy for dengue fever/dengue haemorrhagic fever prevention and control. Report of the Informal Consultation, 18–20 October 1999, World Health Organization, Geneva

DISCUSSION

Fairlie: I am not sure what you are attributing the explosion of dengue fever in Singapore to. You mentioned that one of the factors encouraging spread was urbanization, but Singapore has always been urbanized. It was very successful in the 1960s and 1970s at controlling the mosquito population. Why do you think this explosion in Singapore has occurred in the last few years?

Gubler: It's difficult to know all of the factors involved. The Singapore problem began in the 1980s when Dr Chan Kai-Lok, who devised the program, retired. He had been successful for the better part of 20 years, and then left to work in the University. After this, dengue wasn't taken seriously for a few years until the resurgence began. Singapore has maintained its mosquito population at a relatively low level. Serological surveys indicate that the herd immunity is low. If there are foci of higher mosquito population densities you can have transmission even with low mosquito populations. *Aedes aegypti* is a highly efficient epidemic vector because each mosquito will bite multiple people. If there was one in this room today and it was infected, it would bite four or five of us. It doesn't need to take blood; just probing will inject virus, and all of us would become infected There have been some reorganization issues that may have contributed to the problem. Another possibility is that we know that dengue viruses change genetically and there are some strains with higher infectivity and epidemic potential. Singapore has thousands of migrant workers coming in from around the region, resulting in a constant introduction of new viruses. It is likely that this combination of factors is responsible, although it is not possible to be specific.

Holmes: Dengue also has an enzootic cycle involving primates. Has anything been done actively in this area in recent years? It bothers me that we know there are four serotypes that currently infect humans, but the primate population must harbour diverse strains that potentially could cross to humans like other viruses have.

Gubler: The most active program is by the French, who are monitoring viruses in west Africa. The data suggest that the enzootic cycle has not contributed much to the resurgence of epidemic dengue. No one in Asia is working on this, to my knowledge. Dengue is the only arbovirus we know of that is fully adapted to humans and no longer needs the enzootic cycle, even though it is there.

Holmes: I agree that it doesn't need it, but my concern is that if you look at influenza or West Nile, animal populations harbour viruses that are like dengue which can cross species barriers. Conditions are rife for spread. I wonder whether we should we spend more time trying to survey what is out there in other species that looks like dengue.

Gubler: One hypothesis put out by Al Rudnick 40 years ago was that the virus was moving through different mosquito populations and this could cause changes. No one has really looked at this.

Halstead: For those of you who live here in Singapore, there is an important ferment that needs to be heated up. When I was at the Rockefeller Foundation we sent a consultant around all the universities in southeast Asia to try to determine how much investment was made at the university level in the kinds of scientific skills needed to both cope with dengue and also to understand it. The deficits were alarming. Southeast Asia is likely the ancestral home of the sylvatic dengue viruses. But since Al Rudnick did his studies, no one else has worked on this. Yet we know that wild-caught monkeys from all over southeast Asia have dengue antibodies. The techniques that he pioneered are well known. There seem to be a lot of people trying to save orang utans, but we need people to understand that there are some other very important scientific challenges here. The attributes of these 9000 islands in the Indonesian archipelago makes this is a wonderful place for separate viruses to be isolated.

Gubler: Recently the Indians have described the primate cycle in south India for the first time. The Chinese in Yunnan province have repeatedly claimed that they see dengue virus in some of the primates there. In Vietnam there was some evidence of a sylvatic cycle: a virus was isolated from *Aedes niveus* which is a Finlaya subspecies responsible for sylvatic transmissions in Malaysia. It is likely that sylvatic dengue cycles occur in many areas of the region.

Hombach: Do you see an overall shift in the genetic make-up of the circulating viruses or in their pathogenic properties? Can you see a trend over the years?

Gubler: We recently looked at the DENV-2 and -4 in Puerto Rico over a 20 year period. We set up the virological surveillance system in Puerto Rico in 1981, and

DENV-1 was introduced about the same time. Over a period of 20 years we had constant DENV-4 transmission. Shannon Bennett looked at five sites in the genome and sequenced about 120 viruses over this period. She found that there were small genetic changes occurring on a regular basis as a consequence of genetic drift. Some were clearly beneficial and were selected out, spreading throughout the island; in 1 to 2 years, the whole population of DENV-4 on the island would become a different subtype. In 20 years there were four major clades that emerged, and three of these clades were associated with epidemics, in 1981, 1986 and 1998. As we learn more about these viruses, there is certainly a viral component to epidemic potential, but we still can't answer the question of what role genetic changes in the viruses play in virulence. The dengue viruses haven't been adequately studied and have wrongly been considered to be monolithic: we know that they are not, and that strain variation among the viruses influences both epidemic potential and virulence.

Harris: In Nicaragua we were expecting an epidemic this year due to DENV-3, because we have had other serotypes that haven't been increasing since 1998. True to form, we have this big epidemic, but it is DENV-2 and DENV-1 which have been circulating over the last couple of years. It could be because there are more mosquitoes. We are sequencing those strains.

Gubler: That is very interesting. We saw exactly the same thing in Puerto Rico; we predicted an epidemic in 1986. DENV-2 had just been reintroduced into the island. The epidemic occurred just as we predicted, but it wasn't a DENV-2 epidemic, it was a DENV-4 epidemic. In 1997 DENV-3 was introduced into the island for the first time in 17 years, and we thought there would be an epidemic of DENV-3. There was an epidemic, but it was also DENV-4. We don't understand the dynamics between the serotypes and the population.

Hibberd: In southeast Asia there might be a different scenario. The viruses are endemic. I don't think in this area we need introduction from outside.

Gubler: That's probably true. We know that these viruses move in people, but our data suggest that the dengue viruses that caused the epidemics in Puerto Rico were not introduced, but were the ones that were there for a 20 year period. This also applies to Sri Lanka. The epidemic of 1989/90 was thought to be caused by a newly introduced virus, but analyses suggest that this wasn't the case.

Vasudevan: Also, in Taiwan, viruses are brought in by people travelling to endemic countries.

Gubler: That is true also in Singapore: there is a constant introduction of viruses, but we don't know the relative importance of imported vs. endemic viruses in causing epidemics; probably both play a role.

Evans: So there is no lab that carefully looks at the neutralization scheme based on the pressure of the population.

Gubler: There hasn't been a lot of work in that area.

Evans: If your population puts pressure on the virus, why not have a small neutralization scheme?

Gubler: In Cuba they did see that. Scott Halstead is an author on that paper.

Halstead: Singapore is an island. All the years that Duane Gubler and I have returned to look at this phenomenon, there is very little indigenous transmission. There might be a few cases in a household. Thousands of workers come in from Malaysia each day. If Singapore really cares, it is possible to get early warnings: it is the activity that is occurring in Sumatra, Java and Malaysia that will dictate how much dengue will occur in Singapore. Maybe the *Ae. aegypti* index is creeping up. But there are thousands of viraemic people flooding into this country. If Singapore would set up an interchange of information with the two countries on either side, you could easily identify where the viruses are coming through.

Gubler: Martin Hibberd, you have done some sequencing of the current epidemic: is it DENV-1?

Hibberd: I wish I could answer that question. We have sequences from Indonesia and Malaysia.

Gubler: I hope you are doing the whole genome and not just the envelope.

Hibberd: We are doing the whole genome and we have a supply coming through.

Halstead: We don't know what the herd immunity for dengue is. Singapore isn't very relevant because there are 99% susceptible individuals. Your epidemic isn't controlled by herd immunity but opportunism—a viraemic transient finds a home with *Ae. aegypti*.

Gubler: There is some work by Ooi et al (2006) suggesting that transmission in Singapore has moved out of the home. That is, most of the cases are occurring in adult males in children that are school age or above. There are very few cases in small children and females who stay at home. What is amazing today is the tremendous mobility in southeast Asian cities. People can be all over the city in one day. They are viraemic for 24–48 h before they get sick, so they move the virus around.

Keller: If you look at the increase in dengue in Singapore over the last years, could it have something to do with the growing importance of the Singapore harbour? It is now one of the biggest container terminals with increasing traffic from all over the world. Might there be a correlation?

Gubler: It also correlates well with the increased air traffic. 20 million people a year come through Changi airport.

Gu: There is some evidence that African populations develop less dengue haemorrhagic fever. Is this well characterized?

Gubler: The only data that are really available are those from Cuba, where the rate of dengue haemorrhagic fever in 1981 was higher in Caucasians than in the

African populations. In Cuba there is a tremendous mixture of races and it is hard to separate them, though. The other anecdotal information is that in Africa we see dengue fever but the vascular leak syndrome, i.e. dengue haemorrhagic fever, is very rare.

Halstead: We did a study in Haiti in which US and UN troops had caught dengue. There were 90 strains of dengue virus isolated from these troops. We looked at the rate at which 6 year old children had dengue antibodies by dengue neutralization test. The rate of infection was higher in Haiti than it was in Burma. The protective black gene is a powerful phenomenon. It is amazing how important observations like these get made and then there is no follow-on study. In South America, where populations are largely black, compared with the amount of dengue transmission in the population, the disease is relative mild.

Stephenson: Several people have raised the issue of air travel in terms of the transmission of dengue. There is quite a lot of work underway in other arenas looking at sterilization of aircraft. For a few hundred dollars you can sterilize an aircraft. Would this make any difference, or would transmission occur by so many other routes it is not worth thinking of?

Gubler: First of all, I am not sure you could sterilize an aircraft. However, I don't know of a single incident where airport-associated arboviral disease has occurred through the importation of an infected mosquito. We worry about this with West Nile virus. The problem is that the airlines route aircraft all over the world so they'd have to regularly disinsect the whole fleet. To do this effectively you would need to use a residual insecticide, both in the cargo hold and the passenger compartment.

Stephenson: This is a serious discussion going on primarily with pandemic flu. There are technologies that can disinfect an aircraft for a few hundred dollars. The problem is not the cost, but the turnaround time. There are also remote sensor technologies that can detect febrile people. If you can do both of these, would it make any difference?

Gubler: Yes, it could make a difference. The important thing, however, would be to detect febrile people and prevent them from being exposed to mosquitoes. The international health regulations require countries to control *Ae. aegypti* around airports, although very few with the exception of Singapore, do this today.

Vasudevan: The Taiwanese screen for febrile passengers (Singapore did this only during the SARS crisis).

Halstead: Yes, Taiwan has picked dengue-infected people up this way.

Gubler: We are hoping to start a program like this in Hawaii.

Stephenson: The message I am getting is that this might make a difference.

Halstead: My guess is that most dengue in Singapore is not imported by air travel.

Canard: If viraemic people are the problem, disinsection of a plane wouldn't make any difference. You didn't mention the trade of used tyres around the world: are they disinsected or controlled? I'd have thought this is a huge problem.

Gubler: The used tyre trade has been responsible for moving *Ae. albopictus* around the world in the last 25 years. It is now in Africa. In the USA in the last 20 years we have detected five exotic species of mosquito that have been introduced and become established in the country. At least some of them have come via the used tyre trade. We did disinsection of tyres for a number of years in the USA. It is one of these things that costs a lot of money and when money gets tight it stops. I don't know of any other country that does this. Disinsection of aircraft has been practiced for years. The WHO still recommends it. Many countries did away with it because there was a bigger problem with passengers getting respiratory problems from the sprays.

Harris: They still do it a lot.

Gubler: But they use pyrethroid insecticides that don't work very well.

Malasit: What is the contribution of background flavivirus immunity in different parts of the world to dengue virus resistance? Since most of the antibodies to flaviviruses, including yellow fever, West Nile and Japanese encephalitis viruses, are cross-reactive, is it possible that immunity to one virus influences the protection of another virus?

Gubler: I wish I could give a definitive answer to this important question.

The American tropics are at the highest risk in over 60 years for urban yellow fever epidemics. It is an enigma that we haven't had an epidemic of yellow fever yet. There have been several small outbreaks of urban transmission in Santa Cruz, Bolivia and in Brazil, but no epidemic. If there is an epidemic in the Americas, the virus will move to Asia and the pacific region quickly because of air travel. The question is, what role will the heterotypic antibody, if any, play in preventing transmission in Asia? From a public health standpoint it doesn't matter because the whole region will panic; they will all want vaccine and there isn't enough vaccine available. This underscores the need to control *Ae. aegypti.* There is evidence in monkeys that previous infections with dengue down-regulates the severity of illness associated with wild-type yellow fever infection, but is not protective.

Zinkernagel: In very general terms serotypes would not exist if cross-protective antibodies played any major role. I would rather take the opposite position: while in dengue we might not understand these common low titre ones versus the more serotype-defined specificity, so-called 'cross-protective' antibodies don't exist.

Gubler: It is not cross-protective, but it probably down-regulates the illness and the viraemia.

Zinkernagel: In immunology, we like to use words such as 'regulation' and 'networks'. It is nice thinking, but I consider it to be dirty thinking on clean problems.

Gubler: The only experimental data I can cite date back to 1975. Theiler & Anderson (1975) infected two groups of monkeys with wild-type yellow fever. One group was naïve and the other group had previous dengue infection. All of the dengue-infected monkeys survived while the controls died. This suggests that it modulated the disease, even if it didn't protect against infection.

Zinkernagel: You could think of persistence of low-level infections that would increase interferons and tumour necrosis factor (TNF) as an explanation. Unless these experiments are done properly in a controlled fashion the interpretation is impossible.

Gubler: I agree that timing is everything.

References

Ooi EE, Goh KT, Gubler DJ 2006 Singapore's 35-year experience with vector control: can long term dengue prevention be achieved? Emerg Infect Dis J, in press

Theiler M, Anderson CR 1975 The relative resistance of dengue-immune monkeys to yellow fever virus. Am J Trop Med Hyg 24:115–117

Molecular biology of flaviviruses

Eva Harris, Katherine L. Holden, Dianna Edgil, Charlotta Polacek and Karen Clyde

Division of Infectious Diseases, School of Public Health, 140 Warren Hall, University of California, Berkeley, CA 94720–7360, USA

Abstract. Flaviviruses are enveloped viruses with a single-stranded, 10.7 kb positive-sense RNA genome. The genomic RNA, which has a 5′ cap but no poly(A) tail, is translated as a single polyprotein that is then cleaved into three structural proteins and seven non-structural (NS) proteins by both viral and host proteases. The NS proteins include an RNA-dependent RNA polymerase (NS5), a helicase/protease (NS3), and other proteins that form part of the viral replication complex. Sequences and structures in the 5′ and 3′ untranslated regions (UTR) and capsid gene, including the cyclization sequences, the upstream AUG region, and the terminal 3′ stem-loop, regulate translation, RNA synthesis and viral replication. We have also found that an RNA hairpin structure in the capsid coding region (cHP) influences start codon selection and viral replication of the flavivirus dengue virus (DENV). Peptide-conjugated phosphorodiamidate morpholino oligomers (P-PMOs) were used to further dissect the role of conserved regions of the 5′ and 3′ UTRs; several P-PMOs were shown to specifically inhibit DENV translation and/or RNA synthesis and, hence, are potentially useful as antiviral agents. Regarding the mechanism of DENV translation, we have shown that DENV undergoes canonical cap-dependent translation initiation as well as a non-canonical mechanism when cap-dependent translation is suppressed. Although much remains to be elucidated about the molecular biology of flavivirus infection, progress is being made towards defining the *cis* and *trans* factors that regulate flavivirus translation and replication.

2006 New treatment strategies for dengue and other flaviviral diseases. Wiley, Chichester (Novartis Foundation Symposium 277) p 23–40

Flavivirus life cycle

The flavivirus genus includes the medically important arboviruses dengue virus (DENV), yellow fever virus (YFV), West Nile virus (WNV), Japanese encephalitis virus (JEV), and tick-borne encephalitis virus (TBE), among others. Flaviviruses are enveloped viruses with a single-stranded, 10.7 kb, positive-sense RNA genome that has a type I 5′ cap but no poly(A) tail (Chambers et al 1990). The genomic RNA is translated as a single polyprotein, which is then cleaved into three structural proteins (C, prM/M, E) and seven non-structural (NS) proteins (NS1, NS2A, NS2B, NS3, NS4A, NS4B, NS5) by both viral and host proteases (Fig. 1A). Hydrophobic amino acids preceding the N-termini of prM, E, NS1 and NS4B are

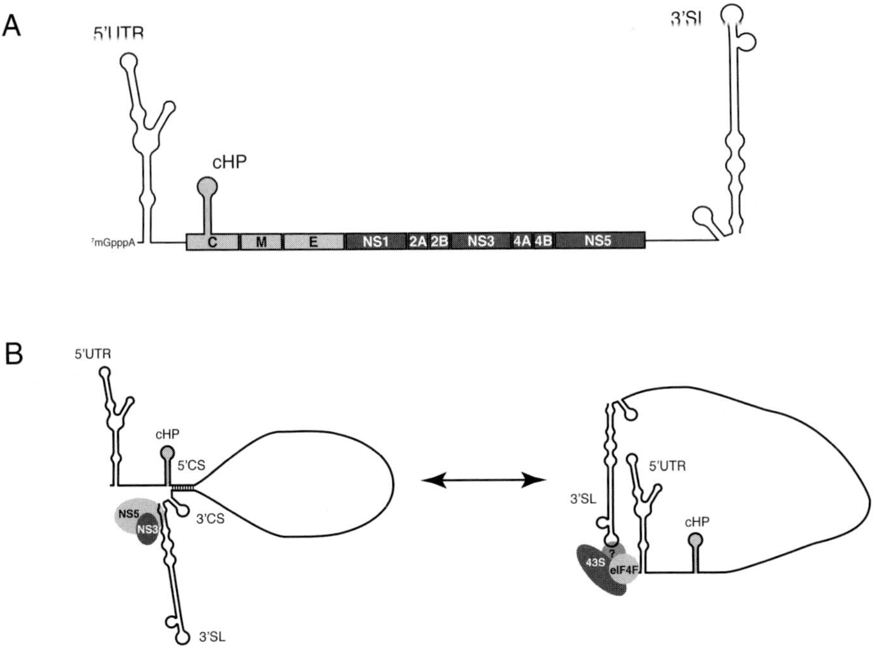

FIG. 1. (A) Schematic diagram of the flavivirus genome. The 2K polypeptide precedes the N-terminus of NS4B. CS, cyclization sequence; 3′SL, 3′ stem-loop. (B) Proposed 5′-3′ end interactions in flavivirus RNA synthesis and translation. cHP, capsid coding region hairpin.

thought to serve as signal sequences for insertion of these proteins into the endoplasmic reticulum (ER) membrane. Cleavage at the N-terminus of the signal sequence for NS4B, which is located in NS4A, generates a 23-amino acid peptide known as 2K (Lin et al 1993). The NS proteins include an RNA-dependent RNA polymerase (RdRp; NS5), a helicase/protease (NS3), and other proteins that form part of the viral replication complex (Khromykh et al 1999, 2000, Lindenbach & Rice 2001). For flaviviruses in general, it is thought that the envelope protein E interacts with a cellular receptor and that viral uptake occurs via receptor-mediated endocytosis (Lindenbach & Rice 2001). Endosome acidification leads to fusion of the viral and endosomal membranes and release of the nucleocapsid into the cytoplasm (Heinz et al 1994a, 1994b). At the ER membrane, the structural proteins and NS1 undergo co-translational translocation and membrane-associated cleavage, whereas the remainder of the NS proteins do not translocate though the ER membrane and remain in the cytoplasm (Falgout & Markoff 1995, Markoff et al 1994). As infection proceeds, virus-induced hypertrophy of intracellular membranes occurs, and vesicle packets thought to be sites of viral replication accumulate (Mackenzie et al 1999). Virus assembly takes place at these membranes, and viral particles pass through the Golgi, where modifications to E and M occur, and are exocytosed via secretory vesicles (Heinz et al 1994a).

Replication of flaviviruses

As with other positive-sense RNA viruses (Gamarnik & Andino 1998), the 5′ terminal region and 3′ UTR of flaviviruses play a key role in regulating viral translation and RNA synthesis (Lindenbach & Rice 2001). Moreover, similar to poliovirus translation (Novak & Kirkegaard 1994), flavivirus translation is coupled to RNA replication (Khromykh et al 1999, 2000). According to a model that has been proposed for the flavivirus Kunjin (KUN), the replication complex is initiated during translation by the binding of NS3 and NS2A to conserved regions in the RdRp, NS5. After completion of translation, the partially assembled replication complex binds to the 3′ UTR of the genome via NS2A (Mackenzie et al 1998), NS3 and NS5, and is transported to the membranous sites of replication (Khromykh et al 1999, 2000). Replication of DENV is important for pathogenesis, and the level of viraemia has been shown to correlate with disease severity (Gubler 1997, Vaughn et al 2000). Furthermore, mutations in the 5′ and 3′ UTRs of DENV have been shown to reduce viral replication in cell culture and mosquitoes, neurovirulence in mice, and viraemia in monkeys (Cahour et al 1995, Men et al 1996, Zeng et al 1998). Like all non-retroviral positive-stranded viruses, replication of the flavivirus genome occurs through a negative-strand replicative intermediate. The 3′ end of this negative-strand flaviviral RNA is critical for priming positive-strand RNA synthesis, and several proteins have been identified that interact with the WNV and DENV minus-strand 3′ end, including TIA-1, TIAR and La autoantigen (Li et al 2002, Shi et al 1996b, Yocupicio-Monroy et al 2003).

The terminal 100 nucleotides of the flavivirus positive-strand RNA 3′ UTR form a conserved stem loop (Brinton et al 1986, Hahn et al 1987), termed the 3′SL. The 3′SL mediates binding to NS5, NS3 (Chen et al 1997), NS2A (Mackenzie et al 1998), elongation factor 1A (Blackwell & Brinton 1997), La (Garcia-Montalvo et al 2004), PTB (De Nova-Ocampo et al 2002) and other host proteins (Blackwell & Brinton 1995) (S.M. Paranjape and E. Harris, unpublished results), based on studies with the flaviviruses WNV, KUN, JEV and DENV. Mutations in the 3′SL reduce translation of DENV reporter RNAs and replicons (Holden & Harris 2004, Holden et al 2006) and negatively impact flavivirus replication and viability (Blackwell & Brinton 1997, Men et al 1996, Tilgner et al 2005, Zeng et al 1998). Additional conserved sequence motifs and predicted secondary structures are found along the length of the flavivirus 3′ UTR (Olsthoorn & Bol 2001, Proutski et al 1997) that play a role in flaviviral RNA synthesis (Alvarez et al 2005b, Lo et al 2003, Tilgner & Shi 2004). These include two dumbbell-like structures that contain the CS2/RCS2 sequences and are predicted to form pseudoknots as well as the terminal five nucleotides of the genomic RNA. Finally, two sets of complementary sequences at the 5′ and 3′ ends of flaviviruses are essential for viral replication: 11-nucleotide cyclization sequences (CS) at the 5′ end of the capsid gene and in the 3′ UTR of the flavivirus genome as well as 16–17-nucleotide upstream AUG region (UAR) sequences found in the 5′ and 3′ UTRs of DENV (Alvarez et al 2005b, Hahn et al 1987, Khromykh et al 2001a, You et al 2001). The CS are required for RNA synthesis (You et al 2001) but do not appear to play a role in viral translation (Edgil & Harris 2006, Lo et al 2003). Thus, some progress has been made in identifying sequences and structures in the 3′ UTR that promote flavivirus translation, RNA synthesis and viral replication.

Flavivirus translation

Viral genomes do not encode translational machinery; therefore, viruses rely on the host cell for protein synthesis (Gale et al 2000). It has been shown that RNA replication of picorna- and flaviviruses is coupled to translation, in that each viral RNA must be translated in order to be replicated (Khromykh et al 1999, 2000, Novak & Kirkegaard 1994). In addition, RNA replication appears to be coupled to viral assembly (Khromykh et al 2001b, Nugent et al 1999). Thus, the same genome must be translated in order to be replicated and must be replicated in order to be packaged into the virion. While the flavivirus 5′ UTRs are more structured than many cellular capped 5′ UTRs, they are shorter and less structured than the uncapped internal ribosome entry site (IRES)-containing 5′ UTRs characteristic of the other members of the *Flaviviridae* family (i.e. hepaci- and pestiviruses). It is thought that flaviviral translation is cap-dependent because the flavivirus genomic RNA contains a m^7G cap at its 5′ terminus and encodes a methyltransferase, RNA 5′-triphosphatase, and putative guanylyltransferase presumed to be involved in capping of this cytoplasmic viral RNA. Consistent with this assumption, the DENV genome can initiate translation via a scanning mechanism (A.-M. Helt and E. Harris, unpublished results), and DENV UTRs do not contain IRES activity (Edgil et al 2006). However, the DENV RNA can also initiate translation via a non-canonical mechanism when translation initiation factors are limiting (see Results section).

5′-3′ end interactions in translation and RNA synthesis

Translation of cellular and viral mRNAs is most efficient when there is an interaction between the 5′ cap and 3′ poly(A) tail (Sachs et al 1997, Tarun & Sachs 1995). Cellular 5′-3′ synergy is usually mediated by the interaction of eIF4G and poly(A) binding protein (PABP), either directly (Tarun & Sachs 1996) or via the bridging protein, PABP-interacting protein (PAIP-1) (Craig et al 1998). Additionally, PABP appears to interact directly with ribosomal subunits, likely via the binding of PABP RNA recognition motifs (RRM) to ribosomal RNA (Proweller & Butler 1996; S. Tarun and A. Sachs, personal communication). Stimulation of translation by the 3′ UTR has been demonstrated for a number of RNA viruses, including rotavirus (Piron et al 1998), poliovirus (Michel et al 2001), brome mosaic virus (Noueiry et al 2003), barley yellow dwarf virus (BYDV) (Guo et al 2001) and, more recently, DENV (Chiu et al 2005, Edgil et al 2003, Holden & Harris 2004). The mechanism of 3′-end enhancement of flavivirus translation has not yet been fully elucidated but likely involves RNA-protein interactions mediated by structures in the 3′ UTR, such as the 3′SL (Fig. 1B). For instance, half of the translational stimulation mediated by the DENV 3′ UTR can been attributed to the action of the 3′SL in the context of both reporter RNAs (Chiu et al 2005, Holden & Harris 2004) and reporter replicons containing the non-structural protein genes in addition to a luciferase reporter (Holden et al 2006, K. Clyde, K. L. Holden and E. Harris, unpublished results). The 'closed loop' model of translation affords several potential advantages. First, it may stabilize the mRNA and translation complex. Second, it may ensure translation only of full-length RNAs that contain both the 5′ and 3′ ends. Third, it may promote efficient 'recycling' of ribosomes and rapid reinitiation of translation on the same strand of RNA (Sachs 2000), which could favour viral translation in the context of competition with translation of cellular mRNA. However, it was recently shown in an *in vitro* system consisting of ribosome-depleted rabbit reticulocyte lysates that stimulation of cellular translation can be mediated by the poly(A) in *trans*, indicating that circularization may not be necessary (Borman et al 2002). High local concentrations of membrane-associated replicating and translating positive sense RNA viruses (Lyle et al 2002, Westaway et al 1997) may allow 5′-3′ end interactions and potentially reinitiation on neighbouring viral RNA molecules as well.

Interactions between the 5′ and 3′ ends of flaviviral genomes also play a critical role in replication of the RNA genome. As mentioned above, hybridization between the 5′ and 3′CS is essential for viral RNA synthesis. Like all positive-sense RNA viruses, the flavivirus genomic RNA undergoes multiple stages of the viral life-cycle that cannot occur simultaneously on the same RNA; for instance, RNA templates undergoing translation cannot be replicated due to the collision of translation and replication complexes proceeding in opposite directions. Regulation of these stages depends on the communication of the viral UTRs. During replication, base-pairing of the CS appears to be necessary for viral RNA synthesis. However, this same interaction is not required for and may even inhibit translation of the viral RNA. The switch between translation and replication of the flavivirus genome is likely to be mediated by architectural remodelling of portions of the genome from that of an RNA-protein bridge between the 5′ and 3′ UTRs to RNA–RNA interactions between the CS and possibly the UARs at the extreme ends of the viral genome (Fig. 1B). In this model, the switch between translation and replication is presumably mediated by viral and/or cellular factors that facilitate communication between the viral UTRs.

Results

We have used several approaches to investigate the regulation of DENV translation and replication, which are summarized below. One set of studies identified a coding region RNA structure that determines translational start site selection and plays a distinct, essential role in viral replication. In another investigation, morpholino oligomers targeting conserved domains in the 5′ and 3′ UTRs were used to dissect the roles of these domains in DENV translation and RNA synthesis. Finally, the mechanism of DENV translation initiation was investigated and found to proceed via both a canonical cap-dependent as well as a non-canonical mechanism.

Regulation of DENV replication by a coding region structural element

RNA *cis*-acting elements are often identified in the UTRs of cellular and viral messages, and several regions of the flavivirus 5′ and 3′ UTRs have been implicated in the regulation of viral translation and RNA synthesis (Alvarez et al 2005a, 2005b, Chiu et al 2005, Holden & Harris 2004, Holden et al 2006, Tilgner & Shi 2004). RNA regulatory elements located in the coding regions are less commonly identified. Using computer algorithms (e.g. *RNAalifold*), we found phylogenetic evidence that a capsid-coding region hairpin structure (cHP) is maintained among mosquito- and tick-borne flaviviruses (Clyde & Harris 2006). In mosquito-borne flaviviruses, the cHP is located 12–16 nucleotides downstream of the first AUG (Fig. 2A), which is in a poor initiation context. Through mutagenic analysis, we demonstrated that the DENV-2 cHP functions in start site selection in human hepatoma (Hep3B) and *Ae. albopictus* mosquito (C6/36) cells, directing first AUG usage proportional to its stability in a position-dependent and sequence-independent manner (Clyde & Harris 2006). This is consistent with a previously proposed mechanism whereby RNA secondary structure 12–15 nucleotides downstream of a start codon in a poor initiation context can enhance recognition of the suboptimal codon (Kozak 1989, 1990) by causing the scanning initiation machinery to pause at the structural element in order to unwind it, allowing the ribosome to remain in contact with a start codon in a poor initiation context (Kozak 1990, 1991).

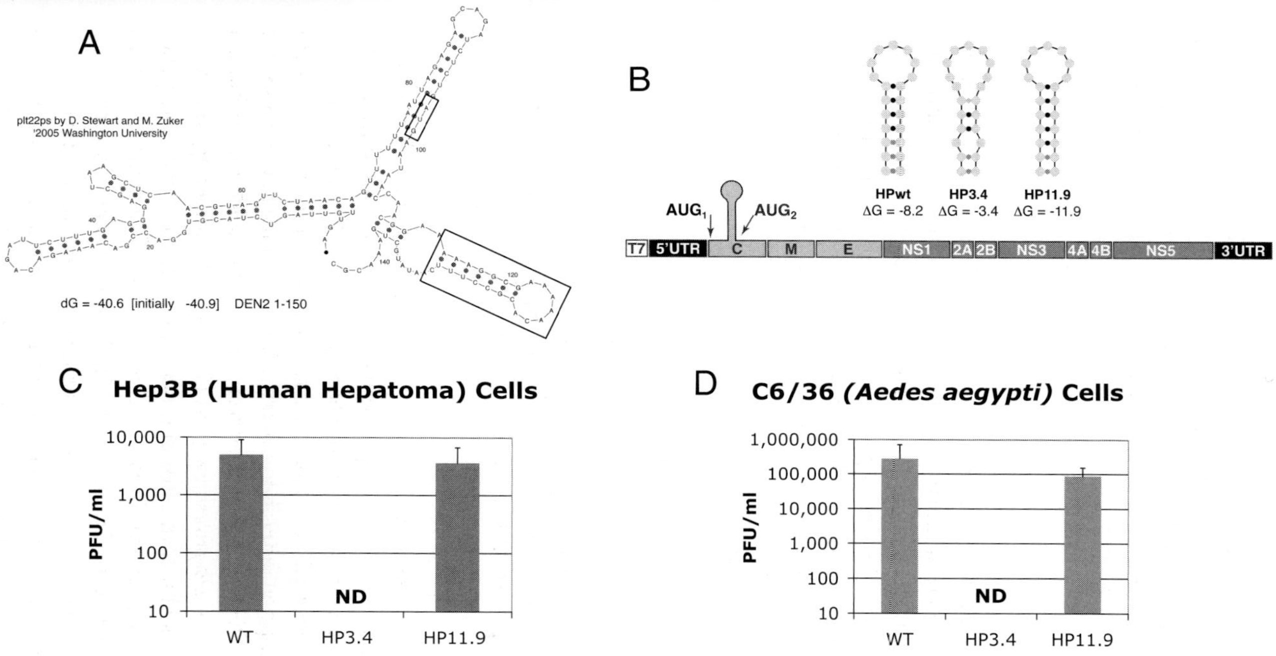

FIG. 2. Regulation of DENV replication by a coding region structural element. (A) The predicted secondary structure of the first 150 nucleotides of DENV-2 showing the cHP. The RNA secondary structure of the first 150 nucleotides of DENV-2 was predicted by *mFold;* the start codon and the cHP are indicated by boxes. (B) Schematic of infectious clone (IC) variants utilized to study the role of the cHP and of the nucleotides that make up the DENV-2 initiation context. The pD2/IC DENV-2 IC was a gift of R. Kinney (Centers for Disease Control and Prevention, Fort Collins, CO). (C) Viral titres from IC-transfected Hep3B cells. *In vitro*-transcribed RNAs were transfected into Hep3B cell monolayers, and viral replication was assessed after 72 h by plaque assay. Titres were normalized to transfection efficiency as determined by qRT-PCR at 2 h post-transfection. Viral titres are expressed as plaque-forming units per ml (PFU/ml) from IC-transfected Hep3B cells. One log reflects the limit of detection of a standard plaque assay. Error bars indicate SD; data derived from four experiments. (D) Viral titres from IC-transfected C6/36 cells. Methods and graph as in C; data are derived from four experiments.

Congruent with the conservation of the cHP among flaviviruses, this *cis* element is necessary for replication of DENV-2 in Hep3B and C6/36 cells. To test the requirement for the DENV-2 cHP in the viral lifecycle, mutations were made in the infectious clone of DENV-2 prototype Thai strain 16681 (pD2/IC) that were predicted to disrupt the cHP structure (Fig. 2B). Viral RNA was *in vitro*-transcribed and transfected into Hep3B and C6/36 cells. Viral replication was assessed after 72 h by plaque assay and normalized to transfection efficiency as measured by quantitative RT-PCR of infectious clone RNA at 2 h post-transfection. In Hep3B and C6/36 cells, disruption of the cHP (pD2/IC-HP3.4) reduced viral replication by at least 2.5 logs and nearly 4 logs, respectively, to levels undetectable by plaque assay (Fig. 2C and 2D). Restoration of the hairpin structure (pD2/IC-HP11.9) restored viral replication to levels similar to wild-type (Fig. 2C and 2D). Similarly, residues at the −3 and +4 positions relative to the start codon, the first start codon and the second in-frame start codon are all required for efficient viral replication in both cell types (data not shown). These data demonstrate that coding region RNA secondary structure plays a regulatory role in translation and replication of DENV and perhaps other flaviviruses.

Analysis of regions of the 5′ and 3′ UTRs involved in DENV translation and RNA synthesis

Viral translation and RNA synthesis of positive-sense RNA viruses is often regulated by the 5′ and 3′ UTRs. Previously, we had demonstrated a role for the DENV-2 terminal 3′SL in stimulating viral translation via deletion analysis in RNA reporter constructs (Holden & Harris 2004). To examine other regions of the DENV-2 UTRs, we targeted conserved domains in the 5′ and 3′ UTRs by complementary peptide-conjugated phosphorodiamidate morpholino oligomers (P-PMOs), as depicted in Fig. 3. The P-PMOs most effective in inhibiting viral replication were those complementary to the DENV 5′ stem-loop (‘5′SL’) and to the DENV 3′ cyclization sequence (‘3′CS’), consistent with recent results with DENV and WNV (Deas et al 2005, Kinney et al 2005). We also identified a third P-PMO, complementary to the top of the 3′ stem-loop (‘3′SLT’), that inhibited DENV replication in BHK cells. Presumably, these P-PMOs interfere with DENV replication by blocking critical RNA-RNA or RNA-protein interactions involved in viral translation and/or RNA synthesis.

To distinguish between effects of DENV-specific P-PMOs on viral translation and/or RNA synthesis, a DENV-2 reporter replicon was constructed (Fig. 4A) using the DENV-2 infectious clone (pD2/IC) as a backbone, with the firefly luciferase gene replacing most of the structural genes. The non-infectious DENV-2 reporter replicon should translate and replicate in a manner similar to wild-type DENV RNA. The DENV-2 reporter replicon was transcribed *in vitro* and trans-

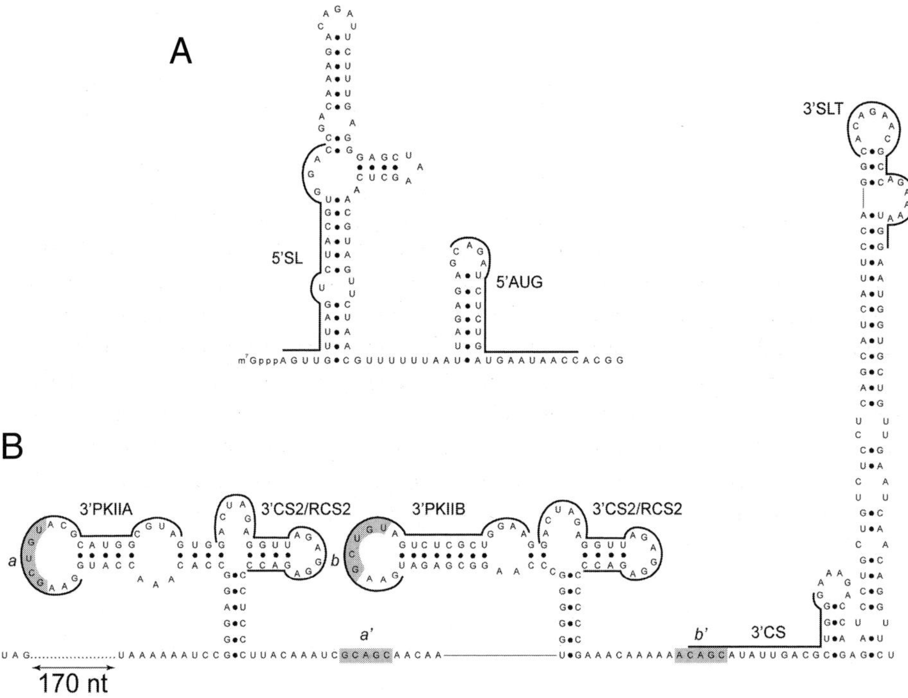

FIG. 3. Schematic diagrams of the targeted locations in the DENV 5′- and 3′-UTRs. (A) Locations of P-PMO target sequences in the DENV-2 5′ UTR. Indicated by lines are target sequences for the DENV 5′SL P-PMO and the 5′ AUG P-PMO. The predicted secondary structure of the DENV-2 5′ UTR was determined using the *mFold* web server (Zuker 2003). (B) Locations of P-PMO target sequences in the DENV 3′ UTR. Indicated by lines adjacent to the 3′ UTR sequence are target sequences for the DENV 3′ PKIIA, 3′ PKIIB, 3′CS2/RCS2, 3′CS, and 3′SLT P-PMOs. The shaded regions indicate the interactions involved in forming the PKIIA and PKIIB pseudoknot secondary structures; the shaded loop sequence is predicted to interact with the shaded region downstream. The secondary structures of the DENV-2 3′ UTR are based on computer predicted secondary structures (Hahn et al 1987, Mohan & Padmanabhan 1991, Olsthoorn & Bol 2001, Proutski et al 1999, Shi et al 1996a) and chemical probing of the terminal 100 nucleotides (Shi et al 1996a). Reprinted with permission from Elsevier (Holden et al 2006).

fected into BHK cells, and luciferase activity was monitored for 96 h post-transfection. Two peaks of luciferase activity were generated by the DENV-2 reporter replicon: an early peak between 4 and 8 h post-transfection and a later peak between 48 and 96 h post-transfection (Fig. 4B). By using mycophenolic acid (MPA), a potent inhibitor of DENV RNA synthesis (Diamond et al 2002), it was determined that the first peak of luciferase activity (4–8 h) correlates with translation of the DENV-2 reporter replicon, whereas the later peak of luciferase activity

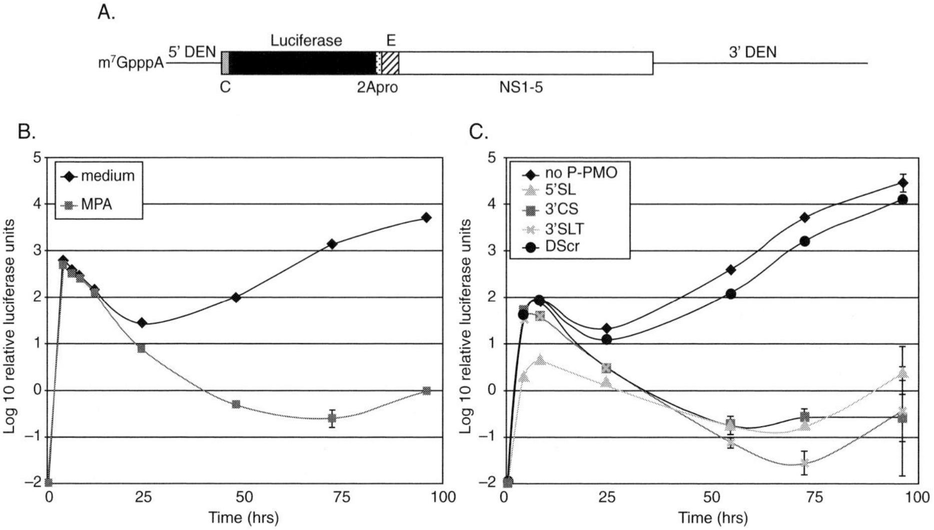

FIG. 4. Effects of the DENV-specific P-PMOs on translation and RNA synthesis of the DENV-2 reporter replicon. (A) Schematic diagram of the DENV-2 reporter replicon. Indicated are the DENV-2 5′ UTR (black line), first 72 nucleotides of C (grey box) fused to the firefly luciferase gene (black box), the FMDV 2Apro (dotted box), the last 90 nucleotides of E (striped box), the entire NS region (white box), and the DENV-2 3′ UTR (black line). (The pD2/IC infectous clone of DENV-2 that was used as the backbone for the replicon was a gift from R. Kinney, Center for Disease Control and Prevention, Fort Collins, CO.) (B) The DENV-2 reporter replicon distinguishes between viral translation and RNA synthesis. Cells were transfected with the DENV-2 reporter mRNA, then treated with medium or MPA (3 μM). Luciferase activity was monitored at 4, 8, 24, 54, 72 and 96 h after transfection. Data shown are representative of three independent experiments. Error bars indicate the SD of duplicate samples. The amount of transfected RNA was not impacted by the presence of the P-PMOs, as determined by quantitative real-time RT-PCR (data not shown). (C) DENV-specific P-PMOs inhibit translation and/or RNA synthesis of the DENV-2 reporter replicon. Cells were treated with the DENV-specific P-PMOs (5′SL, 3′CS, and 3′SLT) or the DScr P-PMO and transfected with the DENV-2 reporter replicon. Luciferase activity was monitored at 4, 6, 8, 12, 24, 48, 72 and 96 h after transfection. Data shown are representative of three independent experiments. Error bars indicate the SD of duplicate samples. The amount of transfected RNA was not impacted by the presence of P-PMO, as determined by quantitative real-time RT-PCR (data not shown). Reprinted with permission from Elsevier (Holden et al 2006).

(48–96 h)—which was inhibited by MPA—is dependent upon RNA synthesis (Fig. 4B). The DENV-2 reporter replicon was then used to distinguish between the effect of the DENV-specific P-PMOs on viral translation and RNA synthesis. Cells were treated with the different P-PMOs before transfection of the DENV-2 reporter replicon and again after RNA transfection; no effect on RNA transfection efficiency was observed, as determined by real-time (RT)-PCR (data not shown).

A 'scrambled' P-PMO ('DScr') was used to control for any non-specific effect of the P-PMOs. The 5′SL P-PMO, which inhibited translation of the DENV reporter mRNA (data not shown), also inhibited translation of the DENV-2 reporter replicon by 95% and consequently, reduced the RNA synthesis peak by over 10 000-fold (Fig. 4C). The 3′CS P-PMO strongly inhibited the appearance of the RNA synthesis peak of the DENV-2 reporter replicon (>10 000-fold) with no significant effect on the first peak of translation, indicating that the 3′CS P-PMO inhibited only viral RNA synthesis (Fig. 4C). The 3′SLT P-PMO reduced translation of the DENV-2 reporter replicon by 40% and inhibited the second peak of luciferase activity that is primarily dependent on RNA synthesis by over 10 000-fold, indicating that it may directly inhibit RNA synthesis in addition to its effects on translation.

Thus, using a novel DENV-2 reporter replicon and a DENV-2 reporter mRNA, we determined that the 5′SL P-PMO inhibited viral translation, the 3′CS P-PMO blocked viral RNA synthesis but not viral translation, and the 3′SLT P-PMO inhibited both viral translation and RNA synthesis. These results show that the 3′CS and the 3′SL domains regulate DENV translation and RNA synthesis and further demonstrate that P-PMOs are potentially useful as antiviral agents.

A non-canonical mechanism of DENV translation initiation

While DENV and other flaviviruses containing capped RNA genomes undergo cap-dependent translation, DENV has been shown to translate efficiently under circumstances in which cap-dependent cellular translation is suppressed through the depletion of the cap-binding protein, eIF4E (Edgil et al 2006). To investigate this phenomenon, we used several approaches to inhibit cap-dependent translation. Initially, the compounds rapamycin, wortmannin and LY294002 were used to suppress cellular translation initiation through the sequestration of eIF4E. The downstream effect of the interaction of rapamycin with the mammalian Target of Rapamycin (mTOR) is the hypo-phosphorylation of the eIF4E-binding protein 1 (4E-BP) (Vanhaesebroeck et al 2001). Similarly, wortmannin and LY294002 inhibit the phosphoinositol-3-kinase (PI3-kinase) pathway, which also results in the hypo-phosphorylation of 4E-BP (Vanhaesebroeck et al 2001). In its hypo-phosphorylated form, 4E-BP binds to and sequesters eIF4E from the eIF4F translation initiation complex, thus inhibiting cap-dependent translation. The eIF4F complex consists of eIF4E, the scaffold protein eIF4G and the helicase eIF4A, and is necessary for recruitment of the 43S ribosomal complex to the 5′-terminus of the mRNA (Gale et al 2000). The impact of treatment with these drugs on translation of cellular cap-dependent mRNAs was compared to their effect on DENV reporter constructs and on viral replication. When cells were incubated with DENV in the presence of rapamycin, LY294002 or wortmannin, viral titres

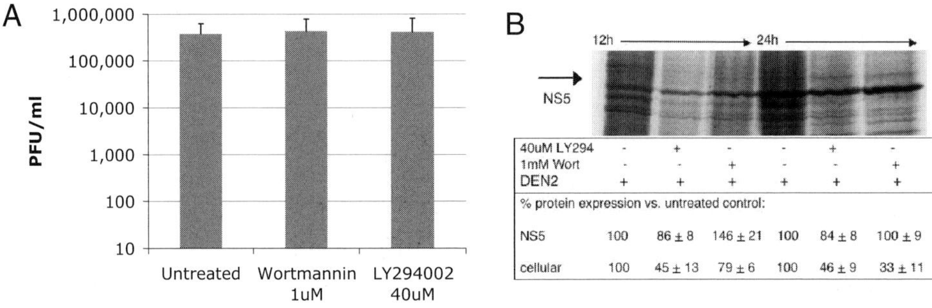

FIG. 5. DENV replication and translation are resistant to inhibitors of cap-dependent translation. (A) DENV replicates in cells exposed to inhibitors for 24 h. BHK cells (2×10^5) were exposed to DENV-2 strain 16681 and simultaneously treated with 40 μM LY294002 or 1 μM wortmannin per well. Cells were incubated for 24 h at 37°C. Cell supernatants were collected and infectious virus titered using BHK21 cells (PFU/mL). The data are expressed as an average of three experiments. Error bars indicate standard deviation. (B) DENV RNA is translated in inhibitor-treated cells. Cells were treated with LY294002 or wortmannin as described above. One hour prior to metabolic labelling (12 or 24 h post-infection), cells were starved in cysteine-methionine-free medium in the presence of inhibitors. Cells were then pulsed with 150 μCi of [^{35}S]cysteine-methionine for 30 min, harvested and analysed by SDS-PAGE. DENV NS5 (arrow) and representative cellular proteins were quantitated as percent protein relative to the untreated control. Data are representative of three experiments.

remained unchanged at 24 h post-infection (Fig. 5A). Furthermore, whereas metabolic labelling of total cellular protein in inhibitor-treated cells revealed an inhibition of cellular protein synthesis 12 h post-infection, translation of DENV proteins, represented by the DENV RNA-dependent RNA polymerase (NS5), was relatively unaffected (Fig. 5B).

Other conditions that inhibit cellular cap-dependent translation, such as siRNA-mediated depletion of eIF4E, also did not affect the yield of DENV progeny from infectious virus, the translation of viral proteins, or the translation of reporter genes flanked by the DENV 5′ and 3′ UTRs (Edgil et al 2006). Additionally, translation of a reporter construct containing both the DENV 5′ and 3′ UTRs required significantly higher concentrations of cap analogue to competitively reduce translation than did constructs containing cellular or viral 5′ UTRs and a poly(A) tail. Importantly, cap-independent translation from both non-functionally capped DENV reporter constructs and infectious DENV RNA was triggered when eIF4E was depleted or sequestered (Edgil et al 2006). Under conditions of reduced eIF4E, this non-canonical DENV translation required both the DENV 5′ and 3′ UTRs for activity and did not proceed via internal ribosome entry but rather by an end-dependent mechanism, as the introduction of stable hairpins at the 5′ end inhibited translation initiation (data not shown). We propose a model

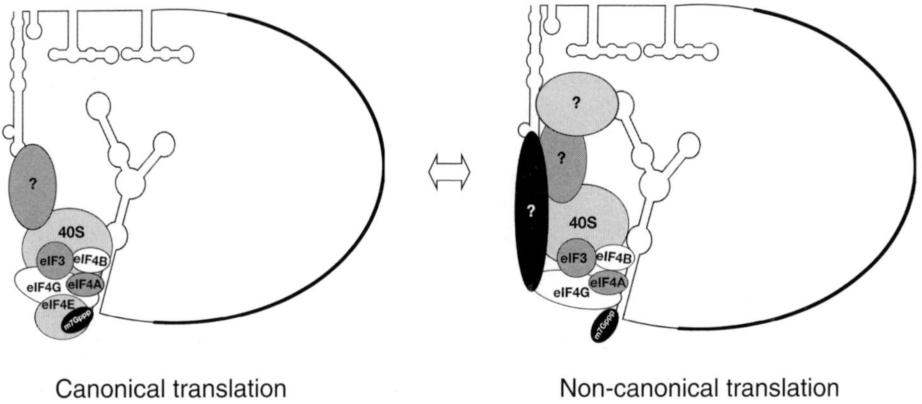

FIG. 6. Model of 5′-3′ interactions in canonical cap-dependent and non-canonical DENV translation initiation. When eIF4E is abundant, a canonical cap-dependent scanning mechanism of translation initiation occurs (left). When eIF4E is limiting, the DENV 3′ UTR interacts with host proteins to deliver and/or stabilize key translation initiation factors at the 5′ UTR in the absence of eIF4E using a non-canonical mechanism of translation initiation (right).

wherein DENV is translated via a canonical cap-dependent mechanism when eIF4E is abundant. However, under cellular conditions where eIF4E is limiting, DENV RNA may undergo a reorganization of viral ribonucleoprotein (RNP) complexes such that RNA structures or sequences in the 3′ UTR interact with higher affinity with protein complexes containing eIF4G and eIF4A to deliver key translation initiation factors to the DENV 5′ UTR (Fig. 6), similar to the mechanism proposed for cap-independent translation of *Luteovirus* RNAs, such as the barley yellow dwarf virus genome (Guo et al 2001). This may be critical for DENV survival in differentiated cell types with low levels of eIF4E (Grolleau et al 1999, Krichevsky et al 1999), such as the dendritic cells and monocyte/macrophages that are the targets of DENV infection in humans (Ho et al 2001, Jessie et al 2004, Wu et al 2000).

Conclusions

In summary, flavivirus replication is regulated by sequences and structures in both the coding and non-coding regions of the viral RNA. Cyclization of the genomic RNA is necessary for RNA synthesis and viral replication and is mediated by the CS and UAR sequences at the 5′ and 3′ ends of the flavivirus RNA. Conserved structures in the 3′ UTR, such as the 3′SL, play a role in both viral translation and RNA synthesis. Additionally, a hairpin structure in the capsid coding region regulates translation start site selection as well as another step in viral replication. As more RNA elements and UTR-binding proteins are identified that modulate viral translation and RNA synthesis, a more detailed understanding of the molecular events in flavivirus replication is being obtained.

Acknowledgments

Funding was provided by the Pew Charitable Trusts (#2617SC) and NIH (AI052324).

References

Alvarez DE, De Lella Ezcurra AL, Fucito S, Gamarnik AV 2005a Role of RNA structures present at the 3′ UTR of dengue virus on translation, RNA synthesis, and viral replication. Virology 339:200–212

Alvarez DE, Lodeiro MF, Ludueña SJ, Pietrasanta LI, Gamarnik AG 2005b Long-range RNA-RNA interactions circularize the dengue virus genome. J Virol 79:6631–6643

Blackwell JL, Brinton MA 1995 BHK cell proteins that bind to the 3′ stem-loop structure of the West Nile Virus genome RNA. J Virol 69:5650–5658

Blackwell JL, Brinton MA 1997 Translation elongation factor-1 alpha interacts with the 3′ stem-loop region of West Nile Virus genomic RNA. J Virol 71:6433–6444

Borman AM, Michel YM, Malnou CE, Kean KM 2002 Free poly(A) stimulates capped mRNA translation in vitro through the eIF4G-PABP interaction. J Biol Chem 277: 36818–36824

Brinton MA, Fernandez AV, Dispoto JH 1986 The 3′-nucleotides of flavivirus genomic RNA form a conserved secondary structure. Virology 153:113–121

Cahour A, Pletnev A, Vazeille-Flacoz M, Rosen L, Lai C-J 1995 Growth-restricted dengue virus mutants containing deletions in the 5′ noncoding region of the RNA genome. Virology 207:68–76

Chambers TJ, Hahn CS, Galler R, Rice CM 1990 Flavivirus genome organization, expression, and replication. Annu Rev Microbiol 44:649–688

Chen CJ, Ku MD, Chien LJ, Hsu SL, Wang YM, Lin JH 1997 RNA-protein interactions: involvement of NS3, NS5, and 3′ noncoding regions of Japanese encephalitis virus genomic RNA. J Virol 71:3466–3473

Chiu WW, Kinney RM, Dreher TW 2005 Control of translation by the 5′- and 3′-terminal regions of the dengue virus genome. J Virol 79:8303–8315

Clyde K, Harris E 2006 RNA secondary structure in the coding region of dengue virus type 2 directs translation start codon selection and is required for viral replication. J Virol 80:2170–2182

Craig AW, Haghighat A, Yu AT, Sonenberg N 1998 Interaction of polyadenylate-binding protein with the eIF4G homologue PAIP enhances translation. Nature 392:520–523

De Nova-Ocampo M, Villegas-Sepulveda N, del Angel RM 2002 Translation elongation factor-1alpha, La, and PTB interact with the 3′ untranslated region of dengue 4 virus RNA. Virology 295:337–347

Deas TS, Binduga-Gajewska I, Tilgner M et al 2005 Inhibition of flavivirus infections by antisense oligomers specifically suppressing viral translation and RNA replication. J Virol 79:4599–4609

Diamond MS, Zachariah M, Harris E 2002 Mycophenolic acid potently inhibits dengue viral infection by preventing replication of viral RNA. Virology 304:211–221

Edgil D, Harris E 2006 End-to-end communication in the modulation of translation by mammalian RNA viruses. Virus Res 119:43–51

Edgil D, Diamond MS, Holden KL, Paranjape SM, Harris E 2003 Differences in cellular infection among dengue virus type 2 strains correlate with efficiency of translation. Virology 317:275–290

Edgil D, Polacek C, Harris E 2006 Dengue virus utilizes a novel strategy for translation initiation when cap-dependent translation is inhibited. J Virol 80:2976–2986

Falgout B, Markoff L 1995 Evidence that flavivirus NS1-NS2A cleavage is mediated by a membrane-bound host protease in the endoplasmic reticulum. J Virol 69:7232–7243

Gale M, Tan S-L, Katze MG 2000 Translational control of viral gene expression in eukaryotes. Microbiol Mol Biol Rev 64:239–280

Gamarnik AV, Andino R 1998 Switch from translation to RNA replication in a positive-stranded RNA virus. Genes Dev 12:2293–2304

Garcia-Montalvo BM, Medina F, Del Angel RM 2004 La protein binds to NS5 and NS3 and to the 5′ and 3′ ends of Dengue 4 virus RNA. Virus Res 102:141–150

Grolleau A, Sonenberg N, Wietzerbin J, Beretta L 1999 Differential regulation of 4E-BP1 and 4E-BP2, two repressors of translation initiation, during human myeloid cell differentiation. J Immunol 162:3491–3497

Gubler DJ 1997 Dengue and dengue hemorrhagic fever: Its history and resurgence as a global public health problem. In: Gubler DJ, Kuno G (eds) Dengue and dengue hemorrhagic fever. CAB International, New York

Guo L, Allen AM, Miller WA 2001 Base-pairing between untranslated regions facilitates translation of uncapped, nonpolyadenylated viral RNA. Mol Cell 7:1103–1109

Hahn CS, Hahn YS, Rice CM et al 1987 Conserved elements in the 3′ untranslated region of flavivirus RNAs and potential cyclization sequences. J Mol Biol 198:33–41

Heinz F, Auer G, Stiasny K et al 1994a The interactions of the flavivirus envelope proteins: implications for virus entry and release. Arch Virol 9:339–348

Heinz F, Stiasny K, Puschner-Auer G et al 1994b Structural changes and functional control of the tick-borne encephalitis virus glycoprotein E by the heterodimeric association with the protein prM. Virology 198:109–117

Ho LJ, Wang JJ, Shaio MF et al 2001 Infection of human dendritic cells by dengue virus causes cell maturation and cytokine production. J Immunol 166:1499–1506

Holden KL, Harris E 2004 Enhancement of dengue virus translation: Role of the 3′ untranslated region and the terminal 3′ stem-loop domain. Virology 329:119–133

Holden KL, Stein D, Pierson TC et al 2006 Inhibition of dengue virus translation and RNA synthesis by a morpholino oligomer to the top of the 3′ stem-loop structure. Virology 344:439–452

Jessie K, Fong MY, Devi S, Lam SK, Wong KT 2004 Localization of dengue virus in naturally infected human tissues, by immunohistochemistry and in situ hybridization. J Infect Dis 189:1411–1418

Khromykh AA, Sedlak PL, Westaway EG 1999 *trans*-Complementation analysis of the flavivirus Kunjin ns5 gene reveals an essential role for translation of its N-terminal half in RNA replication. J Virol 73:9247–9255

Khromykh AA, Sedlak PL, Westaway EG 2000 *cis*-and *trans*-acting elements in flavivirus RNA replication. J Virol 74:3253–3263

Khromykh AA, Meka H, Guyatt KJ, Westway EG 2001a Essential role of cyclization domains in flavivirus RNA replication. J Virol 75:6719–6728

Khromykh AA, Varnavski AN, Sedlak PL, Westaway EG 2001b Coupling between replication and packaging of flavivirus RNA: evidence derived from the use of DNA-based full-length cDNA clones of Kunjin virus. J Virol 75:4633–4640

Kinney RM, Huang CY-H, Rose BC et al 2005 Inhibition of dengue virus serotypes 1 to 4 in Vero cell cultures with morpholino oligomers. J Virol 79:5116–5128

Kozak M 1989 Context effects and inefficient initiation at non-AUG codons in eucaryotic cell-free translation systems. Mol Cell Biol 9:5073–5080

Kozak M 1990 Downstream secondary structure facilitates recognition of initiator codons by eukaryotic ribosomes. Proc Natl Acad Sci USA 87:8301–8305

Kozak M 1991 Structural features in eukaryotic mRNAs that modulate the initiation of translation. J Biol Chem 266:19867–19870

Krichevsky AM, Metzer E, Rosen H 1999 Translational control of specific genes during differentiation of HL-60 cells. J Biol Chem 274:14295–14305

Li W, Li Y, Kedersha N et al 2002 Cell proteins TIA-1 and TIAR interact with the 3′ stem-loop of the West Nile virus complementary minus-strand RNA and facilitate virus replication. J Virol 76:11989–12000

Lin C, Amberg SM, Chambers TJ, Rice CM 1993 Cleavage at a novel site in the NS4A region by the yellow fever virus NS2B-3 proteinase is a prerequisite for processing at the downstream 4A/4B signalase site. J Virol 67:2327–2335

Lindenbach BD, Rice CM 2001 *Flaviviridae*: The viruses and their replication. In: Knipe DM, Howley PM (eds) Fields Virology. Vol. 1 Lippincott Williams and Wilkins, Philadelphia, p 991–1041

Lo M, Tilgner M, Bernard KA, Shi P-Y 2003 Functional analysis of mosquito-borne flavivirus conserved sequence elements within 3′ untranslated region of West Nile Virus by use of a reporting replicon that differentiates between viral translation and RNA replication. J Virol 77:10004–10014

Lyle JM, Bullitt E, Bienz K, Kirkegaard K 2002 Visualization and functional analysis of RNA-dependent RNA polymerase lattices. Science 296:2218–2222

Mackenzie JM, Khromykh AA, Jones MK, Westaway EG 1998 Subcellular localization and some biochemical properties of the flavivirus Kunjin nonstructural proteins NS2A and NS4A. Virology 245:203–215

Mackenzie JM, Jones MK, Westaway EG 1999 Markers for *trans*-Golgi membranes and the intermediate compartment localize to induced membranes with distinct replication functions in flavivirus-infected cells. J Virol 73:9555–9567

Markoff L, Chang A, Falgout B 1994 Processing of flavivirus structural glycoproteins: stable membrane insertion of premembrane requires the envelope signal peptide. Virology 204:526–540

Men R, Bray M, Clark D, Chanock RM, Lai C-J 1996 Dengue type 4 virus mutants containing deletions in the 3′ noncoding region of the RNA genome: analysis of growth restriction in cell culture and altered viremia pattern and immunogenicity in Rhesus monkeys. J Virol 70:3930–3937

Michel YM, Borman AM, Paulous S, Kean KM 2001 Eukaryotic initiation factor 4G-poly(A) binding protein interaction is required for poly(A) tail-mediated stimulation of picornavirus internal ribosome entry segment-driven translation but not for X-mediated stimulation of hepatitis C virus translation. Mol Cell Biol 21:4097–4109

Mohan PM, Padmanabhan R 1991 Detection of stable secondary structure at the 3′ terminus of dengue virus type 2 RNA. Gene 108:185–191

Noueiry AO, Diez J, Falk SP, Chen J, Ahlquist P 2003 Yeast Lsm1p-7p/Pat1p deadenylation-dependent mRNA-decapping factors are required for brome mosaic virus genomic RNA translation. Mol Cell Biol 23:4094–4106

Novak JE, Kirkegaard K 1994 Coupling between genome translation and replication in an RNA virus. Genes Dev 8:1726–1737

Nugent CI, Johnson KL, Sarnow P, Kirkegaard K 1999 Functional coupling between replication and packaging of poliovirus replicon RNA. J Virol 73:427–435

Olsthoorn RC, Bol JF 2001 Sequence comparison and secondary structure analysis of the 3′ noncoding region of flavivirus genomes reveals multiple pseudoknots. RNA 7:1370–1377

Piron M, Vende P, Cohen J, Poncet D 1998 Rotavirus RNA-binding protein NSP3 interacts with eIF4GI and evicts the poly(A) binding protein from eIF4F. EMBO J 17:5811–5821
Proutski V, Gould EA, Holmes EC 1997 Secondary structure of the 3′ untranslated region of flaviviruses: similarities and differences. Nucleic Acids Res 25:1194–1202
Proutski V, Gritsun TS, Gould EA, Holmes EC 1999 Biological consequences of deletions within the 3′-untranslated region of flaviviruses may be due to rearrangements of RNA secondary structure. Virus Res 64:107–123
Proweller A, Butler JS 1996 Ribosomal association of poly(A)-binding protein in poly(A)-deficient *Saccharomyces cerevisiae.* J Biol Chem 271:10859–10865
Sachs A 2000 Physical and functional interactions between the mRNA cap structure and the poly(A) tail. In: Sonenberg N, Hershey JWB, Matthews MB (eds) Translational control of gene expression. Cold Spring Harbor Laboratory Press, Cold Spring Harbor, New York, p 447–465
Sachs AB, Sarnow P, Hentze MW 1997 Starting at the beginning, middle, and end: Translation initiation in eukaryotes. Cell 89:831–838
Shi P-Y, Brinton MA, Veal JM, Zhong YY, Wilson WD 1996a Evidence for the existence of a pseudoknot structure at the 3′ terminus of the flavivirus genomic RNA. Biochemistry 35:4222–4230
Shi P-Y, Li W, Brinton MA 1996b Cell proteins bind specifically to West Nile Virus minus-strand 3′ stem-loop RNA. J Virol 70:6278–6287
Tarun S, Sachs AB 1995 A common function for mRNA 5′ and 3′ ends in translation initiation in yeast. Genes Dev 9:2997–3007
Tarun SZ, Sachs AB 1996 Association of the yeast poly(A) tail binding protein with translation initiation factor eIF4G. EMBO J 15:7168–7177
Tilgner M, Shi PY 2004 Structure and function of the 3′ terminal six nucleotides of the West Nile virus genome in viral replication. J Virol 78:8159–8171
Tilgner M, Deas TS, Shi PY 2005 The flavivirus-conserved penta-nucleotide in the 3′ stem-loop of the West Nile virus genome requires a specific sequence and structure for RNA synthesis but not for viral translation. Virology 331:375–386
Vanhaesebroeck B, Leevers SJ, Ahmadi K et al 2001 Synthesis and function of 3-phosphorylated inositol lipids. Annu Rev Biochem 70:535–602
Vaughn DW, Green S, Kalayanarooj S et al 2000 Dengue viremia titer, antibody response pattern, and virus serotype correlate with disease severity. J Infect Dis 181:2–9
Westaway EG, Mackenzie JM, Kenney MT, Jones MK, Khromykh AA 1997 Ultrastructure of Kunjin virus-infected cells: Colocalization of NS1 and NS3 with double-stranded RNA, and of NS2B with NS3, in virus-induced membrane structures. J Virol 71:6650–6661
Wu SJ, Grouard-Vogel G, Sun W et al 2000 Human skin Langerhans cells are targets of dengue virus infection. Nat Med 6:816–820
Yocupicio-Monroy RM, Medina F, Reyes-del Valle J, del Angel RM 2003 Cellular proteins from human monocytes bind to dengue 4 virus minus-strand 3′ untranslated region RNA. J Virol 77:3067–3076
You S, Falgout B, Markoff L, Padmanabhan R 2001 *In vitro* RNA synthesis from exogenous dengue viral RNA templates requires long range interactions netween 5′- and 3′-terminal regions that influence RNA structure. J Biol Chem 276:15581–15591
Zeng L, Falgout B, Markoff L 1998 Identification of specific nucleotide sequences within conserved 3′-SL in the dengue type 2 virus genome required for replication. J Virol 72:7510–7522
Zuker M 2003 Mfold web server for nucleic acid folding and hybridization prediction. Nucleic Acids Res 31:3406–3415

DISCUSSION

Padmanabhan: We also find in DENV-2 that removal of the nucleotides from the 3′ end affects RNA synthesis in the *in vitro* polymerase assays if an increasing number of nucleotides from five, 10 and 14 nucleotides are removed from the 3′-end of template RNA. This suggests that the terminal nucleotides play an essential role in viral RNA synthesis.

Harris: Pei-Yong Shi has done some work looking at the last six nucleotides in the WNV replicon system. They found that the last two must be C and U, and the next two are also important for RNA replication although the sequence can change as long as base-pairing is maintained. At the fifth position, the sequence is essential, but not at the sixth. Thus, five nucleotides are crucial but six aren't.

Canard: We have the crystal structure of the cap with the methyl transferase. It shows that the nature of the first nucleotide, an adenine, is absolutely crucial.

Gamarnik: We have made replicons, for example with GC at the beginning, which is easier for the cell polymerase, with the thought that we can make more RNA. It doesn't work. You said that the phosphorodiamidate morpholino oligomers (PMO) is targeting the 5′ stem loop, but which region of the loop does this target?

Harris: Essentially, there are two PMOs targeting the 5′ UTR. One starts immediately, with the first 20 nucleotides of the 5′ stem-loop. This completely knocks out activity. Then there is another one that is further downstream, over the AUG.

Screaton: How do you distinguish RNAi from PMOs.

Harris: PMOs are a different sort of structure. They're approximately 20 nucleotides long, with a chemically modified backbone. A lot of the initial PMO work was done by the companies that have created them. The PMO mechanism of action is completely distinct from RNAi.

Neyts: I have a comment on the pentanucleotide motif in the 3′ UTR. We discovered some time ago that in NKV flavivirus this pentanucleotide motif is not CACAG but can have a C or A at position 2 and an A at position 3. We are trying to find whether this has implications for vector specificity. There must be some evolutionary pressure to keep NKV flaviviruses with that particular nucleotide motif.

Harris: There could be a *trans* factor with which it is associating in the mosquito vectors. There is a consensus site for YB-1 binding right in that region as well, which overlaps the pentanucleotide motif and then goes into the base of the stem. This might be a binding site for the YB-1 protein.

Development of novel antivirals against flaviviruses

Chinmay G. Patkar and Richard J. Kuhn[1]

Department of Biological Sciences, Purdue University, 915 West State Street, West Lafayette, IN 47907-2054, USA

Abstract. Dengue virus is responsible for a significant amount of human disease in predominantly tropical areas of the world. Much effort has focused on the development of vaccines against the four serotypes of dengue, and within the next few years a vaccine is anticipated. Less progress has been made at developing antivirals that might reduce disease severity. Recent advances in the structural biology of dengue virus and other flaviviruses have opened new possibilities for the rational design of small molecule inhibitors of virus replication. This chapter describes the structural attributes of the dengue virion and how knowledge of its structure, assembly, and entry mechanisms are guiding new strategies toward the development of compounds that will interfere with the viral replication process.

2006 New treatment strategies for dengue and other flaviviral diseases. Wiley, Chichester (Novartis Foundation Symposium 277) p 41–56

The *Flaviviridae* family comprises a large family of viruses that are causative agents of severe and often fatal disease in humans and agriculturally important animals. *Flaviviridae* consists of three genera; *Flavivirus*, *Pestivirus* (bovine viral diarrhoea virus) and *Hepacivirus* (hepatitis C virus) (Burke & Monath 2001). The *Flavivirus* genus, the largest of the three genera containing more than 70 members, includes viruses such as yellow fever virus (YFV), dengue virus (DENV), West Nile virus (WNV), tick-borne encephalitis virus (TBEV) and Japanese encephalitis virus (JEV); the majority of which are transmitted between vertebrate hosts by insect vectors and cause severe human disease with diverse and complex pathologies that have a significant social and economic impact worldwide (Burke & Monath 2001). YFV has been responsible for numerous outbreaks within the USA in the 1700s and 1800s resulting in thousands of deaths (Foster et al 1998, Gubler 2002). The four serotypes of DENV (1–4) are the most important arthropod-borne human

[1]This paper was presented at the symposium by Richard J. Kuhn, to whom correspondence should be addressed.

viruses. The most serious manifestations of DENV infection, usually precipitated by sequential infection of multiple serotypes, are dengue haemorrhagic fever (DHF) and dengue shock syndrome (DSS), which together cause an estimated 50 million human infections per year (Gubler 2002). The recent introduction and spread of West Nile virus in the USA, a virus previously restricted to the Old World, has further highlighted the public health challenges posed by flaviviruses (Lanciotti et al 1999). The focus of this chapter will be to examine recent efforts to exploit structural biology for the development of new antivirals that will interfere with DENV and other flaviviruses.

The re-emergence of flaviviruses as significant human pathogens has spawned great interest in the development of new vaccines and antiviral agents. The development of a safe and effective vaccine strain was successful in eliminating YFV outbreaks in many countries but it continues to be a serious threat in the developing nations. Inactivated JEV and TBEV have been used as vaccines; however disease resulting from these viruses is still prominent worldwide. Development of vaccines for other flaviviruses has been a painstaking, and as yet an unsuccessful venture. The development of safe and effective vaccines against DENV poses additional problems. The phenomenon of antibody-dependent enhancement (ADE), in which protection against one DENV serotype increases the risk of DHF or DSS when the individual gets infected with another serotype, necessitates the need for the development of equally protective multivalent DENV vaccines (Gubler 2002). Likewise, antiviral drugs that are targeted against all the four serotypes remain to be developed as none are currently available. Candidate therapies include ribavirin, interferon-α2b and anti-virus immunoglobulin. Ribavirin has been shown to be active against many DNA and RNA viruses, including flaviviruses, but the effective concentration in cell culture is very high and it is largely ineffective in animal models (Jordan et al 2000, Morrey et al 2002). Interferon-α2b has been shown to have broad antiviral activity *in vitro* and immunostimulatory effects *in vivo*, and has been considered as a possible therapy for flaviviral encephalitis but clinical trials to date have been slow and not promising (Samuel 2001). Thus, antiviral agents with more potency must be identified. To that end, there has been an ongoing search for antiviral drugs undertaken mostly by random screening of compounds as inhibitors of virus infection. The most widely used antiviral assays were of a low-throughput nature and included monitoring compound inhibition of viral replication through quantification of cytopathic effects or quantification of viral RNA. Advances in our knowledge of the molecular biology of flaviviruses and the construction of flavivirus replicon systems have paved the way for the development of high-throughput screening of viral inhibitors. Also, recent progress in structural studies of flavivirus particles and viral proteins has provided the foundation for structure-based drug design and *in silico* screening of compounds prior to assaying them using virus infection or *in vitro* enzymatic assays. Structure-based drug design and virtual screening connect crystallography and NMR, computational advances in docking algorithms and virtual screening, and traditional techniques of high-throughput screening and combinatorial chemistry to increase the efficiency and speed of drug discovery. Although not initially identified by *in silico* screening, the 'WIN' compounds that act against rhinoviruses and enteroviruses were developed based on successive rounds of structure-based compound refinement. These compounds act by binding into the pocket of a canyon on the floor of the outer coat protein VP1 and stabilizing the outer coat of the virus, thus preventing uncoating of the virus, and hence infection of the host cells (Smith et al 1986, Fox et al 1986). The size of the WIN compounds may be adjusted to obtain the best binding into the canyon. We have undertaken a similar structure-based approach to design and test drugs against flaviviruses, specifically targeting the outer envelope glycoprotein (E protein) of the virus.

Flavivirus structure and assembly

Flaviviruses are small (~50 nm diameter), icosahedral, enveloped viruses containing a single strand of a positive-sense RNA genome. The genome is about 11 000 nucleotides in length, capped at the 5′ end but lacking a 3′ poly-A tract (Lindenbach & Rice 2001). The genome codes for a long polyprotein that is processed into individual proteins by both viral (NS2B-NS3) and cellular proteases. The N-terminal third of the polyprotein comprises the structural proteins—capsid (C), membrane (M, which is expressed as prM, precursor to M) and envelope (E)—that constitute the virus particle, and the rest of the polyprotein gives rise to the replicase proteins—NS1, NS2A, NS2B, NS3, NS4A, NS4B, NS5—that are involved in RNA replication and virus assembly (Lindenbach & Rice 2001). The C protein packages the viral genome and forms the nucleocapsid core. Surrounding this core is a host-derived lipid membrane in which are embedded 180 copies of the glycoproteins, M and E (arranged as 90 dimers), each containing two transmembrane helices (Lindenbach & Rice 2001).

Flaviviruses enter host cells by receptor-mediated endocytosis (Lindenbach & Rice 2001). The E protein contains a cellular-receptor binding site as well as functions as a fusion protein (Hung et al 1999, Crill & Roehrig 2001, Heinz & Allison 2000). No primary receptor for flaviviruses has been clearly established yet, although DC-SIGN, CD14-associated molecules, heparin and other glycosaminoglycans have been proposed to be receptors or attachment molecules (Tassaneetrithep et al 2003). In the endosome, the presence of low pH triggers a fusion event during which the E protein undergoes substantial conformational changes and is converted from a homodimer in the native virus to the fusogenic homotrimer (Allison et al 1995, Stiasny et al 1996). This oligomeric form of the E protein then proceeds to complete the fusion of the viral and host cell membranes. The nucleocapsid core is released into the cytoplasm where the viral genome is uncoated. Viral protein translation and genome replication commence in the cytoplasm. During translation, the polyprotein is translocated into the endoplasmic reticulum (ER) several times by various signal sequences and stop transfer sequences (Lindenbach & Rice 2001). The prM and E proteins are inserted into the lumen of the ER where they form a stable heterodimer (Lorenz et al 2002). Genome replication occurs in membrane bound vesicles where the newly synthesized genome RNA interacts with the C protein. Virus assembly proceeds with the concomitant interaction of RNA, C and envelope proteins promoting the budding of immature particles into the ER (Lindenbach & Rice 2001). The resulting particles, consisting of heterodimers of prM and E, are transported through the *trans*-Golgi network where prM is cleaved by furin to form mature virus particles, consisting of homodimers of E and M proteins (Stadler et al 1997). If prM cleavage by furin is prevented, non-infectious, immature virus particles are released (Elshuber et al 2003).

Flavivirus E protein

The atomic structures of the ectodomain (residues 1–395) of TBEV, DENV-2 and DENV-3 E proteins have been solved using X-ray crystallographic methods and are structurally very similar (Fig. 1) (Rey et al 1995, Modis et al 2003, 2005). The E protein ectodomain, minus the C-terminal ~100 residues consisting of the stem-anchor regions, displays three domains—domain I, the central domain that contains the N-terminus of the protein; domain II, the extended, finger-like 'dimerization' domain that makes most of the intra-dimeric contacts; and domain III, the immunoglobulin-like domain that is thought to be involved in receptor binding. Domains I and II are connected by four polypeptide chains, whereas

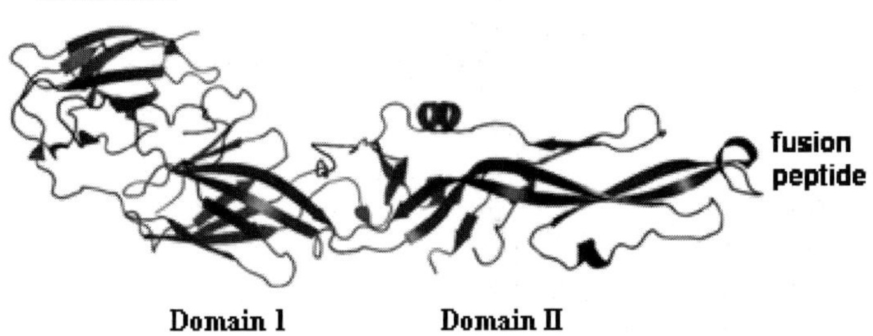

FIG. 1. Structure of the DENV-2 E protein ectodomain, residues 1–395 as determined using X-ray crystallography. The figure shows a monomer of the E protein in the neutral pH conformation. Domains I, II and III are as denoted. The fusion peptide at the distal tip of domain II is also labelled.

domains I and III are connected by a single polypeptide chain. At the distal tip of domain II is a small hydrophobic sequence, the so called fusion peptide, that initiates the fusion process (Allison et al 2001). The E protein structure was found to be remarkably similar to the structure of the E1 protein of Semliki Forest virus (SFV), an alphavirus belonging to the *Togaviridae* family (Lescar et al 2001). Unlike flaviviruses, the alphaviruses separate the functions of receptor binding and membrane fusion into two transmembrane glycoproteins termed E2 and E1, respectively. This distinctive three-domain arrangement of the E protein, shared by fusion proteins from these two viral families, is conserved because of a fundamental structural requirement of the Class II membrane fusion process (Kuhn et al 2002, Lescar et al 2001).

Recently, the structures of native DENV-2 and WNV virus were solved using cryo-electron microscopy (cryo-EM) and image reconstruction methods (Kuhn et al 2002, Mukhopadhyay et al 2003). The atomic structure of the E protein determined by X-ray crystallography can be computationally docked into the cryo-EM density to obtain a 'pseudo-atomic' structure of the virus. Fitting the atomic structure of the E protein into the DENV EM density revealed that there were 90 E dimers lying flat on the viral surface (Fig. 2). Surprisingly, they were not organized into the predicted conventional $T = 3$ icosahedral arrangement. Instead the E proteins can be grouped in sets of three, nearly parallel dimers with these sets arranged to form a 'herringbone' pattern on the viral surface (Kuhn et al 2002). There is a 21° angular difference between domains I and II in the crystal structure of the E protein compared with its position in the fitted structure of the virus particle

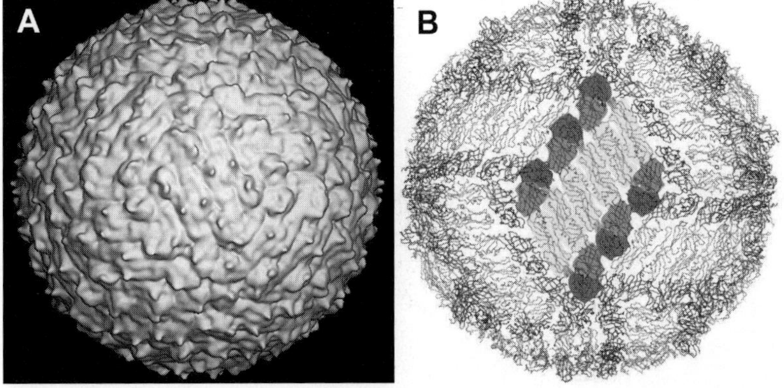

FIG. 2. (A) Surface-shaded view of the cryo-EM reconstruction of mature DENV-2 at 14 Å resolution. (B) The fit of the E protein structure into the cryo-EM density of the mature virus. One raft, consisting of three parallel dimers, is highlighted. The E protein is shaded as in Fig. 1.

(Mukhopadhyay et al 2005, Zhang et al 2005). It has been suggested that this arrangement of E dimers on the viral surface positions the protein in a metastable state that can be released and spring into a fusogenic trimer during entry (Stiasny et al 2001). During the low pH-induced fusion between the viral and cell membranes in the endosome, the E protein undergoes extensive conformational changes and the E homodimers dissociate into monomers and then re-associate to form fusion-competent homotrimers such that their fusion peptides are exposed at the tip of the trimer. The structure of the trimer of E proteins in the putative post-fusion state was solved using X-ray crystallography and shows domain III folded back by about 30 Å toward domain II and rotated ~20°. The domain II of each of the E proteins in the trimer lies parallel to each other with their fusion peptides exposed at the tip of the trimer (Modis et al 2004). The fusion peptides presumably penetrate the endosomal membrane and initiate the fusion process.

The structures of the immature forms of DENV-2 and YFV have also been solved using cryo-EM methods (Zhang et al 2003). The immature particles are dramatically different than the mature native virus. These immature virions are bigger (~60 nm) and present spike-like projections on the surface that are formed by trimers of prM-E heterodimers. Similar to the mature virion, these immature particles lack classical $T = 3$ symmetry. Fitting of the E protein atomic structure into the cryo-EM density revealed that the hinge angle between domains I and II is 6° different from the crystal structure and 27° different than the E fitted in the mature virus (Fig. 3) (Mukhopadhyay et al 2005, Zhang et al 2005). The three prM proteins, which were identified by subtracting the density of E proteins from the

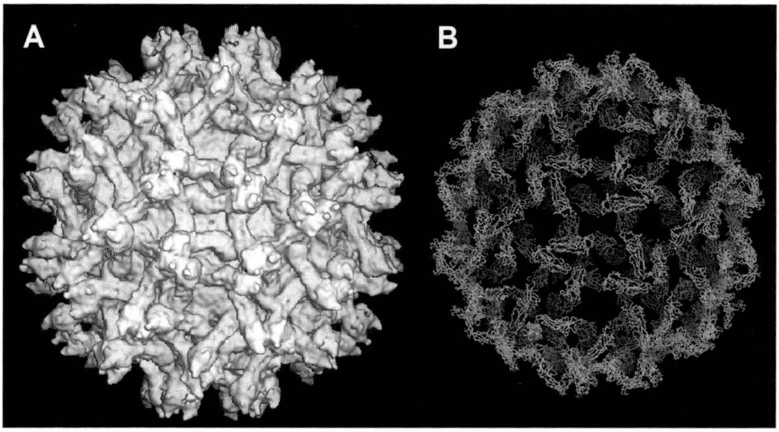

FIG. 3. (A) Surface-shaded view of the cryo-EM reconstruction of immature DENV-2 at 16 Å resolution. (B) Arrangement of the E protein structure in the immature virus particle obtained by fitting the E protein structure into the EM density for the immature particle. The E protein is shaded as in Fig. 1.

total immature density, cover the fusion peptide of the three E proteins in the spike in order to prevent premature fusion. During late maturation steps in the Golgi, prM is cleaved by furin leading to a rearrangement of the E proteins from prM-E heterodimers that are projecting from the viral surface to E homodimers that lie flat on the viral surface. Thus, the E protein undergoes dramatic conformational changes during assembly of the virus particle and during fusion. During this conversion the E protein undergoes a ~30° shift in the hinge region between domains I and II.

The DENV-2 E protein structure solved at Harvard University by Modis and Harrison was determined both with and without the detergent, *n*-octyl-β-D-glucoside (BOG), bound to the E protein (Modis et al 2003). This molecule was found in a hydrophobic pocket that lies in the region between domains I and II that acts as a 'hinge' to promote domain flexing. The E protein has been shown to undergo significant conformational changes in this region during membrane fusion as well as in maturation of the virus particle. Mutations in various flaviviruses that affect virulence and alter the pH threshold required for fusion were mapped to this region between domains I and II (Rey et al 1995). The molecular mechanisms that lead to membrane fusion are not yet completely understood, but the initial events must involve pH-mediated conformational changes resulting in dimer dissociation. The BOG binding may have stabilized the E protein dimer in a certain conformation making the dimer amenable for crystal formation. The DENV-2 E protein structure solved independently at Purdue University by Rossmann and colleagues was similar to the structure solved by Modis and Harrison except that there was a hinge motion between domain I and II of ~10° and lacked the presence of BOG even though it was present in the crystallization buffer (Zhang et al 2005). These data indicate that the flexible hinge region is involved in the numerous conformational changes necessary for membrane fusion and during transition from the immature to the mature viral form. This hinge region thus presents an appealing binding site for compounds that could exhibit anti-viral activity by interfering with the various transitions that the E protein undergoes. Furthermore, sequence analysis of the E proteins from flaviviruses reveals ~40% amino acid identity and these proteins are predicted to be very similar in their overall fold and domain arrangement including the hinge region (Burke & Monath 2001). Thus, there is the possibility for the development of a compound that acts as an inhibitor against a broad spectrum of pathogenic flaviviruses.

Antiviral compound screening

Based on assessment of this data, a comprehensive *in silico* screening of three small compound libraries from the National Cancer Institute against the BOG binding site in the pocket was performed (Zhou & Post, manuscript in preparation). A total of 200,000 ligands were screened by docking them into the BOG binding site. After a rigorous hierarchical and iterative process the lowest-energy conformers were selected. Ultimately, based on visual inspection and consideration of drug-like properties, 23 compounds were selected for biological screening against virus infectivity.

To establish a reliable and robust screen, a luciferase reporter-based assay was developed for investigating the effects of the compounds on virus growth. The YFV genome was modified by inserting a fire-fly luciferase gene at the 3′ end of the genome (in the full-length YFV cDNA plasmid pACNR-FLYF). The expression of the luciferase gene was controlled using an internal ribosomal entry site (IRES) obtained from the encephalomyocarditis virus (EMCV) genome (Fig. 4).

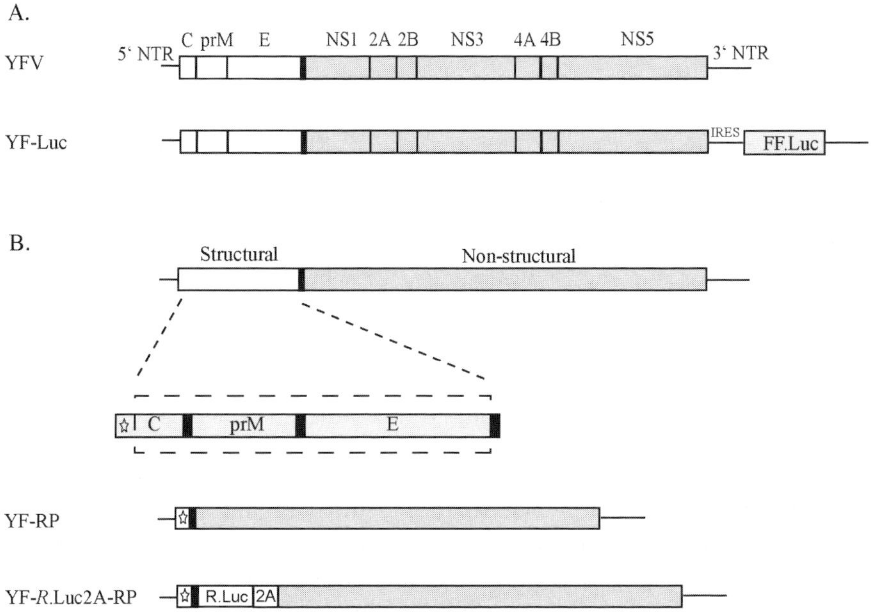

FIG. 4. Schematic representations of the YFV, YF-Luc and YF replicon genomes. Straight lines depict non-coding regions (NTR); boxes depict coding sequences—white (structural proteins), grey (non-structural proteins) and black (signal sequences). (A) Schematic of the YFV genome; genes coding for the structural and non-structural proteins are shown. A fire-fly luciferase gene was inserted at the 3′end of the coding sequence and expressed using an Internal Ribosome Entry Site (IRES) to obtain the genome for the YF-Luc virus. (B) The replicon (YF-RP) was constructed by deletion of the structural protein coding sequence (shown enlarged). The cyclization sequence (☆) was retained to allow genome replication. A *Renilla* luciferase gene was inserted before the signal sequence for NS1. The sequence coding for the autoproteolytic FMDV-2A site was fused to the 3′ end of the luciferase gene and thus, the *Renilla* luciferase could be cleaved off from the growing polyprotein during genome translation.

The virus produced after transfection of *in vitro* transcribed RNA from this cDNA containing the luciferase gene was termed YF-Luc. Growth of YF-Luc could be determined as a measure of the luciferase activity in infected cells. The inhibitory activity of the compounds could be ascertained by the reduction in luciferase activity of YF-Luc infected cells in the presence of active compound. A YFV replicon (YF-RP) was also developed by deleting most of the coding region for the structural proteins from the YFV cDNA and retaining the genes coding for the replicase proteins. After *in vitro* transcription and transfection of RNA, this replicon could independently replicate in cells since it possessed all the signals and could produce the replicase proteins required for replication. However, it could not produce progeny viruses since it lacked genes coding for the structural proteins required for formation of virus particles. A *Renilla* luciferase gene was inserted at the 5′ end of the replicon to yield YF-*R*.Luc2A-RP, and replication of this replicon could now be monitored by measuring luciferase activity in cells transfected with YF-*R*.Luc2A-RP (Figure 4). Thus, we could now test the activity of compounds specifically against YFV genome replication and not virus production using YF-*R*.Luc2A-RP.

Initially, standard cytotoxicity tests were performed using an XTT-based cytotoxicity assay kit (BioVision Inc.) and non-cytotoxic concentrations of the compounds were used for inhibition assays. The assay for screening compounds for activity against YFV growth consisted of infecting baby hamster kidney (BHK) cells with YF-Luc virus at a low multiplicity of infection (MOI = 0.1) followed by addition of the compounds at non-cytotoxic concentrations. A low MOI was employed so as to allow for re-infection of virus and thus gauge the effect of compounds on virus spread. At 36 hours post-infection, cell extracts were taken and luciferase assays performed. From the 23 compounds tested, five showed inhibitory activity against YF-Luc. These compounds that exhibited inhibitory activity against YF-Luc virus were tested for activity against YF-*R*.Luc2A-RP to determine whether the compounds were affecting genomic RNA replication as opposed to virus entry, assembly or exit. This assay required the transfection of BHK cells with *in vitro* transcribed YF-*R*.Luc2A-RP RNA and treatment of transfected cells with the compounds at concentrations that showed inhibitory activity against YF-Luc virus. At 36 hours post-infection, cell extracts were taken and luciferase assays performed. Out of the five compounds tested, four showed inhibition of YF-*R*.Luc2A-RP, indicating that the compounds are inhibiting steps during viral genome replication. Interestingly, three out of these four compounds were found to be active against YF-Luc virus at concentrations lower than those required for inhibition of YF-*R*.Luc2A-RP replication, suggesting that these compounds are affecting some steps other than genome replication.

Assays to determine whether the compounds affect steps during entry of the virus into cells, or during assembly and exit from the infected cells are in progress.

Based on our hypothesis, one would predict that the compounds are mimicking BOG and bind in the hydrophobic pocket on the E protein. This binding might prevent conformational changes that are required of the E protein during fusion. In order to substantiate the idea, cell–cell fusion assays of infected cells in presence of the compounds at low pH will have to be performed. The assay involves treating infected cells with low pH buffer in the presence of the compounds and then assaying for syncytia formation and comparing the efficiency to syncytia formation in untreated cells. However, flaviviruses bud from internal membranes and E protein is not found in substantial amounts on the plasma membrane of infected cells. A more direct fusion assay will be to purify pyrene labelled virus and test fusion efficiency in presence of the compounds in an *in vitro* fusion assay with liposomes. NMR and crystallographic methods can also be pursued to provide direct evidence of compound binding to the E protein. Similar assays will be employed to check whether the compounds have inhibitory activity against other flaviviruses such as DENV and WNV. Furthermore, using similar *in silico* molecule docking techniques using other target sites on the E protein, additional compounds have been identified and will be screened for inhibitors using the assays outlined above.

In summary, we have undertaken a structure-based approach for the rational design of small molecule inhibitors of flaviviruses, specifically directed against the outer envelope glycoprotein. Preliminary assays screening compounds for activity against YFV infection have been performed and the results are encouraging. Further assays to determine the exact site of action of the compounds have been developed and are in progress. Thus, using a powerful combination of structure-based drug design and informative biochemical assays, the goal is to identify novel compounds that selectively inhibit the target site and possess a desirable pharmacological profile that can be developed into potent inhibitors of flavivirus infection.

Acknowledgements

This work was supported by a Public Health Service Program Project Grant (AI55672) from the National Institute of Allergy and Infectious Disease. The work was also funded as part of the NIH NIAID Great Lakes RCE (AI57153).

References

Allison SL, Schalich J, Stiasny K, Mandl CW, Kunz C, Heinz FX 1995 Oligomeric rearrangement of tick-borne encephalitis virus envelope proteins induced by an acidic pH. J Virol 69:695–700

Allison SL, Schalich J, Stiasny K, Mandl CW, Heinz FX 2001 Mutational evidence for an internal fusion peptide in flavivirus envelope protein E. J Virol 75:4268–4275

Burke DS, Monath TP 2001 Flaviviruses. Knipe DM, Howley PM Fields Virology, 4th edn. Lippincott Williams and Wilkins, Philadelphia p 1043–1125

Crill WD, Roehrig JT 2001 Monoclonal antibodies that bind to domain III of Dengue E glycoprotein are the most efficient blockers of virus adsorption to vero cells. J Virol 75:7769–7773

Elshuber S, Allison SL, Heinz FX, Mandl CW 2003 Cleavage of protein prM is necessary for infection of BHK-21 cells by tick-borne encephalitis virus. J Gen Virol 84: 183–191

Foster KR, Jenkins MF, Toogood AC 1998 The Philadelphia yellow fever epidemic of 1793. Sci Am 279:88–93

Fox MP, Otto MJ, McKinlay MA 1986 The prevention of rhinovirus and poliovirus uncoating by WIN 51711: a new antiviral drug. Antimicrob Agents Chemother 30:110–116

Gubler DJ 2002 Epidemic Dengue/Dengue hemorrhagic fever as a public health, social and economic problem in the 21st century. Trends in Microbiol 10:100–103

Heinz FX, Allison SL 2000 Structures and mechanisms in flavivirus fusion. Adv Virus Res 55:231–269

Hung SL, Lee PL, Chen HW et al 1999 Analysis of the steps involved in Dengue virus entry into host cells. Virology 257:156–167

Jordan IBT, Fischer N, Lau JY, Lipkin WI 2000 Ribavirin inhibits West Nile virus replication and cytopathic effect in neural cells. J Infect Dis 182:1214–1217

Kuhn RJ, Zhang W, Rossmann MG et al 2002 Structure of dengue virus: implications for flavivirus organization, maturation and fusion. Cell 108:717–725

Lanciotti RS, Roehrig JT, Deubel V et al 1999 Origin of the West Nile virus responsible for an outbreak of encephalitis in the northeastern United States. Science 286:2333–2337

Lescar J, Roussel A, Wein MW et al 2001 The fusion glycoprotein shell of Semliki Forest virus: an icosahedral assembly primed for fusogenic activation at endosomal pH. Cell 105:137–148

Lindenbach BD, Rice CM 2001 Flaviviridae: The viruses and their replication. D. M. Knipe and P. M. Howley. Fields Virology, 4th edn. Lippincott Williams & Wilkins, Philadelphia p 991–1041

Lorenz IC, Allison SL, Heinz FX, Helenius A 2002 Folding and dimerization of tick-borne encephalitis virus envelope proteins prM and E in the endoplasmic reticulum. J Virol 76:5480–5491

Modis Y, Ogata S, Clements D, Harrison SC 2003 A ligand-binding pocket in the dengue virus envelope glycoprotein. Proc Natl Acad Sci USA 100:6986–6991

Modis Y, Ogata S, Clements D, Harrison SC 2004 Structure of the Dengue virus envelope protein after fusion. Nature 427:313–319

Modis Y, Ogata S, Clements D, Harrison SC 2005 Variable surface epitopes in the crystal structure of dengue virus type 3 envelope glyoprotein. J Virol 79:1223–1231

Morrey JD, Smee DF, Sidwell RW, Tseng C 2002 Identification of active antiviral compounds against a New York isolate of West Nile virus. Antiviral Res 55:107–116

Mukhopadhyay S, Kim B-S, Chipman PR, Rossmann MG, Kuhn RJ 2003 Structure of West Nile virus. Science 303:248

Mukhopadhyay S, Kuhn RJ, Rossmann MG 2005 A structural perspective of the flavivirus life cycle. Nat Rev Microbiol 3:13–22

Rey FA, Heinz FX, Mandl C, Kunz C, Harrison SC 1995 The envelope glycoprotein from tick-borne encephalitis virus at 2Å resolution. Nature (London) 375:291–298

Samuel CE 2001 Antiviral actions of interferons. Clin Microbiol Rev. 55:778–809

Smith TJ, Kremer MJ, Luo M et al 1986 The site of attachment in human rhinovirus 14 for antiviral agents that inhibit uncoating. Science 233:1286–1293

Stadler K, Allison SL, Schalich J, Heinz FX 1997 Proteolytic activation of tick-borne encephalitis virus by furin. J Virol 71:8475–8481

Stiasny K, Allison SL, Marchler-Bauer A, Kunz C, Heinz FX 1996 Structural requirements for low-pH-induced rearrangements in the envelope glycoprotein of tick-borne encephalitis virus. J Virol 70:8142–8147

Stiasny K, Allison SL, Mandl CW, Heinz FX 2001 Role of metastability and acidic pH in membrane fusion by tick-borne Encephalitis virus. J Virol 75:7392–7398

Tassaneetrithep B, Burgess TH, Granelli-Piperno A et al 2003 DC-SIGN (CD209) mediates dengue virus infection of human dendritic cells. J Exp Med 197:823–829

Zhang Y, Corver J, Chipman PR et al 2003 Structures of immature flavivirus particles. EMBO J 22:2604–2613

Zhang YW, Zhang S, Ogata D et al 2005 Conformational changes of the flavivirus E glycoprotein. Structure 12:1607–1618

DISCUSSION

Harris: The pictures of the prM structure are interesting. Why does the virus pass through the prM? It is as if it is going through a whole structure that is capped and then cleaving it, but when it comes around into the entry and fusion, it is a completely different structure that is made to push the fusion peptides into the form for fusion. Yet when you see it in the assembly route, there is fusion and it is capped, but it is presumably not used. Or is it part of the secretory pathway? I'm confused by this.

Kuhn: That's a great question I don't know if I can answer it. Protein folding must take place. On the basis of data from ourselves and others, it is unclear to me what role prM might play in folding. People have been able to express E, and it seems to be reasonably folded in the absence of prM. The early discussion about prM being a chaperone like the alphaviruses is now looking not so clear. In the context of the whole virus particle, with all these pieces lined up in a certain way, then perhaps you do need prM for it to fold correctly. You create a structure that will not fuse prematurely. The driving force behind the rearrangement of this structure is unknown.

Heinz: One of the important functions of the prM protein is to protect the E protein from undergoing those conformational changes already in the acidic *trans*-Golgi network that are necessary for fusion in the endosome. Several years ago we showed that the cleavage of the prM protein also requires an irreversible conformational change in the prM–E complex that is induced by the acidic pH in the *trans*-Golgi. There are two sites in the life cycle of flaviviruses where irreversible conformational changes are induced by acidic pH: one in the *trans*-Golgi for cleavage of the prM protein and the other for fusion in the endosome.

Gamarnik: Which sugar moiety binds to DC-SIGN?

Kuhn: From the cryo-EM, we see that 67 from the icosahedral and the adjacent E come together with DC-SIGN binding between them. The geometry between those two glycosylation units has to be exact, and we don't have this for 153.

Gamarnik: When you say that the fusion peptide is covered by a sugar moiety, is this the 67 or the 153?

Kuhn: The 153.

Screaton: I was interested in the structure you showed of the DC-SIGN. Have you looked to see whether that soluble recombinant protein you made blocks infection?

Kuhn: No.

Screaton: Has that asparagine been mutated?

Kuhn: Work from John Roehrig (Johnson et al 1994) says that that is a conserved glyscosylated site in dengue virus. You can mutate it. DC-SIGN is not absolutely required for entry.

Screaton: Absolutely. I once tried to explore the possibility that DC-SIGN is a molecule for concentrating virus. It will be interesting to see whether the virus you have made that is coated with DC-SIGN would still be infectious, if that were the mechanism. Also, perhaps a virus in which that asparagine is mutated would still be infectious.

Jans: What is your view on the merits of *in silico* screening? It seems to me that one commonly comes up with compounds that inhibit unrelated molecules much better than one's real target.

Kuhn: I'd emphasize that we don't really know what the target is. We have a number of biophysical experiments and we are looking to see whether the compounds are binding to virus and E, and also whether this is actually the important step. The group of people we have is an interesting mix. The structural biologist has concerns about the results from the *in silico* screening but he believes that there is activity, so he likes to think that we have some compounds that work. The medicinal chemist will say that so many of these have not been 'on target'. There is a question of how many positives people have identified by these methods, especially where we are taking a cavity and trying to fit something in there. If it fits in the cavity it must be a strong enough association that it doesn't pop out when these things start moving. It is a coordinated movement among 180 proteins that we are studying.

Keller: Our view on virtual screening is that it is improving, but slowly. Our experience is that virtual screening is quite good at fitting compounds into the active site, but it is bad at telling us which one is a good and which is a bad inhibitor. Novartis hasn't stopped doing high-throughput screening (HTS), which should tell you a lot. We would have stopped this very expensive technology, if virtual screening was a highly successful technique.

Kuhn: From our perspective this is virtually a no-cost screening technique. We don't have the compound libraries for HTS. In interacting with the people who do HTS, we have been trying to get an idea from them about the sorts of compounds that work best. This is hard to do other than by looking at the structure and then interpreting how that molecule might fit in the site.

Keller: It is important that we use virtual screening and try to give feedback to the computational chemists to improve the force fields. Currently, most force fields are neglecting entropic contributions to the free energy of binding. So scientifically it is not very surprising that the success of virtual screening has been limited.

Kuhn: I would just like to see whether any of these compounds are actually there.

Young: I may have missed it, but with the compounds pulled out of the *in silico* screens which were tested in replicon and reporter systems, have you tested those against infectious virus?

Kuhn: Yes, we screened the virus first and then went to the replicon, and separated out RNA synthesis and translation from subsequent assembly and entry steps.

Zinkernagel: What is the symmetry of the neutralizing determinant? Is it twofold or threefold? What does your neutralizing antibody bind to?

Kuhn: We have better data from West Nile virus on this. It depends on the antibody. It seems there is a relationship between neutralizing ability and the organization of where you bind. We have some antibodies that have complete saturation: they have 180 binding sites and we bind 180 FAb fragments with them. We have others where we only bind 60 out of 180, because of steric and other considerations. There seems to be a relationship between how many are bound, where binding occurs and the block.

Zinkernagel: If you go through a range of viruses, for many it is a trimer. Dimers aren't so common.

Kuhn: I agree that this is quite unusual. It is going from a trimer to a dimer, and back to a trimer again. We would say that the antibodies are blocking some of those transitions steps. They aren't necessarily binding trimers or dimers.

Flamand: If I understood correctly from your primary reconstructions of the virus particles there seem to be some regions where the protein shell does not cover the lipid membrane entirely.

Kuhn: In the mature virus there is a complete saturation of the surface. The crystal structure of the E protein dimer shows openings or holes between the two molecules. These are then filled by M, which comes through the holes. The problem is that in our cryo reconstructions we can see the density approaching the hole, but after the hole there are about 15–20 amino acids that we still don't see, and which are probably in different positions depending on which molecule you are looking at. Those holes are filled.

Screaton: Bearing in mind the structure of the transition that occurs, how rigid do you think the coat is? Potentially, there might be some epitopes here for which you only need one antibody to trap a whole virus.

Kuhn: In the cryo-EM, which uses icosahedral averaging to get the highest resolution, when you see these transmembrane domains they are not moving. They are rigid. We get high resolution data from them, indicating they are stuck in place because the icosahedral lattice on the outside must be fairly rigid.

Screaton: So it conceivably could trap a whole virus.

Lescar: It has been suggested that domain III of the flavivirus envelope E protein is a major factor in binding the receptors at the surface of host cells (see e.g. Chu et al 2005). Your data with DC-SIGN shows that domain III is not binding to DC-SIGN.

Kuhn: That's an interesting point. Even if you have the tetramer binding, presumably domain III is still going to be accessible. There is not a lot of surface occupied by DC-SIGN. One possibility is that around the fivefold axis, where there are five domain IIIs that stick up, there are five DC-SIGN binding sites around that perimeter. They might hold it close to the membrane and the true protein receptor is binding to that fivefold axis around five domain IIIs, which still should be accessible.

Stephenson: Going back to the conformational screen, when you selected the compounds from that screen, did you assay those directly or did you first give them to the medicinal chemists to improve them?

Kuhn: We assayed them directly.

Stephenson: You should give them to the chemists first. In a totally different field we have been looking at therapeutic compounds for prions. When we do the conformational screen the compounds coming out are relatively poor. However, when we give them to the medicinal chemists we get dramatic improvements on biological activity.

Kuhn: We need more medicinal chemists!

Harris: I am struck by the dominant neutralizing epitope on West Nile. Things always seem to me to be much clearer in West Nile than in dengue. Is there really a dominant neutralizing epitope in dengue?

Kuhn: I don't know. We have lots of neutralizing antibodies against dengue, and these are against a variety of different sites.

Gu: Some people reported that cholesterol-rich domains are important in Dengue virus entry. What is your opinion on this?

Kuhn: I would like to think that endosomal entry is the dominant entry mechanism.

Gu: It seemed that it is even possible to block the clathrin-mediated receptor endocytosis and still have the dengue virus in the cell.

Kuhn: Perhaps there are different paths for getting into the endosome. The virus has evolved an elaborate and very efficient mechanism for getting into the cell. At some point you will go through a low pH intermediate. It is just such an efficient operation. It may not involve just a single pathway.

References

Chu J, Rajamanonmani R, Li J, Bhuvanakantham R, Lescar J, Ng M-L 2005 Inhibition of West Nile virus entry by using a recombinant Domain III from the envelope glycoprotein. J Gen Virol 86:405–412

Johnson AJ, Guirakhoo F, Roehrig JT 1994 The envelope glycoproteins of dengue 1 and dengue 2 viruses grown in mosquito cells differ in their utilization of potential glycosylation sites. Virology 203:241–249

Entry functions and antigenic structure of flavivirus envelope proteins

Karin Stiasny, Stefan Kiermayr and Franz X. Heinz[1]

Institute of Virology, Medical University of Vienna, Kinderspitalgasse 15, A1095 Vienna, Austria

Abstract. The envelope proteins (E) of flaviviruses form an icosahedral cage-like structure of homodimers that cover completely the surface of mature virions and are responsible for receptor-binding and membrane fusion. Fusion is triggered by the acidic pH in endosomes which induces dramatic conformational changes of E that drive the merger of the membranes. We have identified an alternative trigger that induces the first phase of the fusion process only, but then leads to an arrest at an intermediate stage. These data suggest that the early and late stages of flavivirus fusion are differentially controlled by intersubunit and intrasubunit constraints of the fusion protein, respectively. Details of the molecular antigenic structure of the flavivirus E protein were revealed by the use of neutralization escape mutants as well as recombinant expression systems for the generation of virus-like particles. The experimental data provide evidence that each of the three domains contributing to the external face of the E protein can induce and bind neutralizing antibodies. Broadly flavivirus cross-reactive antibodies, however, primarily recognize a site involving residues of the highly conserved fusion peptide loop which is cryptic and largely inaccessible on the surface of native infectious virions.

2006 New treatment strategies for dengue and other flaviviral diseases. Wiley, Chichester (Novartis Foundation Symposium 277) p 57–73

Flaviviruses are small enveloped viruses that form a genus in the family *Flaviviridae* (Heinz et al 2003) and comprise a number of arthropod-transmitted human pathogens such as yellow fever (YF) virus, the dengue (DEN) viruses, Japanese encephalitis (JE) virus, West Nile (WN) virus, and tick-borne encephalitis (TBE) virus. Their positive-stranded RNA genome (about 11 kb) contains a single long open reading frame encoding three structural proteins (capsid protein, C; precursor of membrane protein, prM; membrane protein, M; envelope protein, E) and seven non-structural proteins (Lindenbach & Rice 2003). Flavivirus assembly takes place at the endoplasmic reticulum and first leads to the formation of immature, prM-

[1]This paper was presented at the symposium by Franz Heinz, to whom all correspondence should be addressed

containing non-infectious particles that are transported through the exocytotic pathway (Lindenbach & Rice 2003). In the *trans*-Golgi network the prM protein is proteolytically cleaved by furin or a related cellular protease (Stadler et al 1997) resulting in the formation of mature infectious virions that are released from infected cells by exocytosis. The structural details of the molecular organization of flavivirus particles were revealed by X-ray crystallography of the E protein (Rey et al 1995, Modis et al 2003, 2005, Zhang et al 2004), NMR spectroscopy and X-ray crystallography of the C protein (Jones et al 2003, Dokland et al 2004), and cryo-electron microscopy of purified mature and immature virions (Kuhn et al 2002, Mukhopadhyay et al 2003, Zhang et al 2003b) as well as recombinant sub-viral particles (Ferlenghi et al 2001). Figure 1 shows schematics of the flavivirus particle organization and the atomic structure of the E protein as present in mature virions. Immature virions are studded with icosahedrically arranged spikes, each of which is composed of three heterodimers of prM and E. The maturation cleavage of prM results in a still poorly understood rearrangement of E proteins at the virion surface leading to the formation of smooth-surfaced particles that carry 90 tightly packed E dimers in a 'herringbone'-like icosahedral arrangement (Kuhn et al 2002, Mukhopadhyay et al 2003) (Fig. 1). In these mature virions the E protein forms a head-to-tail dimer that is oriented parallel to the viral surface and integrated in the membrane at its C-terminus by a double membrane-spanning anchor. A sequence element of about 50 amino acids (the so-called stem; Fig. 1D) connects to the C-terminus of the ectodomain fragment (Zhang et al 2003a) that has been used for determining high resolution structures of the TBE, dengue 2 and dengue 3 virus E proteins by X-ray crystallography (Rey et al 1995, Modis et al 2003, 2005, Zhang et al 2004). Although the crystallized dengue and TBE virus E proteins share only 37% of their amino acids, their overall structures are virtually identical. Each of the monomeric subunits contains three distinct domains (I, II, and III) which are dominated by β-sheet secondary structures (Fig. 1E, F). Most flavivirus E proteins have a single glycosylation site in domain I, in some instances a second site can be present in domain II (Modis et al 2003, 2005) and in certain virus strains the E protein is non-glycosylated (Beasley et al 2004). The E protein is bifunctional; it mediates both attachment to cells and low-pH-triggered membrane fusion after uptake by receptor-mediated endocytosis and is thus the major determinant for the induction of virus-neutralizing antibodies and a protective immunity. Because the flavivirus E and the alphavirus E1 fusion proteins are structurally radically different from the spike-like fusion proteins found in orthomyxo-, paramyxo-, retro-, filo- and coronaviruses, these are now referred to as class II and class I fusion proteins, respectively (Lescar et al 2001). Our work focuses on the molecular mechanism of membrane fusion as a potential target for the development of antiviral agents and the molecular antigenic structure of flaviviruses and its relevance for the induction of a protective immunity.

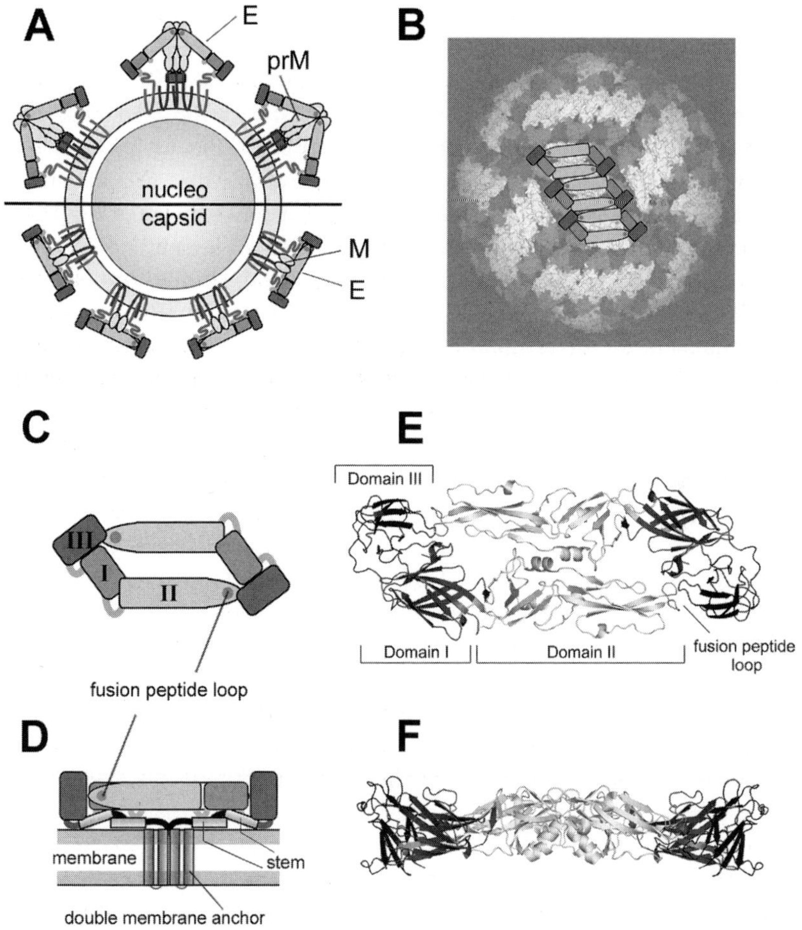

FIG. 1. Structural organization of flavivirus particles. (A) Schematic of immature (top) and mature (bottom) virions. (B) Image of mature virus particle as determined by cryoelectron microscopy (Kuhn et al 2002, Mukhopadhyay et al 2003). One of the rafts consisting of three parallel E homodimers is highlighted schematically using the E dimer representation shown in Fig. 1C. (C–F) Schematics (C, D) and structures (E, F) of E dimers viewed from the top (C, E) and the side (D, F), respectively. The structural element designated 'stem' in D is described in the text. E, F exhibit ribbon diagrams of the atomic structure of a soluble fragment (lacking the stem-anchor region) of TBE virus.

Mechanism of membrane fusion

The fusion of flaviviruses with membranes is strictly dependent on low pH, consistent with the cell entry by receptor-mediated endocytosis. The pH threshold of fusion has been shown to be around pH 6.4, suggesting that this process already occurs in early endosomes. *In vitro* studies on the fusion of TBE virus with liposomes have revealed that this process is extremely efficient, does not require an interaction with specific receptors and proceeds significantly faster than that mediated by class I viral fusion proteins, with $t_{1/2}$ of a few seconds as compared to several minutes, respectively (Corver et al 2000, Melikyan et al 2000, Gallo et al 2001). In contrast to what is known from alphaviruses (Kielian et al 2000), which also possess class II fusion proteins, flavivirus fusion does not have an absolute requirement for the presence of cholesterol or sphingomyelin in the target membrane (Corver et al 2000, Stiasny et al 2003). However, using liposome preparations of different lipid compositions, Stiansy et al (2003) showed that specific interactions involving the 3′-OH group of cholesterol facilitates the binding of the E protein to target membranes as well as the structural transitions necessary for fusion.

Like class I fusion proteins, the class II flavivirus fusion protein exists in a metastable conformation at the surface of mature virions and undergoes dramatic structural changes when it encounters the low pH fusion trigger. As revealed by studies with TBE virus, acidic pH leads to the dissociation of the E dimers (Stiasny et al 1996), the exposure of the internal fusion peptide loop at the tip of domain II (Allison et al 2001, Stiasny et al 2002), and an oligomeric transition into E trimers (the post-fusion structure) (Allison et al 1995) that are energetically more stable than the E dimers (Stiasny et al 2001). The energy released during these transitions is believed to be used for driving the merger of the two opposing membranes. So far it has not been possible to crystallize the full-length form of the E trimer. The soluble forms of TBE and dengue E dimers, however, could be converted into a crystallizable trimeric post-fusion form through the interaction with liposomal membranes at acidic pH (Stiasny et al 2004, Modis et al 2004) and their structures were determined by X-ray crystallography (Bressanelli et al 2004, Modis et al 2004). Apart from the reorientation of the molecule relative to the membrane from horizontal in the virion to perpendicular in the fused membrane, the most prominent difference is a change in the position of domain III relative to the other domains through its movement from the end of the monomer to the side next to domain II (Fig. 2). Although these structures lack the 'stem' element and the two trans-membrane anchors, the arrangement of the three domains and modelling experiments (Bressanelli et al 2004) suggest that the post-fusion structure of E has a hairpin-like organization (similar to that of class I fusion proteins) in which the fusion peptide loop and the membrane anchors are juxtaposed in the fused membranes (Fig. 2C). Based on the atomic structures of E in its dimeric pre- and trimeric post-fusion forms (Rey et al 1995, Modis et al 2003, 2004, 2005, Zhang et al 2004, Bressanelli et al 2004) as well as a number of biochemical and functional studies (Allison et al 1995, 2001, Stiasny et al 1996, 2001, 2002), models of the flavivirus fusion mechanism were proposed consisting of several steps

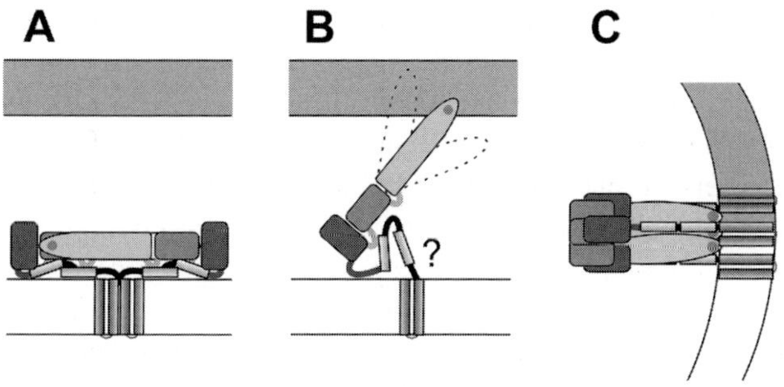

FIG. 2. Schematic diagrams (compare with Fig. 1 C, D) showing the structural changes in the E protein during the fusion process. (A) The dimeric E protein in its native state at the surface of the mature virion. (B) Monomeric E interacting with a target membrane via its fusion peptide loop. Possible hinge movements at the junction between domains II and I are indicated by dotted lines. (C) Formation of the final post-fusion conformation trough relocation of domain III and interactions of the stem-anchor region with domain II, leading to the juxtaposition of the fusion peptide loops and the membrane anchors in the fused membrane.

including (i) the dissociation of the E dimer resulting in the exposure of the previously buried fusion peptide loop, (ii) the attachment of the E protein to a target membrane via the fusion peptide loop, and (iii) the trimerization of E and the formation of a 'hairpin' structure resulting in the merger of the two membranes (Fig. 2).

Because the flavivirus fusion machinery is extremely fast and efficient it has not yet been possible to define structural intermediates of the E protein during its transition into the post-fusion conformation by the use of specifically designed peptide inhibitors, a technology that has been applied successfully to viruses with class I fusion proteins (Earp et al 2005, Matthews et al 2004). In experiments with TBE virus, however, new experimental information about structural intermediates of the flavivirus fusion process was obtained through the exposure of virions to alkaline pH (Stiasny et al 2006a). It was shown that under these conditions the icosahedral envelope organization is opened and the E dimers dissociate into their monomeric constituents (Fig. 2). Similar to what occurs under physiological (acid pH) conditions, the exposure of the fusion peptide and the apparent outward projection of E monomers at alkaline pH lead to their stable attachment to target membranes. Under these conditions, however, the process is arrested at this intermediate stage, and neither fusion activity nor E trimer formation can be observed, suggesting that the domain relocation necessary for hairpin-formation does not occur spontaneously upon dimer dissociation but requires specific protonation events in the monomers.

Molecular antigenic structure

Sites involved in virus neutralization

Since the flavivirus E protein has the dual function of receptor binding and membrane fusion it is the principle target of neutralizing antibodies. A significant degree of information on the binding sites for such antibodies has become available, primarily by the mapping of mutations that lead to escape from neutralizing mouse monoclonal antibodies (Mabs) and, more recently, by the structure determination of protein E–Fab complexes (Oliphant et al 2005) and the specific mutagenesis of E in recombinant forms (Kiermayr et al, in preparation). Figure 3 shows a summary of the positions of single amino acid substitutions found in Mab escape mutants of different flaviviruses. Although there are only limited sets of data for individual flaviviruses, this compilation suggests that the binding of antibodies to sites at the exposed outer surface of each of the three domains can lead to neutralization. Such binding sites can be restricted to individual domains only (e.g. a Fab interacting with domain III of the West Nile virus E protein has been defined in its atomic details), but site-specific mutagenesis experiments with

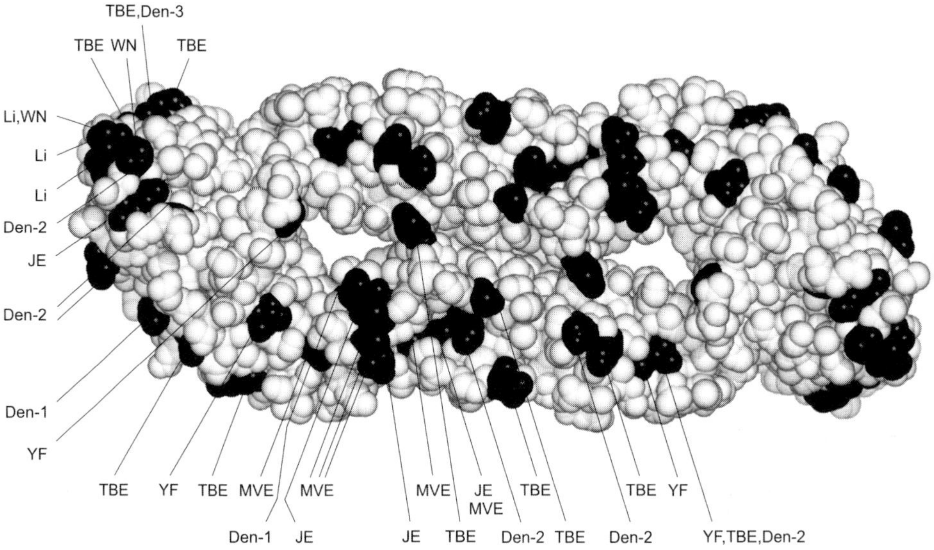

FIG. 3. Surface representation of the TBE virus E dimer. Highlighted in black are the positions of Mab escape mutants identified with different flaviviruses that were assigned to the homologous positions in the TBE virus E protein. Abbreviations are as follows: TBE (tick-borne encephalitis), Den (dengue), JE (Japanese encephalitis), Li (Louping ill), MVE (Murray Valley encephalitis), WN (West Nile), YF (yellow fever).

recombinant subviral particles also provide evidence for subunit-overlapping sites involving amino acid residues from each of the monomeric subunits in the E dimer (Kiermayr et al, in preparation).

Flavivirus cross-reactive sites

Originally, flaviviviruses were grouped together on the basis of antigenic relationships observed in certain assays with polyclonal animal and human immune sera (Calisher et al 1989). These broad cross-reactivities between all of the flaviviruses are characteristic of haemagglutination-inhibition and enzyme immunoassays, whereas neutralization assays are significantly more specific and have allowed the definition of so-called serocomplexes. These comprise more closely related flaviviruses, and cross-neutralization is only observed within but not between such serocomplexes. Molecular details of the sites in E that induce and bind broadly flavivirus cross-reactive antibodies have recently been elucidated in the TBE virus system by the use of cross-reactive Mabs raised against several different flaviviruses as well as by dissecting the antibody populations present in post-infection polyclonal human immune sera (Stiasny et al 2006b). It was shown with low affinity only that the cross-reactive antibodies bind to the surface of native infectious virions and recognize a cryptic site involving conserved amino acids present in the fusion peptide loop at the tip of domain II. Experimentally this site can be made more accessible by dissociating the tightly packed viral envelope into its dimeric E protein constituents, but it also becomes exposed in the course of antigen coating to solid phases (as in certain enzyme immunoassay formats) as well as during the incubation steps in haemagglutination inhibition assays. The information gained so far with monoclonal and polyclonal antibodies suggests that flaviviruses have a single dominant site in their E protein that induces broadly cross-reactive antibodies. This site, however, appears to be only partially accessible at the surface of native and infectious virions and its cryptic nature can explain the lack of cross-neutralization despite extensive cross-reactivities observed in haemagglutination-inhibition and enzyme immunoassays.

Summary and conclusions

The structural information of the flavivirus envelope organization as determined by X-ray crystallography and cryo-electron microscopy combined with the results of biochemical and immunological analyses have provided the basis for a molecular understanding of the mechanism of membrane fusion and lead to new insights into the molecular details of antibody-mediated neutralization and cross-reactivity. This knowledge opens new possibilities for the design of antiviral strategies and can contribute to the development of new flavivirus vaccines.

References

Allison SL, Schalich J, Stiasny K, Mandl CW, Kunz C, Heinz FX 1995 Oligomeric rearrangement of tick-borne encephalitis virus envelope proteins induced by an acidic pH. J Virol 69:695–700

Allison SL, Schalich J, Stiasny K, Mandl CW, Heinz FX 2001 Mutational evidence for an internal fusion peptide in flavivirus envelope protein E. J Virol 75:4268–4275

Beasley DW, Davis CT, Whiteman M, Granwehr B, Kinney RM, Barrett AD 2004 Molecular determinants of virulence of West Nile virus in North America. Arch Virol Suppl 35–41

Bressanelli S, Stiasny K, Allison SL et al 2004 Structure of a flavivirus envelope glycoprotein in its low-pH-induced membrane fusion conformation. EMBO J 23:728–738

Calisher CH, Karabatsos N, Dalrymple JM et al 1989 Antigenic relationships between flaviviruses as determined by cross-neutralization tests with polyclonal antisera. J Gen Virol 70:37–43

Corver J, Ortiz A, Allison SL, Schalich J, Heinz FX, Wilschut J 2000 Membrane fusion activity of tick-borne encephalitis virus and recombinant subviral particles in a liposomal model system. Virology 269:37–46

Dokland T, Walsh M, Mackenzie JM, Khromykh AA, Ee KH, Wang S 2004 West Nile virus core protein; tetramer structure and ribbon formation. Structure (Camb) 12:1157–1163

Earp LJ, Delos SE, Park HE, White JM 2005 The many mechanisms of viral membrane fusion proteins. Curr Top Microbiol Immunol 285:25–66

Ferlenghi I, Clarke M, Ruttan T et al 2001 Molecular organization of a recombinant subviral particle from tick-borne encephalitis virus. Mol Cell 7:593–602

Gallo SA, Puri A, Blumenthal R 2001 HIV-1 gp41 six-helix bundle formation occurs rapidly after the engagement of gp120 by CXCR4 in the HIV-1 Env-mediated fusion process. Biochemistry 40:12231–12236

Heinz FX, Collet MS, Purcell RH et al 2003 Family *Flaviviridae*. In: Van Regenmortel MH, Fauquet CM, Bishop DHL et al (eds) Virus taxonomy. Seventh report of the international committee on taxonomy of viruses. Academic Press, San Diego p 859–878

Jones CT, Ma L, Burgner JW, Groesch TD, Post CB, Kuhn RJ 2003 Flavivirus capsid is a dimeric alpha-helical protein. J Virol 77:7143–7149

Kielian M, Chatterjee PK, Gibbons DL, Lu YE 2000 Specific roles for lipids in virus fusion and exit. Examples from the alphaviruses. Subcell Biochem 34:409–455

Kuhn RJ, Zhang W, Rossmann MG et al 2002 Structure of dengue virus: implications for flavivirus organization, maturation, and fusion. Cell 108:717–725

Lescar J, Roussel A, Wien MW et al 2001 The fusion glycoprotein shell of Semliki Forest virus. An icosahedral assembly primed for fusogenic activation at endosomal pH. Cell 105:137–148

Lindenbach BD, Rice CM 2003 Molecular biology of flaviviruses. Adv Virus Res 59:23–61

Matthews T, Salgo M, Greenberg M, Chung J, DeMasi R, Bolognesi D 2004 Enfuvirtide: the first therapy to inhibit the entry of HIV-1 into host CD4 lymphocytes. Nat Rev Drug Discov 3:215–225

Melikyan GB, Markosyan RM, Hemmati H, Delmedico MK, Lambert DM, Cohen FS 2000 Evidence that the transition of HIV-1 gp41 into a six-helix bundle, not the bundle configuration, induces membrane fusion. J Cell Biol 151:413–423

Modis Y, Ogata S, Clements D, Harrison SC 2003 A ligand-binding pocket in the dengue virus envelope glycoprotein. Proc Natl Acad Sci USA 100:6986–6991

Modis Y, Ogata S, Clements D, Harrison SC 2004 Structure of the dengue virus envelope protein after membrane fusion. Nature 427:313–319

Modis Y, Ogata S, Clements D, Harrison SC 2005 Variable surface epitopes in the crystal structure of dengue virus type 3 envelope glycoprotein. J Virol 79:1223–1231

Mukhopadhyay S, Kim BS, Chipman PR, Rossmann MG, Kuhn RJ 2003 Structure of West Nile virus. Science 302:248

Oliphant T, Engle M, Nybakken GE et al 2005 Development of a humanized monoclonal antibody with therapeutic potential against West Nile virus. Nat Med 11:522–530

Rey FA, Heinz FX, Mandl C, Kunz C, Harrison SC 1995 The envelope glycoprotein from tick-borne encephalitis virus at 2 A resolution. Nature 375:291–298

Stadler K, Allison SL, Schalich J, Heinz FX 1997 Proteolytic activation of tick-borne encephalitis virus by furin. J Virol 71:8475–8481

Stiasny K, Allison SL, Marchler-Bauer A, Kunz C, Heinz FX 1996 Structural requirements for low-pH-induced rearrangements in the envelope glycoprotein of tick-borne encephalitis virus. J Virol 70:8142–8147

Stiasny K, Allison SL, Mandl CW, Heinz FX 2001 Role of metastability and acidic pH in membrane fusion by tick-borne encephalitis virus. J Virol 75:7392–7398

Stiasny K, Allison SL, Schalich J, Heinz FX 2002 Membrane interactions of the tick-borne encephalitis virus fusion protein E at low pH. J Virol 76:3784–3790

Stiasny K, Koessl C, Heinz FX 2003 Involvement of lipids in different steps of the flavivirus fusion mechanism. J Virol 77:7856–7862

Stiasny K, Bressanelli S, Lepault J, Rey FA, Heinz FX 2004 Characterization of a membrane-associated trimeric low-pH-induced form of the class II viral fusion protein E from tick-borne encephalitis virus and its crystallization. J Virol 78:3178–3183

Stiasny K, Koessl C, Lepault J, Rey FA, Heinz FX 2006a Structural and functional dissection of the flavivirus membrane fusion pathway. Submitted

Stiasny K, Kiermayr S, Holzmann H, Heinz FX 2006b Cryptic nature of the dominant flavivirus cross-reactive antigenic site. Submitted

Zhang W, Chipman PR, Corver J et al 2003a Visualization of membrane protein domains by cryo-electron microscopy of dengue virus. Nat Struct Biol 10:907–912

Zhang Y, Corver J, Chipman PR et al 2003b Structures of immature flavivirus particles. EMBO J 22:2604–2613

Zhang Y, Zhang W, Ogata S et al 2004 Conformational changes of the flavivirus E glycoprotein. Structure (Camb) 12:1607–1618

DISCUSSION

Rice: Is the pH 5 treatment necessary to get fusion?

Heinz: Yes. Protonation of specific residues is needed.

Kuhn: If you bring the pH down to 7, can you reform the homodimers? And when you do pH 10 and shift down, do you go back to the native structure and then drive the low pH structure? Or do the monomers trimerize correctly?

Heinz: The monomerization is fully reversible, both in solution and in the context of the whole virus particles. If we take these alkaline pH-treated viruses, bring the pH back to neutral and then look at the oligomeric structure of E, it is all dimeric. Of course, we have to solubilize these particles for analysing the oligomeric state of E. In the electron microscope the back-neutralized virus particles don't look the same as the mature virions. This is understandable, because the mature virions are formed after cleavage of the prM protein in a way which we don't completely understand, unfortunately. However, it is a coordinated process that leads to this herring bone-like organization of E in mature virions. If we resolve this arrangement at alkaline pH the E proteins apparently do not form the very same structure upon back-neutralization.

Harris: Why wouldn't the antibodies that can bind to the fusion peptide be neutralizing too?

Heinz: That is precisely the point. The fusion peptide is one of the most important functional elements in the protein. Why do they not neutralize? The explanation is that these antibodies don't recognize the native virus because these sites are not available at the infectious virion surface. They are, however, accessible in sucrose-acetone-treated antigens as used for haemagglutination-inhibition assays. In addition, in this assay the incubation with the antibodies is done at pH 9, which also leads to the disintegration of the oligomeric structure of E. In ELISA the antigen is coated to a solid phase which also causes a certain denaturation. Cross-reactive antibodies recognize sites that become exposed during the course of these assays. But in neutralization we only measure the interaction with infectious, completely native virus to which these antibodies do not bind.

Rice: Would you expect to see some neutralization if you did the actual infection in the presence of high concentrations of these antibodies?

Heinz: There is one example of this, which is a marginal neutralization at very high concentration. This can be explained in different ways. You might imagine that you had infectious particles where part of the membrane is disintegrated. These antibodies gain access at such a site and in some way impair the infection process, although they cannot completely block it. You could also imagine that the antibody causes some distortion of the native structure through its binding, thus gaining access to internal sites.

Rice: What if you come up with a way of getting the antibody taken up into the endocytic compartment?

Heinz: The pre-requisite for neutralization is that it binds to the native virus.

Fairlie: Have you looked at any other effects on disrupting the monomer/dimer/trimer equilibria, other than pH?

Heinz: We have looked at elevated temperature and urea treatment. In both cases we only got denaturation. We saw no specifically measurable effects. In functional assays we weren't able to induce fusion activity by these alternative methods. Dissociation of the dimer is needed, and any treatment that doesn't allow this at a reasonable temperature will lead to the denaturation of the complex.

Young: You talked about the specific protonation of individual residues, in response to a question about whether a lowering of pH is needed for fusion. Have you started to look at which residues are involved?

Heinz: We don't know which residues those are. There are residues in different regions of the E protein that can be protonated at slightly acidic pH. In the case of the influenza virus haemagglutinin a primary requisite for inducing fusion is the dissociation of the trimeric haemagglutinin 1, which functions like a clamp to hold the haemagglutinin 2 in its metastable conformation. Once you release this clamp the fusion process can proceed to its end. In the case of flaviviruses, the disintegration of the envelope structure and even the dissociation of the dimer is not sufficient. With influenza virus, fusion is a spontaneously occurring process once you release the clamp; in the case of flaviviruses, the triggering of fusion appears to be much more specific.

Flamand: Did you look at the antibody response in patients that recovered from tick-borne encephalitis and their ability to neutralize viruses? Are there differences among individuals?

Heinz: They all have neutralizing antibodies and are protected lifelong.

Flamand: Do their neutralizing antibodies all show similar properties?

Heinz: These sera all contain cross-reactive antibodies. Any serum from a person who has been infected by one of the flaviviruses exhibits broad cross-reactivity, which is related to the recognition of internal structures.

Zinkernagel: Does this cross-reactivity pertain to ELISA assays or to staining infected cell surfaces or neutralization assays?

Heinz: It only relates to haemagglutination inhibition and enzyme immuno assays, not neutralization assays.

Zinkernagel: So it is irrelevant!

Heinz: To a certain degree, yes.

Young: It has been done on infected cell surfaces.

Heinz: The point is that these antibodies are apparently formed because such internal sites are exposed in the course of natural infections. In neutralization assays, however, they remain cryptic.

Halstead: What was your polyclonal antibody? How did you raise it?

Heinz: We used human dengue post-infection sera. The antigen was always tick-borne encephalitis (TBE) virus. We first measured the cross-reactivity of the dengue antibodies with TBE virus coated to the solid phase of an ELISA plate. Then we assessed whether these antibodies were capable of reacting with the native infectious TBE virus in solution.

Halstead: One of the enigmas of antibody-dependent enhancement in dengue is that while you can imagine that antibodies such as the one you described react against fusion domains, antibodies that don't neutralize would in effect form immune complexes and the immune complex would attach to an Fc receptor. Putvatana et al (1984) showed that Japanese encephalitis antibodies don't enhance dengue 2 infections.

Heinz: Yes, enhancement is observed within the dengue sero complex and these antibodies recognize sites that are accessible on the infectious virions. These are not cryptic sites. Enhancement, however, is not observed between viruses of different serocomplexes, such as TBE and dengue virus.

Zinkernagel: With your monoclonals have you tried to compete out your polyclonal sera? With some viruses, such as rabies or rhabdoviruses in general, you can take one of 60 neutralizing monoclonals and you can compete out any neutralizing polyclonal serum from any other species. This says that the neutralizing epitope that is relevant or correlates with so-called serotypes is one unique site on the virus or the infected cell. You can forget about all the rest.

Heinz: We did these sorts of experiments a long time ago and didn't get clear-cut results. But it is known that antibody binding to many different sites at the surface of the E protein can lead to neutralization.

Zinkernagel: Your argument would be that for each serotype there are several (not just one) neutralizing antigenic sites. I would be extremely surprised if that were the case. Biology is not that complicated.

Heinz: Flaviviruses are that complicated! It is a moving target that is recognized by the immune system. You can get sequential infections by different dengue viruses, which differ at their surfaces to different degrees. Some amino acids vary and others are constant, and what is recognized by a Fab footprint is the characteristic face of a given virus.

Zinkernagel: If that were true it would be impossible to define serotypes.

Heinz: It is a question of definition. A serotype is something that can be distinguished in a neutralization assay with polyclonal immune sera. If there is a titre reduction of three steps or whatever, you define it.

Zinkernagel: The serotype is defined by epidemiology. With polio 1, 2 and 3, if you are immune against 1 you are not protected against 2 or 3.

Heinz: To an extent this is also true for flaviviruses. If you have had an infection with yellow fever virus you are not protected against dengue or TBE. If you are vaccinated against TBE you won't be protected against dengue. These are viruses from different serocomplexes, defined by not exhibiting cross-neutralization. Within these serocomplexes, however, at least partial cross-neutralization is seen.

Halstead: It is defined by infection and protection. Dengue 1, 2, 3 and 4 are distinct.

Heinz: But there is partial cross-neutralization and cross-protection.

Halstead: These viruses all cause the same disease, so they have a certain receptor conformation that has to be similar. The concept that any hit on the whole envelope protein is sufficient to define neutralization flies in the face of clinical and epidemiological evidence.

Evans: It makes no sense. If any hit on the envelope is neutralizing and you show that these viruses vary the amino acids on the envelope, then every person would be neutralization dependent. Every virus would be somewhat different. However, there are sero groups.

Heinz: These are two different issues. You have one virus and you ask the question: to which sites on this virus do antibodies have to bind to be able to neutralize the virus.

Evans: The observation made with a single virus may be irrelevant to every other virus.

Heinz: Not every other virus. With TBE we have three subtypes, which differ by 5% of their amino acids in the E protein.

Halstead: I bet people who are immune to one are protected a bit against the others.

Heinz: Yes, there is extensive cross-protection. A 5% difference is not enough to lose this. But if you get to 20% difference, then cross-neutralization and cross-protection begin to be lost.

Jans: You talked about the structure changing with pH. Do you know the accessibility of the fusion peptide at pH 10?

Heinz: It is fully accessible for antibody binding.

Halstead: I believe that CJ Lai has described an antibody directed against the fusion domain, which is in fact neutralizing (Goncalvez et al 2004).

Young: It might be relevant to go back to Gollins and Porterfield's *Nature* paper where they talked about post entry inhibition and neutralization (Gollins & Porterfield 1986). They showed electron microscopy pictures of antibodies interacting with the virus inside the endosome. I guess this is dependent on a high concentration of antibody.

Heinz: Not necessarily. It is only required that the antibody–virus complex is internalized as a complex so that the antibody can inhibit fusion in the endosome.

Young: Unless you have them both going in independently.

Heinz: This is unlikely.

Young: Yes, except for an environment where there is excess antibody.

Heinz: In other virus systems it has been described that antibody–virus complexes are internalized and that the neutralizing activity is exerted somewhere inside the cell. It can be the endosome.

Zinkernagel: The problem is that many of these things can be shown, but we can't yet make reasonable decisions about what is important. We can measure too many things.

Heinz: Most people would agree that neutralizing antibodies are an excellent correlate of protection in the case of flavivirus infections.

Zinkernagel: All these things, such as internalizations in phagolysosomes, have been thought about and can be constructed in selected experimental conditions. But no one knows whether these things work *in vivo*. The easiest way to think about neutralization is to cover the virus up so it cannot dock onto and infect a cell.

Heinz: With respect to defining the functionality of antibodies it is of course important to measure neutralization or passive protection, but then one can try to get a step further and ask about the mechanism of antibody action. Is it preventing binding to receptors or interaction with DC-SIGN, or is it prevention of fusion?

Screaton: Going back to your concept that it doesn't matter where the antibody binds for neutralization, in many ways this is a self-fulfilling prophecy. If you bind 180 antibody molecules on the surface of this tiny virus, there is nothing left that you can see.

Heinz: That is the mechanism of neutralization.

Screaton: Not necessarily. Don't forget immunodominance. You are probably not making nearly as many of these things as you think you are making. You are looking at mouse reagents and immunodominance networks in mice which are likely to be very different than they are in humans. You are probably making only a very small number of these things. This is why you do get these serotype-specific escapes. You are probably not plastering the virus as you think you are. Immunodominance is probably why you get serotype-specific effects here.

Heinz: I agree, and this immunodominance issue is important. But I have a different view of these serotype specific epitopes. I don't think that you have separate structural entities, with structurally distinct serotype epitopes on say dengue 1 and 2. Instead we can look at the surface of the protein as being determined by a specific combination of discontinuous amino acids. Some of them differ to varying degrees in the different flaviviruses. What the antibody sees can be identical, or different at a varying number of spots. The same principal surface can thus define different serotypes.

Halstead: It sounds heretical. Specificity must mean something.

References

Gollins SW, Porterfield JS 1986 A new mechanism for the neutralization of enveloped viruses by antiviral antibody. Nature 321:244–246

Putvatana R, Yoksan S, Chayayodhin T, Bhamarapravati N, Halstead SB 1984 Absence of dengue 2 infection enhancement in human sera containing Japanese encephalitis antibodies. Am J Trop Med Hyg 33:288–294

Goncalvez AP, Purcell RH, Lai CJ 2004 Epitope determinants of a chimpanzee Fab antibody that efficiently cross-neutralizes dengue type 1 and type 2 viruses map to inside and in close proximity to fusion loop of the dengue type 2 virus envelope glycoprotein. J Virol 78:12919–12128

GENERAL DISCUSSION I

Rice: One point that Franz Heinz raised in his paper was the amazing kinetics of fusion that these flaviviruses have. There are very effective fusion inhibitors for other viruses. Franz, does this lessen your enthusiasm for developing these against flaviviruses?

Heinz: I think the whole fusion process is a potential target for antivirals. Compared to HIV, flavivirus fusion is much faster and its inhibition may therefore be more difficult, or may have to be approached in different ways. Flavivirus inhibitors would have much less time to interfere with intermediary structures.

Young: Your question goes to the heart of what Franz said in terms of inhibitors of fusion. It depends on whether you are targeting the transition states or a pre-fusion form. This is what the attempts at *in silico* docking are aiming at. The pre-fusion form exists during the period that the virus is moving between one cell and another.

Kuhn: Since these particles are fusing in an endosome, you have to be sure that if you have any inhibitor, it is already bound to the layers. Otherwise it will have to be in high concentration to be internalized by the endocytosis. The ones that have worked well with HIV will need some kind of tag to hook them to the virus before they will interfere with the wrapping of domain 3 and its helices against domain 2.

Neyts: It is apparently rather expensive to synthesize peptidic fusion inhibitors, such as the HIV fusion inhibitor T-20 or Fuzeon. If one would want to bring peptidic compounds to the clinic for a disease such as dengue, this may be a serious concern.

Rice: In terms of compounds that would fit into a hydrophobic pocket and which would inhibit the required transition, one could envision non-peptidic small molecule inhibitors.

Keller: From my viewpoint what we need are some specific assays which we can use to look at this particular process. I have challenged Subash Vasudevan to develop this sort of assay for looking at this conformational change. This would help if we had to optimize such a compound.

Kuhn: In theory the fusion assay that Franz described can be used as a high-throughput assay for screening compounds.

Fairlie: The fusion assay doesn't give you any information about where the small molecule compounds might be binding. NMR-based drug design would be the best way to go in terms of giving feedback as to which residues in the protein are being modified by small molecules.

Keller: The fusion assay would go some way further than the cellular assay. If you wanted information about the active site, NMR would be fine. I'd be really happy with the fusion assay at this point.

Rice: What has been done by Franz' lab in tick-borne encephalitis (TBE) is a prototype for the kinds of assays we can think about developing for dengue. One advantage that the TBE folks have had is a large panel of antibodies that can be used to look at the exposure of these epitopes under different conditions. I don't know that this exists yet for dengue. NMR would be one way of doing it, but you'd need quite a bit of soluble protein. One of the other things we touched on this morning is the state of the field regarding flavivirus receptors. Are these a potential target for therapy, or are they just too promiscuous?

Keller: If we could find some receptors such as CCL5, which is a G protein-coupled receptor (GPCR) that looks very amenable to inhibition, we would have a good chance. These sugar-like molecules on the surface, such as DC-SIGN, are horrible for drug discovery!

Rice: They are also unproven, in terms of *in vivo* relevance.

Keller: The most important thing from a drug discovery standpoint is that we have perhaps five or six drug targets in the virus. When we look at the attrition in drug discovery process, it is urgent that we have some host targets as well. This will give us a much better chance of finding a drug for dengue.

Rice: Coming from a hepatitis C background, the therapies that people are using today are based on host targets or targets unknown.

Young: One of the other issues we need to look at is when to target infected individuals for therapy. It will be difficult to identify patients early enough and get them appropriate therapeutic treatment to knock down virus infection. This will probably already have happened before the patient presents. Perhaps we should be focusing on how we can prevent progression to more severe disease.

Rice: That is an underlying general topic we need to keep in mind.

Guber: A dengue drug will be effective if you don't have to rely on confirmed cases. You want something very cheap that inhibits viral replication and can be used on the basis of a clinical diagnosis in an epidemic. It probably won't be effective if you have to rely on a confirmed case.

Screaton: I can't conceive of an instance where you are going to be giving a drug prophylactically to a population. This will never happen. If it does, we will run into terrible problems with resistance, just like the Tamiflu story. The realistic target population is people with fever in an infected area.

Heinz: Wouldn't you expect resistance not to be such a problem?

Screaton: It is potentially very serious.

Heinz: If you treat a person with an anti-dengue drug, then the virus tries to escape, but it has to maintain its ability to replicate in the mosquito. There are lots of constraints.

Multiple enzyme activities of flavivirus proteins

R. Padmanabhan*, N. Mueller*, E. Reichert*, C. Yon*, T. Teramoto*, Y. Kono*, R. Takhampunya*¶, S. Ubol¶, N. Pattabiraman†, B. Falgout‡, V. K. Ganesh§ and K. Murthy§

** Department of Microbiology and Immunology and † Lombardi Cancer Center, Georgetown University School of Medicine, Washington DC, USA, ‡ CBER, FDA, Bethesda, MD, USA, § University of Alabama-Birmingham, AL, USA and ¶ Department of Microbiology, Mahidol University, Bangkok, Thailand*

Abstract. Dengue viruses (DENV) have 5′-capped RNA genomes of (+) polarity and encode a single polyprotein precursor that is processed into mature viral proteins. NS2B, NS3 and NS5 proteins catalyse/activate enzyme activities that are required for key processes in the virus life cycle. The heterodimeric NS2B/NS3 is a serine protease required for processing. Using a high-throughput protease assay, we screened a small molecule chemical library and identified ~200 compounds having ≥50% inhibition. Moreover, NS3 exhibits RNA-stimulated NTPase, RNA helicase and the 5′-RNA triphosphatase activities. The NTPase and the 5′-RTPase activities of NS3 are stimulated by interaction with NS5. Moreover, the conserved, positively charged motif in DENV-2 NS3, ^{184}RKRK, is required for RNA binding and modulates the RNA-dependent enzyme activities of NS3. To study viral replication, a variety of methods are used such as the *in vitro* RNA-dependent RNA polymerase assays that utilize lysates from DENV-2-infected mosquito- or mammalian cells or the purified NS5 along with exogenous short subgenomic viral RNAs or the replicative intracellular membrane-bound viral RNAs as templates. In addition, a cell-based DENV-2 replicon RNA encoding a luciferase reporter is also used to examine the role of *cis*-acting elements within the 3′ UTR and the RKRK motif in viral replication.

2006 New treatment strategies for dengue and other flaviviral diseases. Wiley, Chichester (Novartis Foundation Symposium 277) p 74–86

Studies from several laboratories have strongly implicated a complex interplay of viral non-structural proteins as well as cellular proteins, in modulating the process of viral replication (Lindenbach & Rice 2003). Since the title of my presentation is with regard to multiple enzyme activities of flaviviral non-structural proteins, I give below a summary of the work already published from our laboratory as well those reported by others as a prelude to the more recent ongoing research. Our efforts have mainly focused on functional analysis of two viral proteins, NS3 and

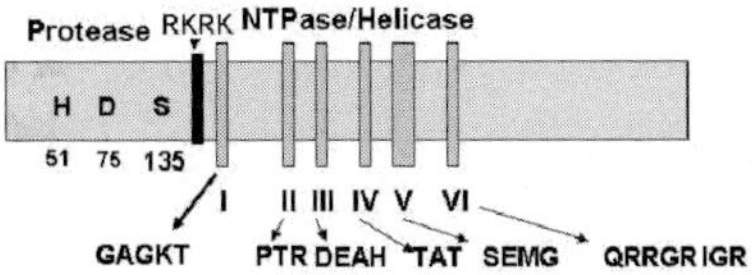

FIG. 1. Multiple domains of NS3. Protease domain is at the N-terminus within 170 amino acid residues. H, D and S are at position 51, 75 and 135. Six conserved motifs comprising the RNA helicase domain are shown as open vertical boxes. Black box is a stretch of basic amino acids.

NS5. NS3 is a paradigm of multifunctional proteins (Fig. 1). The N-terminal (180 amino acid) region is known to have a serine protease domain that requires the NS2B cofactor for protease activity (Chambers et al 1991, Falgout et al 1991, Wengler et al 1991, Zhang et al 1992). The two components of the protease interact and form a stable complex in flavivirus-infected cells (Arias et al 1993, Chambers et al 1993). The NS2B is an endoplasmic resident integral membrane protein and the hydropathy plot of the protein shows a conserved hydrophilic domain of NS2B (NS2BH) flanked by hydrophobic regions (Clum et al 1997). The deletion analysis of NS2B indicated that a conserved hydrophilic domain of ~40 residues (Fig. 2A) is sufficient for protease activity *in vivo* (Falgout et al 1993) as well as *in vitro* (Clum et al 1997). When the His-tagged NS2BH domain linked to serine protease domain is expressed in *Escherichia coli*, it undergoes a *cis* cleavage at the junction of the NS2BH-NS3-pro precursor and the protease can be purified as a complex by metal affinity column. The protease is active in cleaving a radiolabelled NS4B-NS5 precursor substrate or chromogenic and fluorogenic peptide substrates (Leung et al 2001, Li et al 2005, Yusof et al 2000). The crystal structure of the dengue virus (DENV)-2 NS3 protease domain alone in the absence of the NS2BH cofactor domain has been determined (Murthy et al 1999). The structure of the protease domain reveals that the N- and C-terminal β barrels are six-stranded and the catalytic triad consists of His51, Asp75, and Ser135. Molecular modelling of a tetrapeptide into the substrate binding cleft indicates that the protease domain has a shallow substrate binding pocket with no extensive side chain interactions beyond P2 and P2′ residue, suggesting that interaction with the cofactor NS2B results in a dramatic change in conformation that results in more extensive interactions with the substrate beyond P2 and P2′. In fact, the protease activity of NS3-pro is enhanced 10^4-fold against the tripeptide substrate by the interaction with the NS2BH whereas it has negligible effect on hydrolysis of a substrate with a single Arg such as *N*-benzoyl-L-Arg-p-nitroanilide (BAPNA). This structure of the protease domain could explain the effects of mutations of the protease domain previously reported (Valle & Falgout 1998). The crystal structure of the NS3 protease

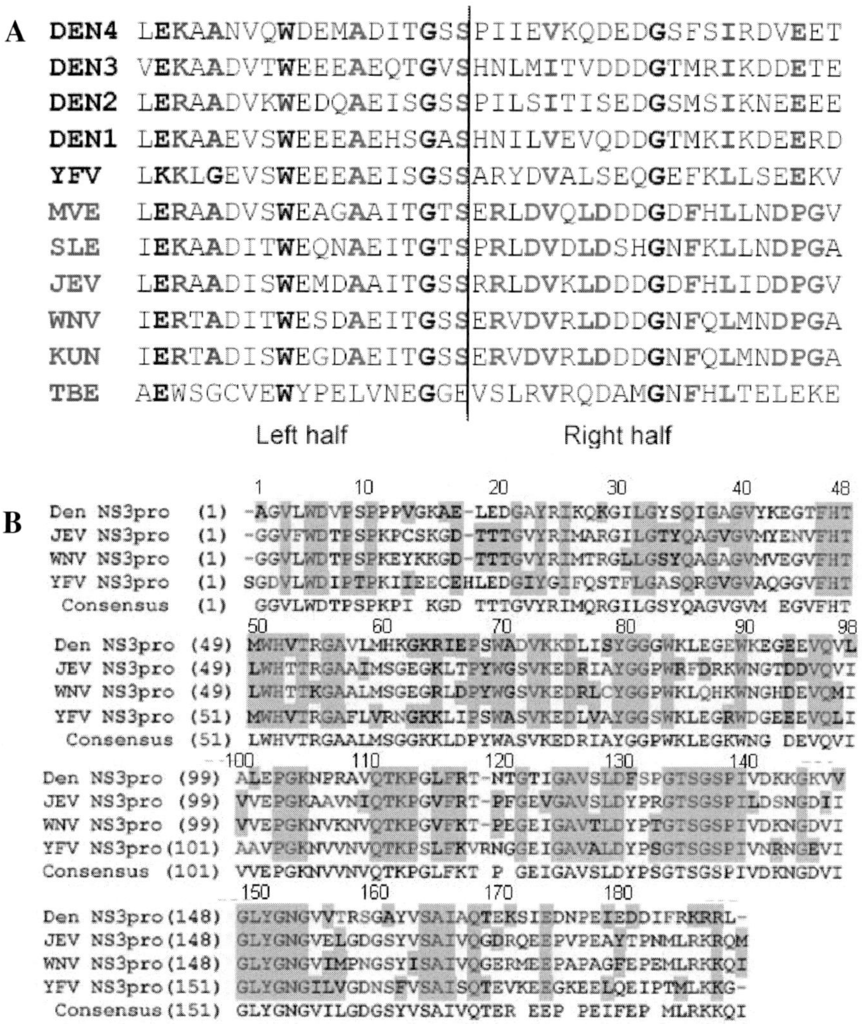

FIG. 2. Sequence alignments of the two-component protease, NS2BH-NS3-pro.

domain in a 2:1 complex with a double-headed mung bean Bowman-Birk inhibitor (MbBBI) is also known (Murthy et al 2000). The MbBBIs are hydrolysed by serine proteases in a standard manner; however, the reaction is completely reversible with no appreciable product release. Both NS3-pro and NS2BH/NS3-pro complex are inhibited by MbBBI using the BAPNA as the substrate with a IC50 value of ~1.8 μM. The structure reveals that the P1 Arg is bound to a bifurcated S1 pocket in a distinct manner (Murthy et al 2000). Although the overall structural features of the NS3-pro/MbBBI complex are in agreement with many other serine protease/inhibitor complexes, there are some differences with respect to interactions made by P1 Arg and Lys with the residues in the S1 pocket of NS3-pro, including the presence of the Arg at P1 in two different conformations and also large conformational changes of some of the residues of NS3-pro that interact with the Arg. However, since the structure of the NS3-pro and MbBBI was obtained in the absence of the NS2BH cofactor the interpretation of these results is made with caution. Based on the bifurcated recognition mode of the P1 Arg by the S1 pocket of the NS3-pro, we reasoned that compounds with terminal biguanidino groups, which could be superimposed on the guanidino groups of the bifurcated Arg side chain, might be possible candidates as selective inhibitors of DENV-2 NS2B:NS3-pro. In fact, of the three compounds tested, only one compound inhibited with a reasonable potency (35 and 44 μM for West Nile, WNV, and DENV proteases, respectively). Two other compounds with a single guanidino group inhibited DENV-2 and WNV proteases with K_i values in the range of 13–23 μM but also inhibited trypsin with K_i values in the 3.2 to 4 μM range (Ganesh et al 2005).

The spacer region between the NS2BH and the NS-pro is not critical for protease activity. An active protease was expressed in *E. coli* in a non-cleavable form by substitution of the natural spacer with a G4-S-G4 linker (Leung et al 2001). Some substrate analogues were found to inhibit the DENV protease in a competitive manner (Leung et al 2001). A comparative study of the substrate specificity of all four DENV NS3-pro was reported using tetra peptide libraries (Li et al 2005). Their results indicated that a strong preference for Arg/Lys at P1 whereas for P2–P4 sites, the order of preference was P2: Arg>Thr>Gln/Asn/Lys, P3: Lys>Arg>Asn>, P4: Nle>Leu>Lys>Xaa. At the prime sites, small and polar residues were preferred at P1′ and P3′, whereas the P2′ and P4′ sites had minimal effect. The N-terminal (P6–P1) cleavage site peptides were also found to inhibit the protease activity in a competitive manner with K_i values in the range of 67–12 μM. The peptides from the P1′-P5′ region had no inhibitory effect (Chanprapaph et al 2005).

The region C-terminal to the protease domain of NS3 has conserved domains found in the DEXH family of NTPases/RNA helicases (Fig. 1). The motif GxGKS/T (domain I) and domain II are required for the ATPase activity of NS3. The NTPase activity of NS3, involved in the hydrolysis of the $\gamma\downarrow\beta\alpha$ phosphoric anhydride bond (shown by an arrow) of NTP has been reported for several flaviviruses including DENV-2 (Li et al 1999, Cui et al 1998, Benarroch et al 2004, Yon et al 2005). Several viral NTPase activities are stimulated by the addition of single-stranded RNA. The presence of conserved RNA helicase motifs in flavivirus NS3 is consistent with its postulated role in viral replication in a key step involving unwinding of the double-stranded RNA replicative form. The RNA helicase activity has been shown for DENV-2 and Japanese encephalitis viruses (Li et al 1999, Utama et al 2000, Benarroch et al 2004, Yon et al 2005). However, the role of the RNA-stimulated NTPase/RNA helicase activity of NS3 in viral life cycle has not been established for any flavivirus. In addition, NS3 has the 5′-RNA triphosphatase activity (5′-RTPase), capable of hydrolysing the $\gamma\downarrow\beta\alpha$ phosphoric anhydride bond (shown by an arrow) of triphosphorylated RNA as shown for full-length DENV-2 NS3 expressed and purified from *E. coli* (Li et al 1999, Benarroch et al 2004, Yon et al 2005). The 5′-RTPase is the first of the four sequential enzymatic reactions that are involved in the addition of 5′-cap to RNA.

The multifunctional NS3 protein exists in a complex with NS5 (Kapoor et al 1995), which itself has two enzyme activities, the 5′-RNA *O*-methyltransferase involved in 5′-capping and the RNA-dependent RNA polymerase required for viral RNA replication in flavivirus-infected cells (Fig. 3). After hydrolysis of the γ-phosphate moiety of triphosphosphorylated RNA (the intrinsic activity of NS3) two additional enzyme activities are required for formation of the 5′-cap, the guanylyltransferase and the two 5′-RNA methyltransferase activities, respectively. Since the flavivirus genomes have type I cap structure at the 5′-end, two steps are involved in the 5′-cap addition: methyl transfer to 7-methylG and to 2′-OH of the first 5′-terminal nucleotide of RNA. The N-terminal domain of NS5 was shown to catalyse the transfer of methyl group from S-adenosylmethionine to 2′-OH and the crystal structure of this domain was reported (Egloff et al 2002). Viruses that replicate in the cytoplasm, in general, provide their own capping machinery and the viral proteins that are involved have multiple functions. Mutational analysis of DENV-2 NS3 indicates that the active site of the NTPase and the 5′-RTPase share one enzymatic function i.e. removal of γ-phosphate moiety of either ATP (NTPase) or the RNA substrate (5′-RTPase) (Bartelma & Padmanabhan 2002, Benarroch et al 2004).

The C-terminal domain of NS5 has the RNA-dependent RNA polymerase (RdRP) activity which is required for RNA synthesis *in vitro* as shown using cell-free systems that utilize endogenous (Grun & Brinton 1986, Chu & Westaway 1987, Uchil & Satchidanandam 2003b) or exogenous viral RNA templates (You & Padmanabhan 1999, You et al 2001, Nomaguchi et al 2003b). According to a current model for viral replication, the synthesis of progeny RNA(+) strands

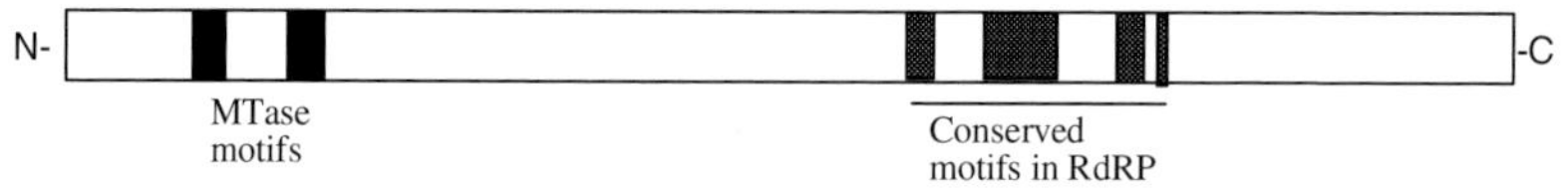

FIG. 3. Flavivirus NS5 is a RNA-dependent RNA polymerase and 2′-*O*-methyltransferase.

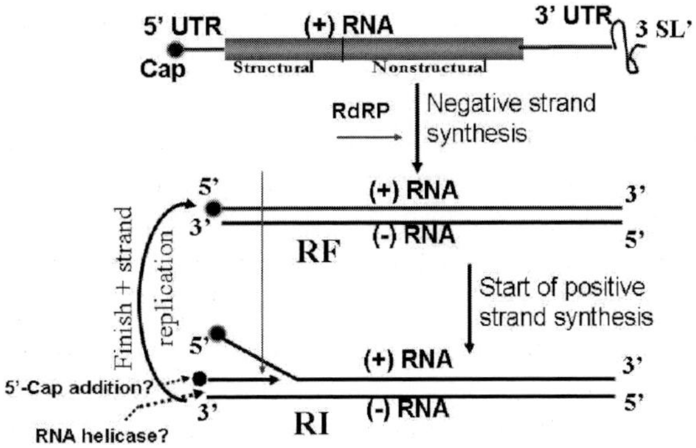

FIG. 4. Replication model for flavivirus RNA. The simplistic model is based on studies on the intracellular forms of DENV-2 and KUNV-infected cells. Three forms were identified: replicative intermediates (RI), replicative form (RF) and virion RNA (vRNA). The viral replicase and the vRNA (+) template are in association with membranes. The synthesis of progeny vRNA occurs in a semiconservative and asymmetric manner on a recycling RF (middle structure) and RI (bottom) templates.

occurs via asymmetric and semiconservative replication on a template of dsRNA as a replicative intermediate (RI) or replicative form (RF), which serve as recycling templates (Fig. 4). This *in vitro* assay is useful to detect the intracellular forms of DENV-2 replicative RNA when infected cells were treated with inhibitors of viral replication. The exogenous subgenomic RNA template-dependent and template-specific replication system established in our laboratory has been useful in defining the requirements for minus and plus strand RNA synthesis. The cell lysate system gave the first evidence for long range interaction involving the conserved self-complementary 5′- and 3′-cyclization (CYC) motifs (You & Padmanabhan 1999) as well as the 5′- and 3′-stem-loop structures that are required for (−) strand RNA synthesis. Physical interaction between the 5′- and 3′-ends was shown by psoralen-UV cross-linking (You et al 2001) and by atomic force microscopy (Alvarez et al 2005). Recombinant DENV-2 or WNV NS5 alone expressed and purified from *E. coli* can synthesize (−)RNA *in vitro* (Ackermann & Padmanabhan 2001, Nomaguchi et al 2003a, 2003b) which also requires a functional interaction between the 5′- and 3′-ends (You & Padmanabhan 1999, Khromykh et al 2001, Corver et al 2003, Lo et al 2003).

Experimental procedures

Our laboratory previously described the first *in vitro* flavivirus protease assay (Yusof et al 2000) which is now used in high-throughput screening to identify inhibitor compounds that are described below and reported by the Novartis group elsewhere (Li et al 2005). A high throughput assay for the NTPase activity of NS3 was also reported recently (Yon et al 2005). The *in vitro* RNA-dependent RNA polymerase assays that utilize exogenous viral RNA templates and either lysates from DENV-2-infected mammalian or mosquito cells or purified NS5 polymerase has been described previously (You et al 2001, You & Padmanabhan 1999, Ackermann & Padmanabhan 2001). The *in vitro* assays that utilize endogenous membrane-bound viral replicase associated with the three intracellular forms of viral RNA are as described by others (Chu & Westaway 1987, Uchil & Satchidanandam 2003a). The construction of DENV-2 replicon RNA encoding the Renilla luciferase and its use in studying the role of *cis*-acting elements and trans-acting factors will be described elsewhere.

Results and discussion

Viral proteases

In our quest to identify compounds that inhibit DENV-2 and WNV proteases, we have launched a high-throughput screening endeavour using our *in vitro* protease assay. In the initial screen of ~32K small molecule compounds, ~200 compounds having ≥ 50% inhibition were identified from which compounds having similar core structures were grouped. From this list, eleven compounds that could be grouped into three core structures were selected for further analyses. The analyses of their inhibitory properties using both *in vitro* and cell-based assays and their cytotoxicity assays are in progress.

Characterization of the NTPase/RNA helicase and 5′-RNA triphosphatase activities of NS3

To understand the role of RNA-stimulated NTPase/RNA helicase and 5′-RTPase activities of NS3 in viral replication, we expressed the full length NS3 in *E. coli* with an N-terminal His tag and purified the protein in a soluble form. We characterized the RNA-stimulated NTPase and 5′-RTPase activities of NS3 and showed that the interaction of DENV-2 NS3 with NS5 stimulated both activities. The NS5-mediated stimulation of the NTPase activity reached a plateau at 1:1 stoichеometry indicating that the NS3/NS5 complex is the functional unit that is involved in modulating the NS3 activities. A previous study from our laboratory indicated that the RNA helicase activity of the N-terminal-truncated NS3 (NS3Δ160) was very low that required 2.7 μM of purified NS3Δ160 protein to unwind less than 5% of a 29 bp RNA duplex, whereas the hepatitis C virus NS3 and the bovine diarrhoea virus p80 RNA helicases required 0.1–1 pmol

(10–100 nM) of enzyme for unwinding similar substrates. We also reported that the RNA-stimulated NTPase activity of (NS3Δ160) was abolished by mutation of the positively charged motif, ^{184}RKRK → ^{184}QNGN (Li et al 1999).

In this recent study, the RNA helicase activities of full length NS3 containing the protease domain (NS3FL) either with or without the presence of the NS2BH at the N-terminus, were analysed. The results showed that the presence of protease domain enhanced the NTPase, the 5′-RTPase, and the RNA helicase activities associated with the C-terminal region of NS3 suggesting that the overall folding of the NS3 molecule is important for its function. The mutant NS3 containing the ^{184}RKRK → ^{184}QNGN motif had severely reduced basal NTPase and the 5′-RTPase activities which could not be enhanced by the addition of NS5 as well as significantly reduced RNA helicase activity. Since all RNA-dependent enzyme activities of NS3 were affected by this mutation, we examined the binding of the wild-type and the mutant NS3 to single-stranded RNA. We found that the binding of NS3 to single-stranded RNA was abolished by this mutation. This positively charged motif is conserved among several flavivirus NS3 proteins examined (Fig. 5) suggesting that this motif could play an important role in the virus life cycle.

Construction and characterization of DENV-2 subgenomic replicon encoding Renilla luciferase

We have constructed the DENV-2 replicon that expresses the *Renilla* luciferase in pRS424 yeast-*E. coli* shuttle vector. This replicon will allow us to quantify the DENV-2 replication in our analyses of *cis*-acting elements and trans-acting factors required for replication. The choice of pRS424 vector is to introduce site-directed mutations into the replicon by *in vivo* recombination in yeast.

DEN1	IEDEVF**RKR**NLTIMDL
DEN2-NGC	IEDDIF**RKRK**LTIMDL
DEN2-16681	IEDDIF**RKRR**LTIMDL
DEN3	LEEEMF**KKR**NLTIMDL
DEN4	VDEDIF**RKKR**LTIMDL
KUN	FEPEML**RKK**QITVLDL
JEV	YTPNML**RKR**QMTVLDL
WNV	FEPEML**RKK**QITVLDL
MVEV	PEML**KKR**QLTVLDL
MVEV-1-51	YNPEML**KKR**QLTVLDL
YF-17D	EIPTML**KKG**MTTVLDF
YF-ASIBI	EIPTML**KK**GMTTILDF
YF-IVORY	EIPTML**KK**GMTTILDF

FIG. 5. Sequence alignment of positively charged motifs of flavivirus NS3.

Conclusions

In summary, the flavivirus enzymes, the viral proteases, the NTPase/RNA helicases, the 5′-RTPase, 5′-RNA methyltransferases, and RNA-dependent RNA polymerases are excellent targets for development of antiviral therapeutics because specific inhibitors can be identified using the *in vitro* assays that are amenable for high throughput screen. Since the key pathways of flaviviral life cycle are conserved, potential for development of antiviral drugs that would interfere with more than one flavivirus is promising. With concerted multidisciplinary efforts, development of antiviral drugs against flaviviruses can become a reality in the near future.

Acknowledgement

The work described here was supported by grants from NIH/NIAID (AI32078, AI54776 and AI45623).

References

Ackermann M, Padmanabhan R 2001 De novo synthesis of RNA by the dengue virus RNA-dependent RNA polymerase exhibits temperature dependence at the initiation but not elongation phase. J Biol Chem 276:39926–39937

Alvarez DE, Lodeiro MF, Luduena SJ, Pietrasanta LI, Gamarnik AV 2005 Long-range RNA-RNA interactions circularize the dengue virus genome. J Virol 79:6631–6643

Arias CF, Preugschat F, Strauss JH 1993 Dengue 2 virus NS2B and NS3 form a stable complex that can cleave NS3 within the helicase domain. Virology 193:888–899

Bartelma G, Padmanabhan R 2002 Expression, purification, and characterization of the RNA 5′-triphosphatase activity of dengue virus type 2 nonstructural protein 3. Virology 299:122–132

Benarroch D, Selisko B, Locatelli GA et al 2004 The RNA helicase, nucleotide 5′-triphosphatase, and RNA 5′-triphosphatase activities of Dengue virus protein NS3 are Mg^{2+}-dependent and require a functional Walker B motif in the helicase catalytic core. Virology 328:208–218

Chambers TJ, Grakoui A, Rice CM 1991 Processing of the yellow fever virus nonstructural polyprotein: a catalytically active NS3 proteinase domain and NS2B are required for cleavages at dibasic sites. J Virol 65:6042–6050

Chambers TJ, Nestorowicz A, Amberg SM, Rice CM 1993 Mutagenesis of the yellow fever virus NS2B protein: effects on proteolytic processing, NS2B-NS3 complex formation, and viral replication. J Virol 67:6797–6807

Chanprapaph S, Saparpakorn P, Sangma C et al 2005 Competitive inhibition of the dengue virus NS3 serine protease by synthetic peptides representing polyprotein cleavage sites. Biochem Biophys Res Commun 330:1237–1246

Chu PW, Westaway EG 1987 Characterization of Kunjin virus RNA-dependent RNA polymerase: reinitiation of synthesis in vitro. Virology 157:330–337

Clum S, Ebner KE, Padmanabhan R 1997 Cotranslational membrane insertion of dengue virus type 2 NS2B/NS3 serine proteinase precursor is required for efficient processing in vitro. J Biol Chem 272:30715–30723

Corver J, Lenches E, Smith K et al 2003 Fine mapping of a cis-acting sequence element in yellow fever virus RNA that is required for RNA replication and cyclization. J Virol 77:2265–2270

Cui T, Sugrue RJ, Xu Q et al 1998 Recombinant dengue virus type 1 NS3 protein exhibits specific viral RNA binding and NTPase activity regulated by the NS5 protein. Virology 246:409–417

Egloff MP, Benarroch D, Selisko B, Romette JL, Canard B 2002 An RNA cap (nucleoside-2′-O-)-methyltransferase in the flavivirus RNA polymerase NS5: crystal structure and functional characterization. EMBO J 21:2757–2768

Falgout B, Pethel M, Zhang YM, Lai CJ 1991 Both nonstructural proteins NS2B and NS3 are required for the proteolytic processing of dengue virus nonstructural proteins. J Virol 65:2467–2475

Falgout B, Miller RH, Lai C-J 1993 Deletion analysis of dengue virus type 4 nonstructural protein NS2B: Identification of a domain required for NS2B-NS3 protease activity. J Virol 67:2034–2042

Ganesh VK, Muller N, Judge K et al 2005 Identification and characterization of nonsubstrate based inhibitors of the essential dengue and West Nile virus proteases. Bioorg Med Chem 13:257–264

Grun JB, Brinton MA 1986 Characterization of West Nile virus RNA-dependent RNA polymerase and cellular terminal adenylyl and uridylyl transferases in cell-free extracts. J Virol 60:1113–1124

Kapoor M, Zhang L, Ramachandra M et al 1995 Association between NS3 and NS5 proteins of dengue virus type 2 in the putative RNA replicase is linked to differential phosphorylation of NS5. J Biol Chem 270:19100–1916

Khromykh AA, Meka H, Guyatt KJ, Westaway EG 2001 Essential role of cyclization sequences in flavivirus RNA replication. J Virol 75:6719–6728

Leung D, Schroder K, White H et al 2001 Activity of recombinant dengue 2 virus NS3 protease in the presence of a truncated NS2B co-factor, small peptide substrates, and inhibitors. J Biol Chem 276:45762–45771

Li H, Clum S, You S, Ebner KE, Padmanabhan R 1999 The serine protease and the RNA-stimulated NTPase domains of dengue virus type 2 converge within a region of 20 amino acids. J Virol 73:3108–3116

Li J, Lim SP, Beer D et al 2005 Functional profiling of recombinant NS3 proteases from all four serotypes of dengue virus using tetrapeptide and octapeptide substrate libraries. J Biol Chem 280:28766–28774

Lindenbach BD, Rice CM 2003 Molecular biology of flaviviruses. Adv Virus Res 59:23–61

Lo MK, Tilgner M, Bernard KA, Shi PY 2003 Functional analysis of mosquito-borne flavivirus conserved sequence elements within 3′ untranslated region of West Nile virus by use of a reporting replicon that differentiates between viral translation and RNA replication. J Virol 77:10004–10014

Murthy HM, Clum S, Padmanabhan R 1999 Dengue virus NS3 serine protease. Crystal structure and insights into interaction of the active site with substrates by molecular modeling and structural analysis of mutational effects. J Biol Chem 274:5573–5580

Murthy HM, Judge K, DeLucas L, Padmanabhan R 2000 Crystal structure of Dengue virus NS3 protease in complex with a Bowman-Birk inhibitor: implications for flaviviral polyprotein processing and drug design. J Mol Biol 301:759–767

Nomaguchi M, Ackermann M, Yon C, You S, Padmanbhan R 2003a De novo synthesis of negative-strand RNA by dengue virus RNA-dependent RNA polymerase in vitro: nucleotide, primer, and template parameters. J Virol 77:8831–8842

Nomaguchi M, Teramoto T, Yu L, Markoff L, Padmanabhan R 2003b Requirements for West Nile virus (−)- and (+)-strand subgenomic RNA synthesis in vitro by the viral RNA-dependent RNA polymerase expressed in Escherichia coli. J Biol Chem 279:12141–12151

Uchil PD, Satchidanandam V 2003a Architecture of the flaviviral replication complex: protease, nuclease and detergents reveal encasement within double-layered membrane compartments. J Biol Chem 278:24388–24398
Uchil PD, Satchidanandam V 2003b Characterization of RNA synthesis, replication mechanism, and in vitro RNA-dependent RNA polymerase activity of Japanese encephalitis virus. Virology 307:358–371
Utama A, Shimizu H, Morikawa S et al 2000 Identification and characterization of the RNA helicase activity of Japanese encephalitis virus NS3 protein. FEBS Lett 465:74–78
Valle RP, Falgout B 1998 Mutagenesis of the NS3 protease of dengue virus type 2. J Virol 72:624–632
Wengler G, Czaya G, Farber PM, Hegemann JH 1991 In vitro synthesis of West Nile virus proteins indicates that the amino-terminal segment of the NS3 protein contains the active centre of the protease which cleaves the viral polyprotein after multiple basic amino acids. J Gen Virol 72:851–858
Yon C, Teramoto T, Mueller N et al 2005 Modulation of the nucleoside triphosphatase/RNA helicase and 5′-RNA triphosphatase activities of dengue virus type 2 nonstructural protein 3 (NS3) by interaction with NS5, the RNA-dependent RNA polymerase. J Biol Chem 280:27412–27419
You S, Padmanabhan R 1999 A novel in vitro replication system for Dengue virus. Initiation of RNA synthesis at the 3′-end of exogenous viral RNA templates requires 5′- and 3′-terminal complementary sequence motifs of the viral RNA. J Biol Chem 274:33714–33722
You S, Falgout B, Markoff L, Padmanabhan R 2001 In vitro RNA synthesis from exogenous dengue viral RNA templates requires long range interactions between 5′- and 3′-terminal regions that influence RNA structure. J Biol Chem 276:15581–15591
Yusof R, Clum S, Wetzel M, Murthy HM, Padmanabhan R 2000 Purified NS2B/NS3 serine protease of dengue virus type 2 exhibits cofactor NS2B dependence for cleavage of substrates with dibasic amino acids in vitro. J Biol Chem 275:9963–9969
Zhang L, Mohan PM, Padmanabhan R 1992 Processing and localization of Dengue virus type 2 polyprotein precursor NS3-NS4A-NS4B-NS5. J Virol 66:7549–7554

DISCUSSION

Rice: Do you need all four of those basic amino acids to preserve the RNA binding activity?

Padmanabhan: We are doing this experiment now.

Rice: You mentioned that it was defective for replication if you put it in the replicon. Do you know what step in replication is compromised—translation, minus or plus strand RNA synthesis?

Padmanabhan: No. We do not know that at present. We have to quantify the minus and plus strands in the replicon-transfected cells to know at which step the RNA-binding of NS3 is important.

Canard: Have you looked at the interaction between an NS3 and NS5 using Biacore? Do you measure an interaction?

Padmanabhan: NS3 and NS5 interact was shown by us using Biacore, but we have not determined the dissociation constant for the NS3/NS5 interaction for the wild-type and mutant NS3 proteins yet.

Vasudevan: The NS5 protein is expressed in *E. coli*. Have you tried phosphorylating it and then looking at NS3/NS5 interactions?

Padmanabhan: No, but that would be a good experiment to do.

Satchidanandam: If you put your compounds which inhibit NS3 activity on infected cells and look at the RNA synthesis activity, what happens?

Padmanabhan: We haven't done this yet. We first intend to do this with the replicon. We have not done an infectivity assay either with the West Nile virus.

Rice: Have you been able to map the sites of phosphorylation, and are they close to this basic region?

Padmanabhan: No. We have not mapped the phosphorylation sites on NS5. Moreover, we do not know which cellular kinase is involved in phosphorylation of NS5 in dengue virus-infected cells.

Canard: I read in your abstract that polymerase is found in the nucleus. This has been known for a while. Why is NS5 going there? Do we know if NS5 goes into the nucleus with the viral RNA or just alone to do other things?

Satchidanandam: We have seen both NS3 and NS5 inside the nucleus. They are co-localized at membrane-bound sites attached to the inner nuclear membrane. We can detect de novo RNA synthesis there. NS5 is also distributed towards the nucleoplasm. However, we don't find RNA-synthesizing activity all over, except where it co-localizes with NS3. I don't know why two sites are needed for making viral RNA.

Canard: There are certainly strange things going on in the complex. If you look at the methyltransferase activity of the methyltransferase domain alone you will find good activity, but if you use the full-length NS5 you don't see much, whereas you find normal polymerase activity. There is an access problem of the capped RNA within the methyltransferase when it is in full-length NS5.

Padmanabhan: The complex of NS3 and NS5 may have guanylyltransferase and/or 7-methyl guanine methyltransferase activity. Membrane localization may also play a role. With Semliki Forest virus of the alphavirus family, the interaction of nsP1 with membrane lipids has been shown to be important for capping (Salonen et al 2005). This may also be true for flavivirus.

Fairlie: One of the things that you showed is that the helicase activity rose with the NS2B/NS3. Is the protease activity also enhanced?

Padmanabhan: When we use the NS2B/NS3 full-length construct, there is protease activity catalysing the auto-proteolysis of the protein. Clearly, the folding of the protease domain enhances the helicase activity, especially when the auto-proteolysis is blocked by mutation of histidine 51 to alanine (Yon et al 2005).

Gamarnik: It has been previously reported that NS3 has an internal cleavage site (Teo & Wright 1997). After the protease cleaves itself does the RNA binding domain stays in the N- or C-terminal of the protein?

Padmanabhan: I am not sure whether the internal cleavage of NS3 has any biological significance. It may be an artefact of *in vitro* systems. In my opinion it doesn't have any regulatory role.

Gamarnik: Where is the RNA binding domain of NS3?

Vasudevan: The RNA binding is in the helicase domain. The ATPase is the interphase between first and second subdomains; the RNA binding involves subdomain 3.

Kuhn: I thought your binding is around 184. That is N-terminus of the internal cleavage.

Padmanabhan: I am sorry that I misunderstood Andrea Gamarnik's question. Richard Kuhn is correct. The positively charged motif, RKRK, starts at position 184. It is between the protease domain and the first of the conserved helicase motifs and also at the N-terminal to the internal cleavage site. I am not sure whether the internal cleavage of NS3 has any biological significance

Kuhn: Cleavage is occurring on the helicase, on an α helix.

Gamarnik: The RNA binding domain is in the helicase, then?

Rice: One has to be a little careful here. We are talking about basic residues in a region between the protease and helicase. Mutagenesis can modulate RNA binding, but this doesn't necessarily mean that those four residues are directly involved in RNA binding. Domain 3 of the helicase is probably involved in interaction with the RNA substrate.

Padmanabhan: Also the RNA helicase works by binding to a single-stranded RNA first, and then it moves and binds a double-stranded RNA.

Rice: We don't want to get into this!

Padmanabhan: We don't really know the role of these four basic amino acids. They seem to be important for single-stranded RNA binding *in vitro*, as well as being important for replication in a replicon-based assay.

Kuhn: Do you know whether this stimulation of the helicase with NS5 requires binding?

Padmanabhan: We don't know, but we assume it requires binding to NS3. We need to look at the mutants.

References

Salonen A, Ahola T, Kaariainen L 2005 Viral RNA replication in association with cellular membranes. Curr Top Microbiol Immunol 285:139–173

Teo KF, Wright PJ 1997 Internal proteolysis of the NS3 protein specified by dengue virus 2. J Gen Virol 78:337–341

Yon C, Teramoto T, Mueller N et al 2005 Modulation of the nucleoside triphosphatase/RNA helicase and 5′-RNA triphosphatase activities of dengue virus type 2 nonstructural protein 3 (NS3) by interaction with NS5, the RNA-dependent RNA polymerase. J Biol Chem 280:27412–27419

Towards the design of flavivirus helicase/NTPase inhibitors: crystallographic and mutagenesis studies of the dengue virus NS3 helicase catalytic domain

Ting Xu*[1], Aruna Sampath†[1], Alex Chao†, Daying Wen†, Max Nanao§, Dahai Luo*, Patrick Chene‡, Subhash G. Vasudevan† and Julien Lescar*[2]

* *School of Biological Sciences, Nanyang Technological University, 60, Nanyang Drive Singapore 637551,* † *Novartis Institute for Tropical Diseases, 10 Biopolis Road, Chromos Building, Singapore 138670,* ‡ *Novartis Institute for Biomedical Research, Oncology Department, CH-4002 Basel, Switzerland and* § *EMBL, Grenoble Outstation, Grenoble, 38000, France*

Abstract. Infectious diseases caused by flaviviruses are important emerging public health concerns and new vaccines and therapeutics are urgently needed. The NS3 protein from flavivirus is a multifunctional protein with protease, helicase and nucleoside 5′ triphosphatase activities (NTPase). Thus, NS3 plays a crucial role in viral replication and represents an interesting target for the development of specific antiviral inhibitors. We have solved the structure of an enzymatically active fragment of the dengue virus NTPase/helicase C-terminal catalytic domain in several related crystal forms. The structure is composed of three domains, bears an asymmetric distribution of charges and comprises a tunnel large enough to accommodate single strand RNA. A concave face formed by domains 2 and 3 is proposed to bind a nucleic acid duplex substrate. Comparison of the various copies of dengue and yellow fever virus NS3 NTPase/helicase catalytic domains reveals mobile regions of the enzyme. Such dynamic behaviour is likely to be coupled with directional translocation along the single strand nucleic acid substrate during strand separation. We used structure-based site directed mutagenesis to identify regions of the enzyme that are crucial for its ATPase or nucleic acid duplex unwinding activity.

2006 New treatment strategies for dengue and other flaviviral diseases. Wiley, Chichester (Novartis Foundation Symposium 277) p 87–101

[1] These two authors have contributed equally to this work.
[2] This paper was presented at the symposium by Julien Lescar, to whom correspondence should be addressed.

Dengue fever is a viral infection transmitted by *Aedes* mosquitoes to humans with tens of million cases occurring annually. Dengue virus is an increasingly important problem for countries in tropical and subtropical regions with several billion people living in areas where one of the four serotypes of dengue virus, DENV-1–4, circulate. Severe forms of the disease called dengue haemorrhagic fever and dengue shock syndrome occur in a number of cases and new vaccines and therapeutics are urgently needed to combat this disease. Other important human pathogens such as yellow fever, Japanese encephalitis and West Nile virus also belong to the *Flaviviridae* family of enveloped viruses. The dengue virion contains a single stranded, positive sense RNA genome of approximately 11 kb which is translated into a large polyprotein during the infectious life cycle. Cellular and viral proteases process the polyprotein into three structural (C, prM and E, a class II viral fusion protein [Bressanelli et al 2004, Lescar et al 2001]) and seven non-structural proteins (NS1, NS2A, NS2B, NS3, NS4A, NS4B and NS5) (see Lindenbach & Rice 2001 or Kuhn & Rossmann 2005 for reviews).

NS3 is a multifunctional protein of 618 amino acids, endowed with protease, helicase, NTPase and 5′-terminal RNA triphosphatase activities. Its N-terminal 180 amino acids comprise a serine protease domain, and its C-terminal domain is thought to be involved in viral genomic replication by disrupting secondary structure elements or duplexes of RNA, a putative prerequisite for efficient viral RNA replication by the RNA-dependent RNA polymerase NS5. The region spanning residues 180–618 of the Den NS3 comprises two amino-acid sequence motifs named Walker A: GK(S/T), and Walker B: DEx(D/H) that are found in nucleotide binding proteins coupling NTP hydrolysis with directional movement, nucleic acid duplex destabilization, RNA processing and DNA recombination and repair (Walker et al 1982, Caruthers & MacKay 2002). Functionally, the helicase and NTPase activities of the NS3 protein have been characterized for several members of the *Flaviviridae* including hepatitis C virus (HCV) (Utama et al 2000), dengue virus (Benarroch et al 2004), West Nile virus (WNV) (Borowski et al 2001), yellow fever virus (YFV) (Warrener et al 1993) and Japanese encephalitis virus (JEV) (Utama et al 2000). Dengue virus with impaired helicase activity is not able to replicate, demonstrating the importance of NS3 in the *Flaviviridae* life cycle (Matusan et al 2001). In the helicase structures determined so far, a 'core' α/β structural motif of about 150 amino acids (having structural similarity with the RecA protein involved in homologous DNA recombination (Story & Steitz 1992) is visible as a tandem that has probably arisen through gene duplication. Amino acids from the most conserved motifs Walker A and B belong to the amino-terminal α/β domain and interact with the NTP substrate and Mg^{2+} respectively. Helicases are enzymes that convert chemical energy obtained through ATP hydrolysis into mechanical energy used to separate a nucleic acid duplex into its individual strands. Dengue NS3 provides a simple model to understand this activity at the atomic level. Here we summarize recent crystallographic and mutagenesis data on the dengue virus serotype 2 NTPase/helicase domain and suggest possible routes for the design of compounds with antiviral activity.

Materials and methods

Methods for the cloning, purification and structure determination have been described (CCP4 1994, Xu et al 2005). In brief, the catalytic domain of the DENV-2 helicase (Den NS3: 171–618) and the full length NS3 protein were expressed in *E. coli* as fusion proteins with at their N-terminus the thioredoxin and a $(His)_6$ tag followed by an enterokinase cleavage site. Crystals of dengue NS3: 171–618 were grown by the hanging drop vapour diffusion method over wells containing 0.1 M MES pH 6.5, 0.2 M $(NH_4)_2SO_4$, 14% polyethylene glycol 8000 and 10 mM $MnCl_2$. For data collection, crystals were soaked in the same solution containing 25% glycerol for cryoprotection, mounted and cooled to 100 K in a nitrogen gas stream. Diffraction intensities were recorded on beamline ID14-4 at the E.S.R.F. (Grenoble, France). Integration, scaling, merging of the intensities, model building and refinement were carried out as described (Xu et al 2005). Mutagenesis was performed with standard PCR methods using the full length NS3 protein.

Results and discussion

Functional characterization

The helicase catalytic domain of NS3 displays strand displacement activity using both dsDNA and dsRNA as substrates (Xu et al 2005). We compared the unwinding activity of the truncated NS3 protein with the full length NS3 protease/helicase (NS3FL). Interestingly, NS3FL shows a significantly higher unwinding activity compared to the helicase domain. In the absence of a structure for the full length NS3 protein, how the N-terminal protease domain of dengue NS3 influences the helicase activity is currently unknown. We determined the kinetic parameters for ATP hydrolysis for both NS3: 171–618 and NS3FL. Both enzymes have similar turnover numbers ($k_{cat} = 6.9\,s^{-1}$ and $5.9\,s^{-1}$, for NS3: 171–618 and NS3FL respectively). NS3: 171–618 has a higher affinity for ATP (K_m 33 ± 3 μM) than NS3FL (K_m 297 ± 34 μM). Thus, the NS3 fragment we crystallized is enzymatically active both in NTP hydrolysis and duplex unwinding. Enzymatic activities are likely to be further regulated by additional protein-protein interactions in the context of the viral replication complex.

Overall structure

The structure of dengue NS3 helicase (Fig. 1) comprises three domains of equal dimensions. Domain I (residues 181–326) and domain II (residues 327–481) show little sequence identity with each other, but are structurally similar. They feature a large central six-stranded parallel β-sheet, flanked by four α-helices. Domain III (residues 482–618) is composed of four approximately parallel α-helices (α_1'', α_3'', α_4'', α_7''), surrounded by three shorter helices (α_2'', α_5'', α_6''), and augmented by two

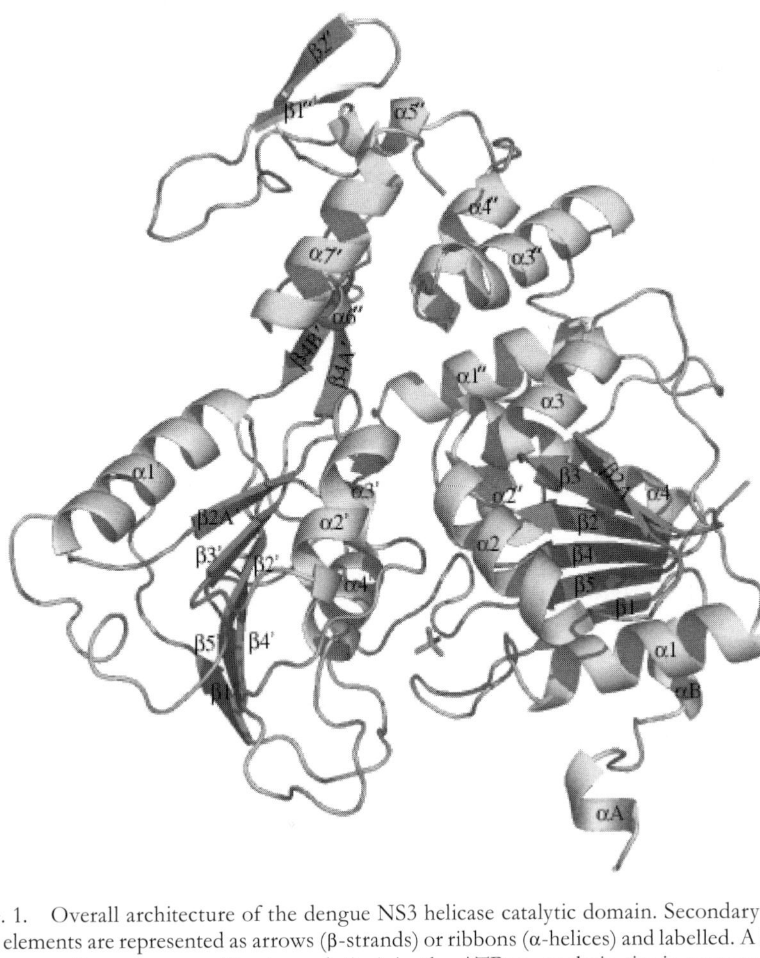

FIG. 1. Overall architecture of the dengue NS3 helicase catalytic domain. Secondary structure elements are represented as arrows (β-strands) or ribbons (α-helices) and labelled. A bound sulfate ion (from the crystallization solution) in the ATPase catalytic site is represented as sticks.

anti-parallel β-strands largely exposed to the solvent. A long β-hairpin (β_{4A}', β_{4B}') extends from domain II into domain III. A tunnel runs across the centre of the most basic face of the protein. A side-by-side comparison between the Den NS3: 171–618 and the HCV NS3 helicase structures highlights the conservation of the tandem core structure while the third terminal domain bears no structural similarity (Fig. 2), an observation consistent with the pattern of amino-acid sequence identities between the individual domains of den NS3 and their counterparts in HCV (Xu et al 2005).

A superposition with the yellow fever virus helicase (Wu et al 2005) reveals possible hinge motion between domain III and the RecA tandem repeat, as well as movements in helices α1′ and α7′ (Fig. 3). Recently, residues 303–618 of dengue NS3 were shown to bind to the RNA dependent RNA polymerase NS5 (Brooks et al 2002). This interaction might involve the C-terminal domain III of dengue NS3. Indeed, a major challenge for future investigations consists in the precise mapping of the intermolecular interactions that take place within a functional viral replication complex.

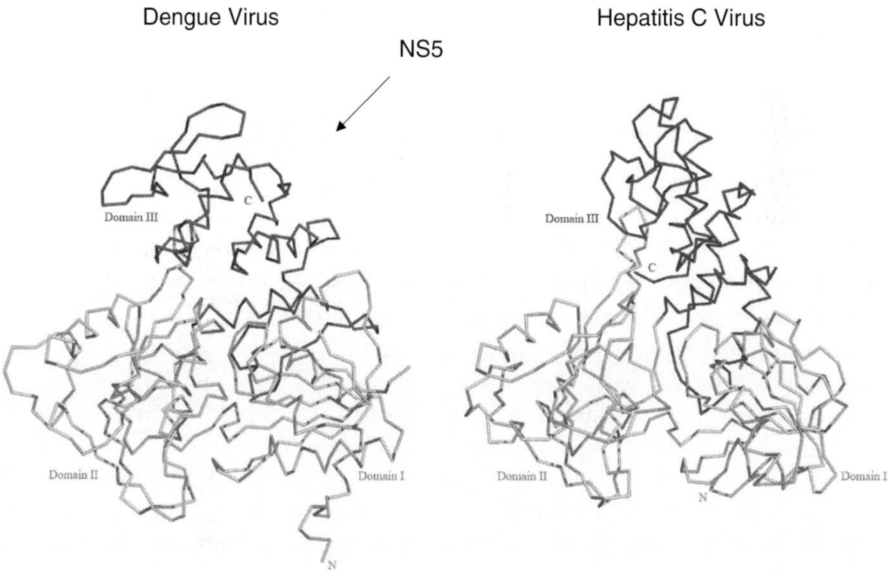

FIG. 2. Side by side view of dengue virus NS3 (left) and HCV helicase catalytic domains (right) (Kim et al 1998) displayed in the same orientation. Only the α-carbon atoms traces are shown. Their structurally conserved RecA-like domains I and II are shaded in grey. Their domains III bear no significant structural similarity. The putative binding site of NS5 is indicated.

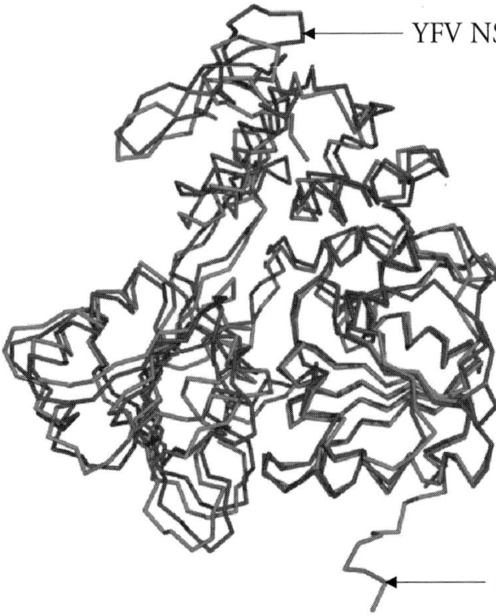

FIG. 3. Comparison of flavivirus helicase catalytic domains. The Cα traces of yellow fever virus and dengue virus helicases were superimposed, giving a r.m.s deviation of 2.3 Å for 415 residues.

The NTPase active site

The NTP substrate is primarily bound through ionic interactions through its phosphate moieties whilst the ribose and base bulge out from the NTP binding cavity (Wu et al 2005, Xu et al 2005). As a result, there are few structural constraints for their recognition by the enzyme, an observation consistent with a lack of specificity for the base. This is also consistent with the capacity of NS3 to hydrolyse short oligo-ribonucleotides regardless of the nature of the base moiety (Benarroch et al 2004). We obtained a crystal form (Table 1), with a Mn^{2+} and a sulfate ion from the crystallization buffer located in the ATP binding pocket. The cation makes hydrogen bonds with the carboxylate group of Glu285 of the DExH motif II and is further coordinated by six water molecules (Fig. 4). A comparison of the orientations of the residues in this cavity with the yellow fever virus helicase bound to ADP (Wu et al 2005) shows that a sulfate ion is located at a position close to the γ-phosphate of an ATP substrate. Thus, this structure might resemble a product state that would follow the release of ADP from the NTP binding pocket.

TABLE 1 Refinement statistics (Mn^{2+} complex)

Resolution range (Å)	20.0 – 2.75
Intensity cutoff (F/σ[F])	0.
No of reflections: completeness (%)	97.7
Used for refinement	24536
No of non hydrogen atoms	1243
Protein	3140, 3480
missing residues[a]	(38, 11)
SO_4^{2-}	3
Water molecules	229
Mn^{2+}	2
Rfactor[b] (%)	20.6
Rfree[c] (%)	26.4
Rms deviations from ideality	
Bond lengths (Å)	0.0071
Bond angles (°)	1.401
Ramanchandran plot	
Residues in most favoured regions (%)	84.0
Residues in additional allowed regions (%)	16.0

[a]Values are given for molecule 1 and 2, respectively.
[b]$Rfactor = \Sigma \|\|F_{obs}\| - \|F_{calc}\|\| / \Sigma \|F_{obs}\|$.
[c]Rfree was calculated with 5% of reflections excluded from the whole refinement procedure.

Nucleic acid binding sites

A tunnel lined with a number of basic residues runs approximately through the centre of the structure. Based on previous structural work on the HCV and PcrA helicases (Yao et al 1997, Kim et al 1998, Velankar et al 1999) we propose the following working model for a nucleic acid substrate binding by dengue NS3: 171–618 (Fig. 5). Translocation in the $3' \rightarrow 5'$ direction along a single-stranded nucleic acid tail of about 6–8 (ribo)-nucleotides, trapped in the tunnel, would result from interdomain movements triggered by the hydrolysis of a nucleotide at the NTP binding site. The duplex portion of the substrate would contact the concave surface between domains II and III (Fig. 5).

Structure-based site directed mutagenesis

Based on the 3D structure, we carried out site-directed mutagenesis to probe the function of a number of residues for ATPase activity and duplex unwinding. Fourteen mutations were introduced within the helicase domain of the NS3 protein and each mutated protein was tested for its ATPase, RNA helicase and RNA binding activities. A mapping of residues identified in this study that are critical

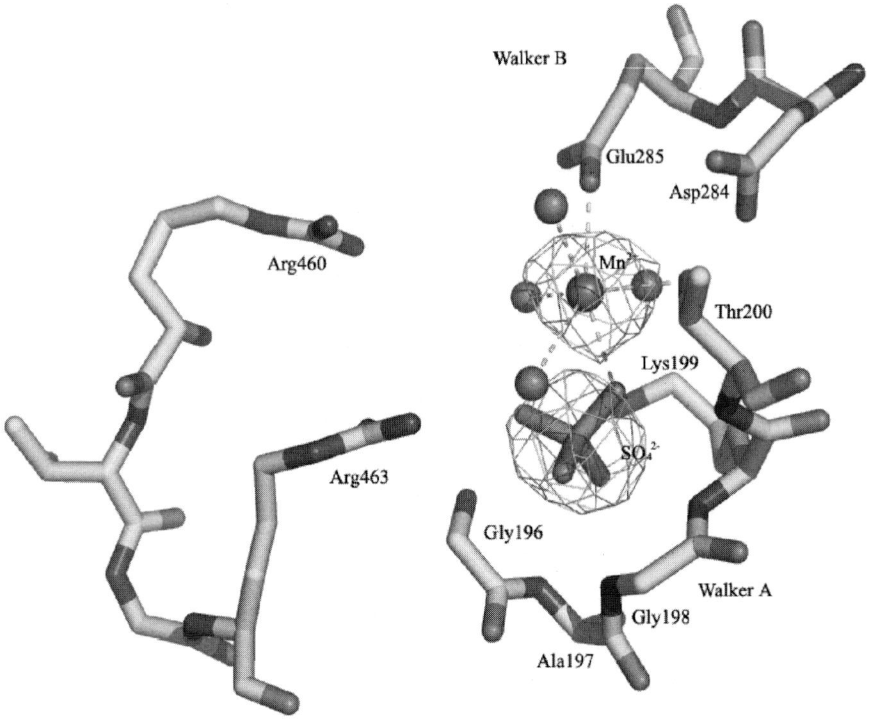

FIG. 4. Close-up view of the ATP binding pocket of dengue NS3: 171–618 (sticks) with a bound Mn^{2+} and a sulfate ion from the crystallization solution. Residues are labelled. The divalent Mn^{2+} (represented as a dark grey sphere) is coordinated by residues Glu285 and four water molecules.

for enzymatic activity is shown in Fig. 6. We identified four residues which selectively disrupt either the ATPase (Arg460, Arg463) or helicase (Ile365, Arg376) activity of the full-length protein, when individually mutated to alanine (Fig. 6). The sole substitution into alanine of Lys396 which is located at the surface of domain II resulted in a NS3 protein having lost both enzymatic activities. We propose that residue Ile365 which is found at the tip of domain II and lines the putative single-strand RNA binding tunnel at a position closed to the fork (Fig. 5), is crucial for translocation of the enzyme along the nucleic acid substrate.

Conclusion

We reported two mutations (Ile365Ala, Arg376Ala) for which the helicase activity is completely abolished despite an intact ATPase activity. Interestingly, a shallow

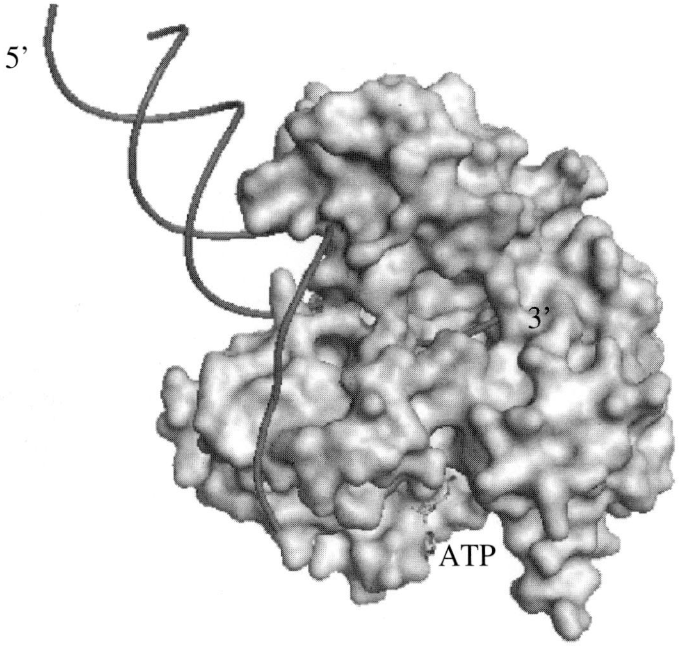

FIG. 5. Surface representation of the Den NS3 helicase domain in the same orientation as in Fig. 1. The putative interaction between the enzyme and a nucleic acid substrate is shown. The model was generated in part by superposition with the HCV helicase in complex with a dU_8 oligonucleotide (PDB code: 1A1V; Kim et al 1998).

pocket is located in the vicinity of Ile365, lined by residues protruding from domain II (Thr 408, Asp 409, Leu 443) and domain III (Asp 603, Leu 605). This previously unidentified region could be a regulatory site controlling structural transitions between closed and open forms of the enzyme. Taken together, our results point to the helicase activity for the design of antiviral therapeutics: compounds binding to allosteric pockets might achieve both specificity and activity by hindering enzyme movements required for nucleic acid strand separation. Ongoing structural work on the NS2B/NS3 protease, NS3 helicase domain, NS5 methyltransferase and polymerase catalytic regions hold much promise for the future of rational drug design against flaviviruses.

Our hope is that the development of specific compounds with antiviral activity against flaviviruses will parallel the successes that recently followed structural determinations of key viral proteins from HCV or HIV.

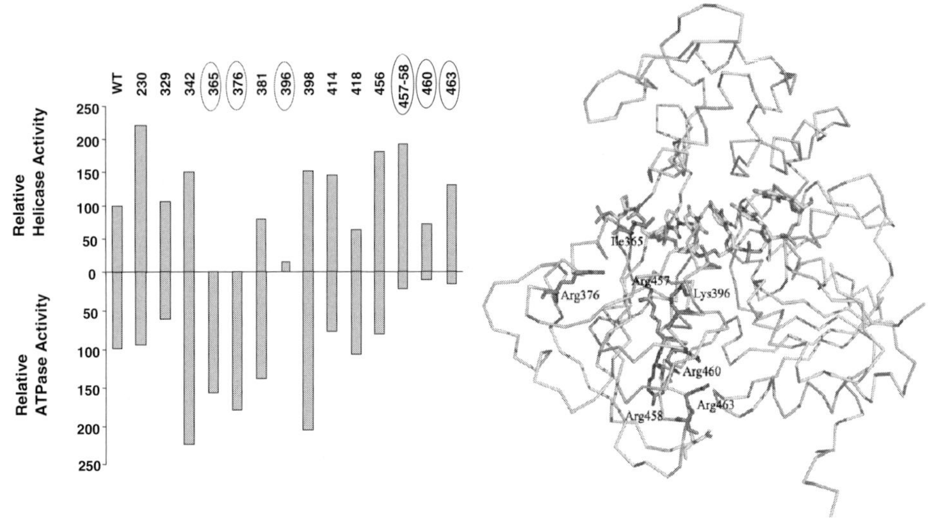

FIG. 6. Results of the site-directed mutagenesis study. Left panel shows the residual ATPase (lower histogram) and helicase activity (upper histogram) for each residue which was mutated into alanine in our study. The activities are expressed as a percentage of wild-type which was arbitrarily set to 100%. Residues which drastically disrupt either of the two enzymatic functions when substituted by alanine are circled. Right panel shows the mapping of these residues onto the 3D structure of NS3 helicase.

Acknowledgements

Financial support via Grants from N.T.U. (RG29/05), the Singapore Biomedical Research Council (03/1/21/20/291 and 02/1/22/17/043) and the Singapore National Medical Research Council (NMRC/SRG/001/2003) to J.L. laboratory is acknowledged as well as provision of excellent beam-time and support by the E.S.R.F. (Grenoble, France).

References

Benarroch D, Selisko B, Locatelli GA, Maga G, Romette JL, Canard B 2004 The RNA helicase, nucleotide 5′-triphosphatase, and RNA 5′-triphosphatase activities of Dengue virus protein NS3 are Mg^{2+}-dependent and require a functional Walker B motif in the helicase catalytic core. Virology 328:208–218

Borowski P, Niebuhr A, Mueller O et al H 2001 Purification and characterization of West Nile virus nucleoside triphosphatase (NTPase)/helicase: evidence for dissociation of the NTPase and helicase activities of the enzyme. J Virol 75:3220–3229

Bressanelli S, Stiasny K, Allison S et al 2004 Structure of a flavivirus envelope glycoprotein in its low-pH-induced membrane fusion conformation. EMBO J 23:728–738

Brooks AJ, Johansson M, John AV, Xu Y, Jans DA, Vasudevan SG 2002 The interdomain region of dengue NS5 protein that binds to the viral helicase NS3 contains independently functional importin beta 1 and importin alpha/beta-recognized nuclear localization signals. J Biol Chem 277:36399–36407

Caruthers JM, McKay DB 2002 Helicase structure and mechanism. Curr Opin Struct Biol 12:123–133
Collaborative Computational Project Number 4 1994 The CCP4 suite: programs for protein crystallography. Acta Crystallogr D Biol Crystallogr 50:760–763
Kim JL, Morgenstern KA, Griffith JP et al 1998 Hepatitis C virus NS3 RNA helicase domain with a bound oligonucleotide: the crystal structure provides insights into the mode of unwinding. Structure 6:89–100
Kuhn RJ, Rossmann MG 2005 Structure and assembly of icosahedral enveloped RNA viruses. Adv Virus Res 64:263284
Lescar J, Roussel A, Wien MW et al 2001 The fusion glycoprotein shell of Semliki Forest Virus: an icosahedral assembly primed for fusogenic activation at endosomal pH. Cell 105:137–148
Lindenbach BD, Rice CM 2001 Flaviviridae: The viruses and their replication. In: Fundamental virology. Lippincott-Raven, Philadelphia, p 589–639
Matusan AE, Pryor MJ, Davidson AD, Wright PJ 2001 Mutagenesis of the Dengue virus type 2 NS3 protein within and outside helicase motifs: effects on enzyme activity and virus replication. J Virol 75:9633–9643
Story RM, Steitz TA 1992 Structure of the recA protein-ADP complex. Nature 355:374–376
Utama A, Shimizu H, Hasebe F et al 2000 Role of the DExH motif of the Japanese encephalitis virus and hepatitis C virus NS3 proteins in the ATPase and RNA helicase activities. Virology 273:316–324
Velankar SS, Soultanas P, Dillingham MS, Subramanya HS, Wigley DB 1999 Crystal structures of complexes of PcrA DNA helicase with a DNA substrate indicate an inchworm mechanism. Cell 97:75–84
Walker J E, Saraste M, Runswick MJ, Gay NJ 1982 Distantly related sequences in the alpha- and beta-subunits of ATP synthase, myosin, kinases and other ATP-requiring enzymes and a common nucleotide binding fold. EMBO J 1:945–951
Warrener P, Tamura JK, Collett MS 1993 RNA-stimulated NTPase activity associated with yellow fever virus NS3 protein expressed in bacteria. J Virol 67:989–996
Wu J, Bera AK, Kuhn RJ, Smith JL 2005 Structure of the Flavivirus helicase: implications for catalytic activity, protein interactions, and proteolytic processing. J Virol 79:10268–10277
Xu T, Sampath A, Chao A et al 2005 Structure of the dengue virus helicase/NTPase catalytic domain at a resolution of 2.4 Å. J Virol 79:10278–10288
Yao N, Hesson T, Cable M et al 1997 Structure of the hepatitis C virus RNA helicase domain. Nat Struct Biol 4:463–467

DISCUSSION

Canard: How do you place the polymerase?

Lescar: The only mapping that was done was before we got the structure.

Vasudevan: It was with the yeast two-hybrid system. Most likely it is in domain III.

Canard: So, if I understand correctly, the helicase is behind the polymerase, and the helicase is unwinding after the polymerase. The polymerase is synthesizing a double-stranded RNA, and the helicase is behind, unwinding.

Kuhn: Remember, the region is believed to be between the two domains of NS5. The polymerase, RDRP, could be wrapped around.

Canard: I have a problem with this. It can be ahead, unwinding double stranded RNA to produce single-strand RNA. Or it can be behind, unwinding, but the polymerase has just turned round.

Vasudevan: This is only a model! The single-stranded RNA could well be going up towards NS5, using that as a template. It could be sticking up and you would still get replication.

Canard: NS3 could be ahead. It is known that there is a lot of double-stranded RNA in the cell. If this were the case, there wouldn't be any double-stranded RNA. A Danish group in Aarhus have assayed double-stranded RNA from many different viruses. They find a lot of double-stranded RNA from flaviviruses.

Rice: What is the evidence that it has anything to do with the elongation reaction at all?

Vasudevan: The whole thing is part of a complex. In the virus, the non-structural proteins appear to form a replication complex.

Rice: That's what people say. I am wondering what the evidence is.

Satchidanandam: NS3 has to go ahead of the polymerase, in order to unwind the double stranded RF template. In all flaviviruses, there is always a 10- to 20-fold excess of plus strands over minus strands inside the infected cell. It also appears that all the plus strand has to be synthesized only from RF templates as there are no free negative strands available in the cell to serve as a template for plus strand synthesis. Thus, the role of NS3 is pivotal for making viral genomic RNA (plus strands).

Young: It is a form that is expressed by every single RNA virus that transcribes its genome. David Baltimore put together that classification in the 1970s. For every RNA virus there is a significant accumulation of double-stranded RNA.

Gamarnik: We really don't know when the helicase activity is necessary during the RNA replication process.

Padmanabhan: NS3 does not seem to stimulate the NS5 polymerase activity, at least in our hands.

Satchidanandam: With your recombinant NS5 protein you can do both primed and *de novo* synthesis *in vitro*, with your short templates. There you don't seem to need NS3.

Padmanabhan: It is only negative-strand synthesis, so it doesn't have all the secondary structures.

Satchidanandam: You are not using double-strand RNA as your template. Your templates are all single stranded.

Rice: If you do use double-stranded RNA, say with a single-stranded tail on a primer, is the polymerase capable of making a strand? I suspect you will find the answer is yes.

Gamarnik: It is likely that you need the helicase activity.

Rice: That would be wonderful.

Kuhn: Do you know what role that first N-terminal helix plays.

Lescar: It might be an artefact.

Kuhn: Our structure doesn't have that. Our structure lacks helicase activity until you extend it a few amino acids, when it probably would be comparable to what you are showing. We have to go a few amino acids beyond what we have the structure for, so I wonder whether this actually has some kind of function.

Lescar: In our crystal form (Xu et al 2005), the N-terminal helix of the dengue NS3 helicase catalytic domain (residues 171–618) is in contact with another helicase molecule present in the asymmetric unit. I was tempted to say that it was an artefact due to crystal packing forces, but we have no definitive evidence since this segment is absent in the slightly shorter yellow fever virus NS3 helicase structure solved by Wu et al (2005). If we use the full length protease-helicase 3D structure from HCV (Yao et al 1999) as a template, and superimpose the dengue helicase domain onto it, then this rather naive modelling doesn't work well because a much longer linker would then be needed in order to connect the helicase and protease domains from Dengue virus NS3 protein. This suggests that the N-terminal α-helix is probably not present in the native structure of the full-length NS3 bifunctional enzyme.

Rice: What are people's thoughts on developing helicase inhibitors? This is one of the more complicated enzymatic reactions from a screening perspective.

Vasudevan: We are developing this as one of the targets here at NITD. We haven't done a full high-throughput screen yet, but we have just developed an assay that could be used in such a screen. This will be done by the end of 2005.

Keller: The HCV helicase has been a very difficult target for drug discovery. I think it will also be rather challenging for dengue. We have easier targets, but we are going to give the helicase a good shot. No one has developed any antivirals yet which inhibit a helicase.

Rice: The trick is the primary screen and then the counter screens.

Vasudevan: The I365A mutant that Julien Lescar mentioned suggests that we may be able to find allosteric helicase inhibitors.

Kuhn: Whoever reviewed our NIH RCE grant suggested that we shouldn't go after any helicase targets!

Rice: Boehringer Ingelheim and ViroPharma had a joint program and screened hundreds of thousands of compounds.

Vasudevan: In dengue there aren't many targets, so this is well worth going after at this stage.

Rice: What kind of approaches are people taking to try to figure out what the helicase is doing in virus replication? What do we need to do in this area? I was just wondering what the viral replication structural biologists are doing to try to build up a higher-order understanding of how these various subunits fit together in an RNA amplification machinery.

Vasudevan: The only person who has published data in this area, looking at the enzyme in the replication complex is Vijaya Satchidanandam; everyone else has taken a reductionist approach and looked at expressed proteins.

Satchidanandam: I would say any drug that is targeting a component of the replicase is going to have problems accessing the enzyme. This is because of the membrane organization of the replicase complex.

Gamarnik: We don't know whether both RNA strands are made inside this replication complex involving cellular membranes. Polymerase inhibitors such as nucleoside analogues work on other viruses that have similar membranous structures, thus they must have access to the viral enzyme.

Satchidanandam: The flaviviruses are particularly membrane enclosed. The replicase and the double-stranded templates are completely within membrane structures. The RNAi work that is being done only works when the construct is transfected before infection. It never works if you put it in after infection.

Keller: It is not surprising that you can't get RNAi in, but with a small molecule compound it should be possible. I wouldn't worry too much about this. We can modulate properties of compounds quite well, and we can study how well they penetrate these systems. The question about the protein complex is a fascinating one. In inflammation we also have these protein complexes, and I don't think anyone has a good understanding of how they work.

Young: There are data by Khromykh and colleagues who found that in complementation experiments it only worked in *cis*—they couldn't complement in *trans* (Liu et al 2003). We are probably looking at a dynamic interaction between nucleic acid-protein and protein-protein, and it could be that the interactions are only occurring as these nascent molecules are actually folding. If we take post-folding molecules and stick them together they are just not going to work.

Satchidanandam: Haven't several groups tried doing RNAi for flavivirus infection?

Rice: It is not clear that you need to target an established replication complex in order to have effective antiviral activity. If you inhibit the formation of new complexes, that may be sufficient.

Canard: People working in companies have told me that they have screened several hundred thousand compounds for helicase inhibitory activity, and haven't found any that work. It could be that there is an engine performing ATP hydrolysis, and it is dynamic because movement is required for strand separation. In my opinion, this site is moving a lot when the ATP is being consumed. We have an idea of the structure, but it may be very different during helicase activity. Perhaps another site that is more static would be better to target for inhibition.

Kuhn: What goes along with that is that perhaps the helicase activity is not what we should be assaying for. Perhaps we need a polymerase assay.

Vasudevan: Or a replicon based assay.

Kuhn: That gets more complicated.

Padmanabhan: You can just use the stimulation of ATPase by polymerase.

Kuhn: You might interfere with interaction with the polymerase.

References

Liu WJ, Sedlak PL, Kondratieva N, Khromykh AA 2002 Complementation analysis of the flavivirus Kunjin NS3 and NS5 proteins defines the minimal regions essential for formation of a replication complex and shows a requirement of NS3 in cis for virus assembly. J Virol 76:10766–10775

Wu J, Bera AK, Kuhn RJ, Smith JL 2005 Structure of the flavivirus helicase: implications for catalytic activity, protein interactions, and proteolytic processing. J Virol 79:10268–10277

Xu T, Sampath A, Chao A et al 2005 Structure of the Dengue virus helicase/nucleoside triphosphatase catalytic domain at a resolution of 2.4 Å. J Virol 79:10278–10288

Yao N, Reichert P, Taremi SS, Prosise WW, Weber PC 1999 Molecular views of viral polyprotein processing revealed by the crystal structure of the hepatitis C virus bifunctional protease-helicase. Structure 7:1353–1363

Finding new medicines for flaviviral targets

Thomas H. Keller, Yen Liang Chen, John E. Knox, Siew Pheng Lim, Ngai Ling Ma, Sejal J. Patel, Aruna Sampath, Qing Yin Wang, Zheng Yin and Subhash G. Vasudevan

Novartis Institute for Tropical Diseases (NITD), 10 Biopolis Road, #05-01 Chromos, 138670 Singapore

Abstract. With the incidence of dengue fever increasing all over the world, there is an urgent need for therapies. While drug discovery for any disease is a long and difficult process with uncertain success, dengue fever poses an additional complication in that most of the target patient population is young and lives in developing countries with very limited health care budgets. Recent progress in drug discovery for dengue and an analysis of approaches toward hepatitis C virus (HCV) therapeutics suggest that NS5 polymerase is the most promising target for dengue. Moreover such inhibitors may be useful for several other flaviviral diseases. NS3 proteases will be more challenging targets, especially if oral delivery is desired. Recent work has shown that potent inhibitors can be designed readily, but optimization of pharmacokinetic parameters will probably be a long an arduous task, especially since the primary binding pockets prefer to bind basic amino acids. NS3 helicase can also be considered a viable drug target for flaviviral diseases. It has however proved to be a challenging for HCV and selectivity issues versus human helicases must be overcome.

2006 New treatment strategies for dengue and other flaviviral diseases. Wiley, Chichester (Novartis Foundation Symposium 277) p 102–119

With the incidence of dengue fever increasing all over the world, there is an urgent need for therapies. While drug discovery for any disease is a long and difficult process with uncertain success, dengue poses an additional complication in that most of the target patient population lives in developing countries with very limited health care budgets. In this article we will discuss the prospects for antivirals to treat dengue and other flavivirus infections and suggest the best avenue to achieve success in drug discovery.

Platform approach

During the work on ATP-competitive inhibitors for kinases, it became evident that there were strong synergies in assays, medicinal chemistry, drug design etc.

between different kinase projects and that it would be advantageous to attack the many drug targets in this class in a coordinated fashion (ter Haar et al 2004). As a consequence directed compound collections were created, which represented likely inhibitor scaffold. At the same time structural biology was emphasized, so that the interactions of any compound with the target compounds could be rapidly analysed, and structure–activity relationships would emerge. In addition a large panel of assays was set up to examine the selectivity of any emerging compound.

Such synergies clearly exist for flaviviral diseases. While it is currently unclear whether it will be possible to create drugs that have antiviral activity against several flaviviruses or whether different chemical entities are needed for dengue and for example Japanese encephalitis, there is no question that synergies especially in molecular biology, biochemistry and chemistry can be exploited to attack several of these viral diseases. NITD is following this avenue in all of our drug discovery efforts.

Target product profile

At the start of a drug discovery effort for any disease, it is important to consider the need of a potential patient. In dengue we have at least two potential points of intervention as shown in Fig. 1: an antiviral drug that would be used to reduce the viral load in the early stages of the disease or a modulator of host targets that would prevent or treat dengue haemorrhagic fever (DHF). Both approaches have potential challenges and pitfalls. While the effective window of opportunity for using an antiviral drug may be short, designing an immune-modulator poses the difficult

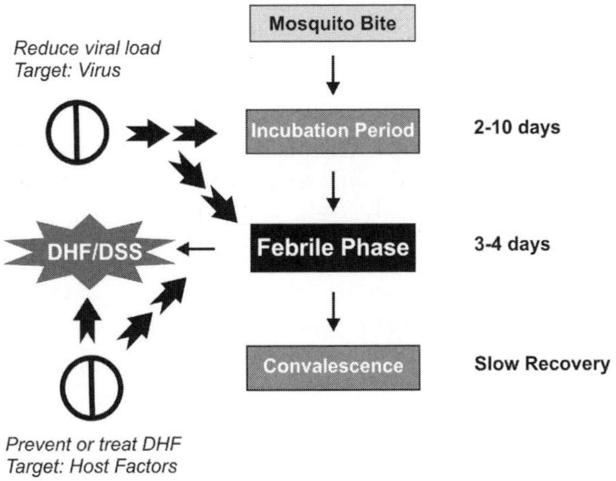

FIG. 1. Potential intervention strategy for dengue antivirals or modulators of host targets.

challenge of finding a treatment without unwanted side effects or toxicity. This latter problem is not made easier by our lack of detailed understanding of the pathogenesis profile of the disease and the absence of a disease model.

NITD has decided to initially focus on an antiviral drug whose target product profile is shown in Table 1. Since dengue epidemics occur mainly in developing countries, a cheap oral drug that neither induces nor inhibits liver enzymes and that can be given once a day would be ideal.

The influence of a drug on the cytochrome P450 enzymes is a very important consideration for flaviviral infections as rural patients in tropical countries often are afflicted with other diseases. In such situations it would be extremely important that any dengue drug does not interact with other medications. Furthermore the aim to keep the cost of goods as low as possible would suggest that a potent drug with a long half-life would be desirable.

The decision to aim for an oral drug has an important influence on the choice of drug targets, as such drugs must be potent inhibitors of the chosen target protein, but more importantly possess the physicochemical properties to ensure good oral pharmacokinetics (Lipinski & Hopkins 2004). It is therefore essential to choose targets that are likely to be inhibited by drug-like molecules, since the success of a lead optimization often depends on the ability of chemists to produce compounds with good pharmacokinetics. With these considerations a target portfolio has been developed for the viral enzyme activities that are essential for its replication in infected cells.

TABLE 1 Dengue drug: target product profile (TPP)

	Minimal product profile	*Added value*
Route of Administration	Oral	
Frequency of Dosing	Once a day	
Clinical Efficacy	1. Reduces symptoms 2. Reduces incidence of severe dengue disease 3. Active against all 4 serotypes	Also effective in other flaviviral diseases
Patient Population	1. Patients with severe Dengue symptoms. 2. High risk groups 3. Must be indicated for children ages 5 and up	1. Children younger than 5 years 2. Prophylactic use
Others	Stable to heat/humidity; long shelf life; low/reasonable costs of goods; ease of formulation/low cost exipients	

NS3 protein

Dengue NS3 protein is the second largest viral protein of the dengue virus with four distinct enzymatic activities: protease, RNA helicase, NTPase and RNA triphosphatase. It is a highly conserved protein among the serotypes 1–4 with 98% sequence similarity and 67% identity. It is also highly conserved across the flaviviridae family with more than 60% similarity and 50% identity.

NS3 protease

An obvious drug target in this protein is the viral serine protease domain in the N-terminal part. The enzymatic activity of this protein, which is dependent on at least 40 amino acids of NS2B, is vital for the post-translational proteolytic processing of the polyprotein precursor and is essential for viral replication and maturation of infectious dengue virions (Falgout et al 1991).

Leung et al (2001) have shown that the dengue protease domain fused to 40 amino acids of NS2B through a flexible protease resistant linker (CF40-Gly-NS3pro185) was soluble and catalytically active. A suitable substrate was identified by Li et al (2005) using a tetrapeptide library containing over 130 000 substrates. Not surprisingly this study found a preference for Arg at both P1 and P2 positions, the best substrate for all four serotypes of dengue having the sequence Bz-Nle-Lys-Arg-Arg-AMC (AMC = 7-Amido-4-methylcoumarin). The substrate profiling suggested that P1 and P2 are important both for binding and turnover, while changes in P3 mainly influence binding and P4 amino acids are essential for efficient catalysis.

With all the tools in hand we chose an assay based on dengue virus (DENV)-2 CF40-Gly-NS3pro185 with the fluorogenic substrate Bz-Nle-Lys-Arg-Arg-AMC as our high-throughput screening (HTS) assay. The result of the first HTS campaign was unfortunately not very encouraging. Despite screening over 1 000 000 compounds we were only able to identify a number of promiscuous inhibitors, which either formed aggregates or did not show useful structure–activity relationships (SAR). This outcome was not a total surprise, since serine proteases are notoriously difficult targets for HTS. Because of the importance of the target for our drug discovery effort, we are currently running a second HTS campaign with a slightly different assay format using full length DENV-2 NS3 in parallel with West Nile CF40-Gly-NS3pro185.

Traditionally most drug discovery projects for prototypical serine proteases like thrombin (Gustafsson et al 2004), elastase or HCV NS3 (Llinas-Brunet et al 1998) have started from inhibitors based on substrate peptide sequences. Such a peptidomimetic approach can often produce very potent inhibitors in a short period of time, however turning these early leads into drugs can be extremely challenging.

The failure of our first HTS made such an approach however necessary, since it allowed us to rapidly map the active site and assess the feasibility of a project based on structure-aided design. As a first step we synthesized a number of peptides with electrophilic warheads (Fig. 2). These warheads interact with the active site serine of the enzyme to form transition state analogues which are very convenient starting points for inhibitor design.

Table 2 shows a selection of peptides that were tested on NS3 protease enzymes from dengue 2, West Nile and yellow fever virus. While the sample is rather small, the data is useful to illustrate a few general trends in SAR that we observed during our work on over 70 peptides (Z. Yin, unpublished work 2005). A non-covalent, charged warhead such as a carboxylic acid theoretically could provide binding efficiency via electrostatic interactions. A product based peptide acid was reported as a competitive inhibitor of the hepatitis C virus NS3-4A serine protease with a K_i of 0.6 μM (Llinas-Brunet et al 1998). In our hands, carboxylic acid **1** failed to show any activity against Dengue serine protease. Simple amides are known to be relatively inert towards certain serine proteases and can serve as substrate-like enzyme inhibitors (Brady et al 1995). However tetrapeptide amide **2** showed only activity at high micromolar concentration.

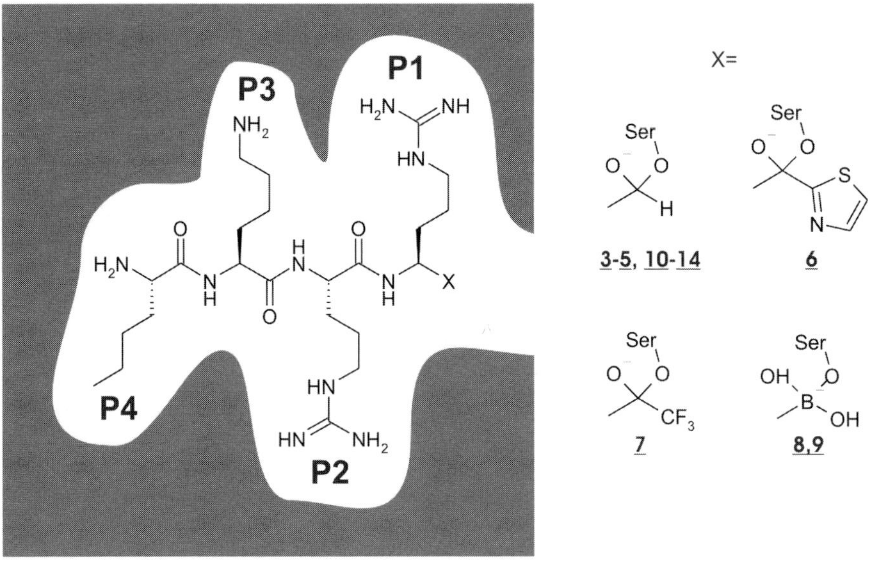

FIG. 2. Binding mode of transition state inhibitors **3–14**. Binding pockets of side chains are labelled P1 to P4 (see text). Transition state inhibitors contain an electrophilic carbonyl group or boron atom which forms a covalent but reversible bond to the active site serine.

TABLE 2 Comparison of peptide activity on flavivirus NS3 proteases[a]

	Inhibitors	K_i (μ*M*)[b] *Dengue 2*	K_i (μ*M*)[b] *Yellow fever*	*Ki* (μ*M*)[b] *West Nile*
1	Bz-Nle-Lys-Arg-Arg-OH	>500	nd	nd
2	Bz-Nle-Lys-Arg-Arg-NH2	>500	nd	nd
3	Bz-Nle-Lys-Arg-Arg-H	5.8	0.36	4.10
4	Bz-Lys-Arg-Arg-H	1.30	0.34	0.75
5	Bz-Arg-Arg-H	10.00	0.74	2.15
6	Bz-Nle-Lys-Arg-Arg-Thiazole	36.60	10.50	13.20
7	Bz-Nle-Lys-Arg-Arg-CF3	0.70	0.38	0.30
8	Bz-Nle-Lys-Arg-Arg-B(OH)2	0.04	0.05	0.03
9	Bz-Lys-Arg-Arg-B(OH)2	0.20	0.04	0.04

[a] Preparation of the enzyme and assay conditions are described in Li et al 2005
[b] K_is are averages of at least two independent determinations
nd, not determined

Surprisingly several of the standard serine protease warheads do not work very well for the examined NS3 proteases. Inhibitors incorporating an α-keto heterocycle moiety like **6** proved to be less active than aldehyde **3**. These heterocyclic groups gave potent inhibitors with the related serine proteases elastase, chymase and thrombin (Ni & Wagman 2004). We surmise that their lack of activity may be due to some steric restriction in the active site of the enzyme, which prevented optimal binding. In our hands α-keto amides with a deprotected arginine at P1 were unstable and therefore useless as probes.

Boronic acids, trifluoroketone and aldehydes gave the best results. These peptides have helped us to get invaluable insight into the behaviour of the enzymes and allowed us to make rapid progress towards X-ray crystal structures of NS3 proteases. A detailed discussion of the SAR of these peptides is beyond the scope of this article, however we currently can say that the shorter peptides (e.g. Table 2, compounds **5** and **9**) still maintain appreciable inhibition which bodes well for drug discovery (Z. Yin unpublished work 2005).

Table 3 shows an examination of the importance of the different binding pockets (P1–P4) in dengue 2 NS3 protease. The alanine scan suggested that P1 and P2 are very important for inhibitor binding, which is in good agreement with the substrate profiling studies. A slight surprise is that the replacement of the arginine at P2 causes a greater decrease in potency than the one at P1. The inhibitor residues at P3 only contribute a small amount of binding energy, while the norleucine at P4 can be eliminated without any affect on inhibitor potency. The latter result matches our substrate studies in which a suboptimal substitution at P4 maintained K_m but displayed sevenfold decrease in k_{cat} [substrates Bz-nKRR-ACMC ($k_{cat} = 1.39\,s^{-1}$) and Bz-TKRR-ACMC ($k_{cat} = 0.20\,s^{-1}$)] (Li et al 2005).

TABLE 3 Effect of alanine scan on dengue inhibition[a]

	Inhibitors	K_i *(μM)*[b] *dengue 2*
10	Bz-Nle-Lys-Arg-Arg-H	5.8
11	Bz-Nle-Lys-Arg-*Ala*-H	193
12	Bz-Nle-Lys-*Ala*-Arg-H	>500.0
13	Bz-Nle-*Ala*-Arg-Arg-H	22.1
14	Bz-*Ala*-Lys-Arg-Arg-H	5.3

[a] Preparation of the enzyme and assay conditions are described in Li et al 2004.
[b] K_is are averages of at least two independent determinations.

Moreover, they are in agreement with the recently published data that pointed out marked destabilization of the enzyme-inhibitor interactions in the presence of a small chain residue such as Ala or Ser at P4 (Chanprapaph et al 2005).

The design of inhibitors and virtual screening would be greatly facilitated by structural information. Unfortunately the currently available crystal structures (Murthy et al 2000) lack the NS2B cofactor and are therefore of uncertain value for drug discovery. As a consequence, an intensive effort was mounted to generate protein structures of different NS3 protease constructs in the presence of inhibitors, so far without conclusive results for the dengue NS3 protease. As a substitute, a homology model based on an unpublished X-ray crystal structure of West Nile NS3 protease (N. Schiering, personal communication 2005) was generated and used to rationalize the SAR. Figure 3 shows a snapshot of the homology model with the docked peptide Ac-Lys-Arg-Arg-H (J. Knox, personal communication 2005).

Our homology model suggests that the active site of dengue NS3 protease is rather flat with few distinct pockets which could be used for inhibitor binding. As discussed previously this situation makes drug discovery on this target rather challenging and is probably also the reason why the HTS did not produce a good lead. Nevertheless, because of the importance of this target for flavivirus drug discovery we will continue to pursue all avenues of lead discovery.

NS3 helicase

The NTPase/ helicase domain of dengue NS3 protein is located within the C-terminal end and contains seven conserved motifs associated with SF2 class of RNA NTPases and helicases (Gorbalenya et al 1989, 1993). NS3 helicase activity is essential for unwinding dsRNA, an intermediate step during viral RNA synthesis. The unwinding mechanism is an energy-dependent reaction driven by ATP

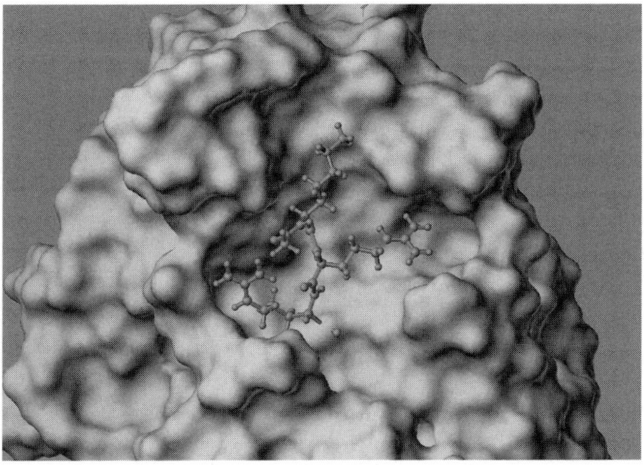

FIG. 3. Homology model of Dengue 2 CF40-Gly-NS3pro185 in complex with Ac-Lys-Arg-Arg-H.

hydrolysis. Mutational analysis of dengue 2 NS3 protein within and outside the helicase motifs have shown that absence of helicase activity correlates directly with lack of viral replication (Matusan et al 2001). NS3 helicase activity is potentially an interesting antiviral target, however so far these domains have proven to be challenging targets for modern drug discovery.

It is important to consider that helicases are involved in a variety of cellular processes involving nucleic acid metabolism, e.g. replication, transcription, translation, DNA recombination and repair. Because of the conserved nature of helicase motor function, inhibitors targeting helicases can be potent inhibitors of host protein. To effectively target dengue viral helicase, it is very important to select an inhibitor which specifically inhibits dengue viral helicase activity. In this direction, we are pursuing both a structure based approach as elaborated by the previous presentation, as well as a screening based approach. In the latter, we have developed a FRET (fluorescence resonance energy transfer)-based helicase unwinding assay (Fig. 4) in high-throughput format which will be used to evaluate compound libraries. Since helicase unwinding activity is driven by ATP hydrolysis, the FRET-based assay can select inhibitors which affect both the ATPase activity and RNA unwinding activity. We have also developed a secondary assay which detects ATPase activity which will help determine mode of action of inhibitors. In addition to HTS, the recently resolved crystal structure of the catalytically active helicase domain (Xu et al 2005) will permit fragment based screening (Rees et al 2004) as well as *in silico* high-throughput docking to identify potential helicase inhibitors.

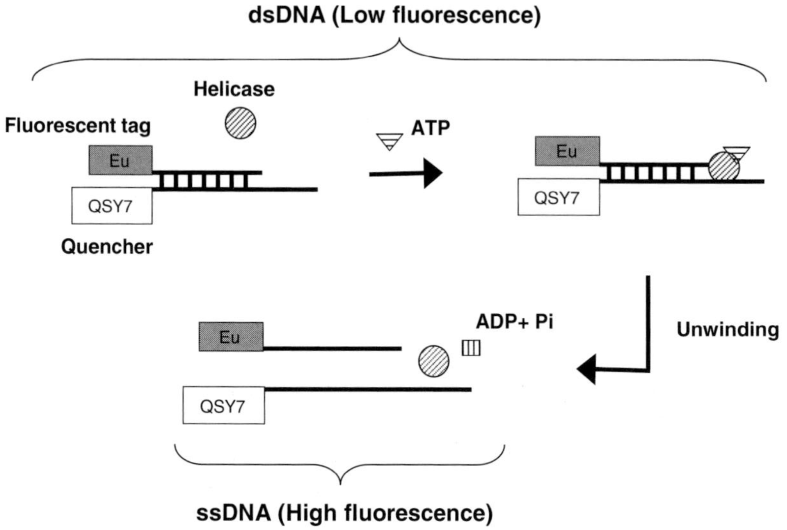

FIG. 4. FRET-based helicase unwinding assay. The measured fluorescent signal depends on the interaction between the fluorescent tag and the quencher. Spatial separation of the two DNA strands through ATP driven unwinding increases the signal. An inhibitor would block the unwinding activity of the helicase.

NS5 polymerase

Dengue NS5 protein is the largest viral protein in the flavivirus genome and contains two enzymatic activities: an S-adenosyl methionine transferase (SAM) and an RNA-dependent RNA polymerase (RdRp) domain. The NS5 polymerase is the most conserved protein across dengue serotypes, with more than 75% sequence identity and it is essential for viral replication in all flaviviridae. Mutations within the GDD motif of this protein result in non-viable virus.

Polymerases have been the most useful target class for antiviral drug discovery. Numerous compounds targeting this class of enzymes, such as famciclovir (herpes simplex virus), tenofovir (HIV) or adefovir (hepatitis B) are in clinical use. While the above drugs work against different subtypes of polymerase enzymes, the first compounds inhibiting hepatitis C (HCV) RdRp have entered clinical trials or are in preclinical development. It is currently unclear whether these compounds will also inhibit the replication of flaviviruses, but scientists at Merck (Olsen et al 2004) have reported that compound **A** (a polymerase inhibitor of the nucleoside class) has antiviral activity on a number of members of flaviviridae (Table 4). We have examined 7-deaza-2′-‘C-methyl adenosine (**A**) and one of its analogues (**B**) in our in-house cellular assay (Q.Y. Wang, personal communication) and confirmed their

TABLE 4 Cellular activity of A[c] against Flaviviridae (μM)[b]

	BVDV[a]	*West Nile*	*Dengue 2*	*Yellow fever*
EC_{50}	0.3	4.5	15	15
CC_{50}	>50	250	>320	>320

[a] Bovine diarrhoea virus.
[b] Data from Olsen et al (2004).
[c] 7-Deaza-2′-C-methyl-adenosine.

TABLE 5 Cellular activity of A and B against dengue 2[e] (μM)

	EC_{50}[d]	CC_{50}[d]
A[a]	20.1	>100
B[b]	5.3	>100

[a] 7-Deaza-2′-C-methyl-adenosine.
[b] 2′-C-methyl-adenosine.
[c] TSV01.
[d] Values are averages of at least two independent determinations.
[e] See description of assay in text.

activity on the dengue 2 virus (Table 5). The NITD cell-based assay is performed in a 96-well plate format, with BHK21 (BHK = baby hamster kidney) cells that are infected by dengue 2 virus in the presence of inhibitors. After 2 days of incubation, virus load is measured by quantifying the amount of viral envelope protein produced using ELISA.

Even though all of the compounds discussed above were derived from nucleosides, we have decided as a first priority to focus on non-nucleoside inhibitors for flavivirus RdRp. The reason for this strategy originates in the target product profile. Because all nucleosides are prodrugs which are converted in cells to the active triphosphate, the space for chemical innovation is restricted. Furthermore all work is done in cellular systems, which makes the tasks of optimizing parameters that dictate PK and tolerability slow and difficult. Because of the absence of structural information, we will follow a two-pronged strategy to find a lead. Taking advantage of recent developments in mass spectrometry, we have identified compounds that bind to the dengue NS5 protein. This is a very convenient and quick way to find potential chemical starting points, since there is no need for a sophisticated assay. A drawback of this technique is that not all hits will have the desired effect on the function of the protein. Nevertheless we have already identified very promising inhibitors of dengue 2 NS5 polymerase using this technology. These hits are currently going through a standard hit-to-lead process.

For the HTS campaign we have developed a Scintillation Proximity Assay (SPA) (Fig. 5) (Ferrari et al 1999), which will be used for the screening of the Novartis compound and natural product collections during the second half of 2005. In order to follow our platform strategy for this target, cellular assays to test the resulting compounds on other flaviruses are currently being set up in the NITD BSL-3 laboratory.

Conclusions

Dengue and other flaviviruses have probably somewhere between four and six drug targets in their genome. History and drugability arguments suggest that NS5 polymerase is most likely to yield a drug in the near future, since chemical starting points are available and the optimization should be reasonably straight forward. It is likely that inhibitors of this target class will have broad spectrum activity and will therefore also be useful for other flaviviral diseases.

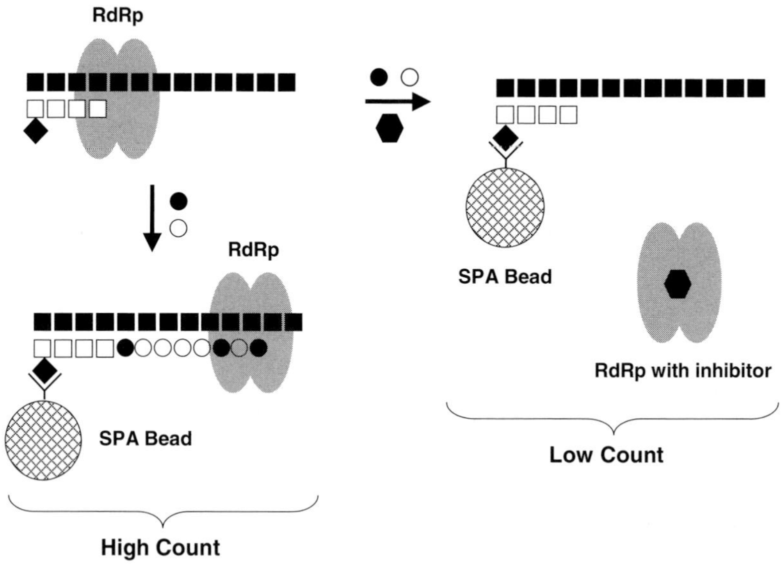

FIG. 5. Scintillation proximity assay for screening of polymerase (RdPp) inhibitors (⬢). A poly C template (■) together with an oligo G_{20} primer (□) is incubated with RdPp in the presence of GTP (●) and [^{3}H]GTP (○). Incorporation of radioactivity into the primer strand is measured using SPA beads which are attached to the primer through Streptavidine (◆).

NS3 protease is also a target that looks promising. However the optimization of a lead will be considerably more challenging than for the polymerase. The task will be especially hard if no suitable lead can be found by HTS and a peptidomimetic approach needs to be adopted. In such a case a multi-year effort will be necessary to optimize pharmacokinetics, especially oral bioavailability.

NS3 helicase, while attractive from a biological standpoint, has proven to be a difficult target for HCV drug discovery. If HTS can identify a lead, then the most challenging aspect of such a project would be to find a compound with sufficient selectivity versus host helicases. This problem should be surmountable with a concerted effort in lead optimization.

Two other potential targets in the flavivirus genome, the SAM domain and E protein, have not been discussed in this short overview. While for both targets structural information is available, we currently do not have enough information to judge whether they are attractive drug targets.

Overall, drug discovery for flaviviral diseases has made great strides in the last two years and there is cause for optimism. It is, however, important to remember that timelines for both preclinical and clinical developments are long and that therefore, even under best conditions, a drug for dengue will not be available to the patients before the start of the next decade.

References

Brady SF, Sisko JT, Stauffer KJ et al 1995 Amide and α-keto carbonyl inhibitors of thrombin based on arginine and lysine: synthesis, stability and biological characterization. Bioorg Med Chem 3:1063–1078

Chanprapaph S, Saparpakorn P, Sangma C et al 2005 Competitive inhibition of the dengue virus NS3 serine protease by synthetic peptides representing polyprotein cleavage sites. Biochem Biophys Res Commun 330:1237–1246

Falgout B, Pethel M, Zhang YM et al 1991 Both nonstructural proteins NS2B and NS3 are required for the proteolytic processing of dengue virus nonstructural proteins. J Virol 65:2467–2475

Ferrari E, Wright-Minogue J, Fang JWS et al 1999 Characterization of soluble Hepapatis C virus RNA-dependent RNA polymerase expressed in Escherichia coli. J Virol 73:1649–1654

Gorbalenya AE, Koonin EV 1993 Helicases: amino acid sequence comparisons and structure-function relationships. Curr Opin Struct Biol 3:419–429

Gorbalenya AE, Kunin EV, Donchenko AP et al 1989 Two related superfamilies of putative helicases involved in replication, recombination, repair and expression of DNA and RNA genomes. Nucleic Acids Res 17:4713–4730

Gustafsson D, Bylund R, Antonsson T 2004 Case history: A new oral anticoagulant: the 50-year challenge. Nat Rev Drug Disc 3:649–659

Leung D, Schroder K, White H et al 2001 Activity of recombinant dengue 2 virus NS3 protease in the presence of a truncated NS2B co-factor, small peptide substrates, and inhibitors. J Biol Chem 276:45762–45771

Li J, Lim SP, Beer D et al 2005 Functional profiling of recombinant NS3 proteases from all four serotypes of dengue virus using tetrapeptide and octapeptide substrate libraries. J Biol Chem 280:28766–28774

Lipinski D, Hopkins A 2004 Navigating chemical space for biology and medicine. Nature 432:855–861

Llinas-Brunet M, Bailey M, Fazal G et al 1998 Peptide-based inhibitors of the hepatitis C virus serine protease. Bioorg Med Chem Lett 8:1713–1718

Matusan AE, Pryor MJ, Davidson AD et al 2001 Mutagenesis of the Dengue virus type 2 NS3 protein within and outside helicase motifs: effects on enzyme activity and virus replication. J Virol 75:9633–9643

Murthy HMK, Judge K, DeLucas L et al 2000 Crystal structure of Dengue virus NS3 protease in complex with a Bowman-Birk inhibitor: implications for flaviviral polyprotein processing and drug design. J Mol Biol 301:759–767

Ni ZJ, Wagman AS 2004 Progress and development of small molecule HCV antivirals. Curr Opin Drug Disc Dev 7:446–459

Olsen DB, Eldrup AB, Bartholomew L et al 2004 A 7-deaza-adenosine analogue is a potent and selective inhibitor of hepatitis C virus replication with excellent pharmacokinetic properties. Antimicrob Agents Chemother 48:3944–3953

Rees DC, Congreve M, Murray CW et al 2004 Fragment-based lead discovery. Nat Rev Drug Disc 3:660–672

ter Haar E, Walters WP, Pazhanisamy S et al 2004 Kinase chemogenomics: targeting the human kinome for target validation and drug discovery. Mini Rev Med Chem 4:235–253

Xu T, Sampath A, Chao A et al 2005 Structure of the Dengue virus helicase/nucleoside triphosphatase catalytic domain at a resolution of 2.4 ANG. J Virol 79:10278–10288

DISCUSSION

Young: You had a series of substrates and you made substitutions in P1–P4. You identified P2 as being perhaps more important than P1. P1 has that relatively large pocket that it can bind into, whereas P2 is more open. Why is this?

Keller: Our data tell us that P2 is more important. We know this goes a bit against what Pad Padmanabhan has shown in his crystal structure. I can't tell you all the data we have, but some of it has been published in Yin et al (2006a,b). We consistently see that P2 is more important than P1. We can discuss what this means for the structure.

Padmanabhan: One of the compounds you showed still had two basic residues. Do you know where it is cleaving?

Keller: Our reviewers have pointed this out! We believe our data are correct.

Padmanabhan: You mutated the P1 arginine to alanine in the sequence (from –P3 to P1, -Lys-Arg-Arg-H to -Lys-Arg-Ala-H), then alanine becomes the P1′ and the P2 arginine now becomes a P1, and the cleavage could occur between Arg and Ala. This is a possibility that you can't rule out.

Keller: We believe that we can rule this out since we know that the aldehyde binds to the serine to form a tetrahedral transition state mimic (see Erbel et al 2006).

Fairlie: One of the interesting things about that molecule is that all the other aldehydes have an arginine at P1. We know that arginine cyclyses back onto the aldehyde, so that in solution they are not really aldehydes. In the second compound, you have an alanine at P1, which is therefore going to behave more like an aldehyde, and less like a combination of hydrated forms.

Keller: We can see this equilibrium between the aldehyde and the cyclized form but we think it is so fast that it doesn't really have an impact on the K_i. We have also made some peptides with rigid arginine analogues which cannot cyclize and we get very similar results.

Padmanabhan: If we have the structure of the NS2B/NS3 complex, that will solve a lot of questions.

Hibberd: Could you comment on the relative merits of some host targets. Do you see these as a more difficult problem?

Keller: The problem with host targets is usually selectivity. This is less of an issue with some of the viral targets. If you tell us which host targets will be useful then we can evaluate them. There are so many candidates it is hard to say. In general, modulating the immune system is just more difficult than killing the virus.

Padmanabhan: If you compare the K_i of these two peptide inhibitors of DENV-2 NS3 protease complex (Bz-Nle-Lys-Arg-Arg-H and Bz-Nle-Lys-Arg-Ala-H) in which the arginine in one is changed to alanine, the K_i is increased from 5.8 μM to 193. This would also suggest that P1 Arg is also important.

Keller: I am not saying that P1 is not important. I am just saying that our data suggest P2 is more important than P1. I can't say how much more important.

Rice: Can you tell us any more about your polymerase inhibitor? Does it show cross-inhibitory potential?

Keller: It is active in the cell at about 20 μM. Otherwise we can't say anything about it. We found it two months ago and it looks good. It has the right spectrum of potency. I am convinced that we find a lot of non-nucleoside hits from HTS and it will be a matter of choosing the right compound for lead optimization.

Canard: If you are interested in fragment-based screening, it is better to know where it binds for this reason.

Keller: My bias is that we will have enough leads from HTS without fragment-based screening. We would only go to these more sophisticated ways of finding leads if you don't have anything from HTS.

Padmanabhan: Have you tried to see whether some of the compound hits inhibit the interaction between the cofactor and the protease domain? You used the NS2B/NS3 protease for these assays. Some of these compounds might inhibit the interaction between this and the cofactor.

Keller: I doubt it. We have the warheads (e.g. aldehydes), and without the warheads they are inactive. This suggests to me that they interact with the serine in the active site to form a tetrahedral transition state analogue.

Padmanabhan: Do they inhibit by competitive inhibition?

Fairlie: They are all competitive inhibitors.

Keller: We have done Lineweaver-Burk plots and they are clearly competitive.

Screaton: Using a peptide like this, one of the things that would worry me is that you get the specificity, but what about all the endogenous proteases?

Keller: You are right, this is miles from a drug. What we have done this for is to understand the protein a little bit better. After this we will start looking for better affinity by changing the peptide and removing amide bonds. In the end you would even like to remove the aldehydes if possible. You can do a lot, but it takes years of work; this is just a first step but it has given us a lot of information about the protein. We know much more about NS3 now than a year ago.

Farrar: After two or three years of learning much more do you remain as optimistic as you were two or three years ago about these compounds going through to preclinical testing?

Keller: That's a nasty question! Since I knew at the beginning that serine proteases are very tough targets, my opinion has not changed.

Farrar: In the industry generally, how does starting with your six potential targets and ending up with one compound in preclinical trials, fit with other areas of drug discovery?

Keller: That's the normal attrition rate.

Farrar: But you didn't have many at the start.

Keller: Since there are not that many targets for dengue we will look more carefully. If this had been another indication, we would have by now abandoned the NS3 protease, but here we will look more carefully. We always need more targets.

Vasudevan: Even for one target, like NS5, you can have different classes of compounds.

Rice: NS5 is not one target. It is probably 20 targets or more.

Keller: I'm not trying to make everyone pessimistic, it is just a reality check. Even with a good compound the chances of failure are considerable.

Young: There are strong data showing the role of certain cytokines in leading on to vascular leak and so on. What about cytokines as potential targets?

Keller: With cytokines you'd probably have to go in with antibodies or soluble receptors. With small molecular weight compounds you could try to target the signal transduction pathways post-receptor or pre-expression, and this has been done. For Singapore, an antibody would be a useful drug, but not for developing countries because it is too costly to produce.

Screaton: One of the functions of thalidomide is as a cytokine blocker with quite broad specificity.

Harris: Are you interested in working on other potential targets that come out of the literature?

Keller: Subhash Vasudevan's task is to work through the biology and identify targets.

Vasudevan: Alpha glucosidase is one target for which there are some known inhibitors.

Keller: This might be an interesting target but selectivity concerns will be large. It wouldn't be my choice for dengue.

Canard: Do we have an idea why double-strand RNA isn't chopped off like in the process of silencing RNAs? Would this be a target if there is a natural resistance mechanism for this?

Rice: In terms of the RNAi, I don't think people have a good idea. It could be sequestered in vesicles. You can certainly artificially deliver double-stranded RNAs and selectively target the viral RNA for degradation.

Canard: So there is no specific mechanism.

Rice: Not that I am aware of. Some people have overexpressed the capsid proteins of these viruses, which can be a non-specific RNA-binding protein, and shown that this will antagonize the RNAi pathway.

Canard: What about the interferon results?

Rice: There are quite a few papers appearing where people have used various kinds of biological readouts to look at whether or not one or more of the flavivirus proteins can antagonize interferon (IFN). I think there are several small non-structural proteins that have been implicated.

Gu: The problem is that many of these IFN suppression studies are done *in vitro*. Has anyone tested why interferon therapy doesn't work with dengue *in vivo*?

Vasudevan: Maria Guzman and colleagues tried this in Cuba in 1981 during an outbreak. They claimed some sort of success.

Rice: I thought the results had been mixed in the clinic.

Halstead: I remember the trial, but not the results. It wasn't pursued.

Halstead: There are high levels of γ and α IFN in the early stages of dengue infection. I don't know what this means; most infectious diseases cause IFN production. If it were administered early in the course of disease, perhaps it would have some beneficial effect. It is pretty toxic.

Fairlie: In any case, it is too expensive.

Farrar: We tried it in JE and saw no benefit. The damage may have already been done in the neurons.

Halstead: Someone needs to think a little more creatively about this resistance gene that lies fallow. Black people don't get the severe form of dengue. There is some evidence of attenuated disease. This phenomenon appears to be related to the vascular permeability syndrome. There is an interesting origin to this. Clinically, yellow fever and dengue haemorrhagic fever (DHF) day by day are almost identical diseases. The best hypothesis for this resistance gene is that it is actually a yellow fever resistance gene. There is a lot of anecdotal evidence that yellow fever has lower mortality in black people than whites. It is likely that west African black people are genetically resistant to yellow fever. In Africa, the sylvatic yellow fever virus is fully pathogenic for humans.

Screaton: There is an interesting polymorphism in caspase 10. There is a correlation between caspase 10 mutations and susceptibility to septic shock in black people.

Farrar: The study of host genetics has been limited until now by our ability only to study single genome polymorphisms. We have to move to much more of a fishing trip to look at the whole haplotypic structure through the most important classes of immune response genes, and move away from attempts to identify candidate genes. Sometimes fishing trips are very important, and the technology is there to do this.

Halstead: If you follow the hypothesis here, there is a big difference in Africa between west Africa, central Africa and east Africa. There has never been a yellow fever epidemic in east Africa. People in east Africa would therefore not be expected to have a resistance gene.

Holmes: The strains of yellow fever that circle in east Africa are different from those in west Africa. This complicates things greatly.

Malasit: I would like to raise another issue. The data in the literature come from the study of patients who have been admitted to hospital, but this is just the tip of the iceberg. We need to study the larger cohort of infected patients who aren't admitted into hospital. Unless this is done, it will be difficult to verify targets and the genetic search will be affected.

Rice: You need the right samples to do this kind of analysis.

Farrar: In the setting where everyone has been exposed, this may be less important than for a rarer disease. We have a common disease in terms of seroconversion. The whole population has been exposed and most of them have not got disease, so in dengue we may be lucky in terms of looking at targets.

Hibberd: This comes back to the point that we are looking at a severity question. Also, it is important to realize that that the virus has something to do with it. We need to approach from both sides.

Evans: I don't think you'll find those targets by looking at populations. I think you'll find them by looking at virology. We need to understand the replication complex and the proteins that are involved. With regard to race, African Americans in the USA have much less severe respiratory syncytial virus (RSV) than whites. It is a completely pulmonary pathology, so this may be a general phenomenon. It makes a lot more sense to me to look at the proteins that associate with the capping complex, for example, than to go on general fishing expeditions.

Farrar: We did some work a long time ago showing that susceptibility to shock depends on age. When you are very young you are much more susceptible to shock. This probably has to do with ability to cope with plasma leak. It may have nothing to do with virus or immune response genes. It may be a physiological issue.

Hibberd: The whole genome approaches that we are now all capable of doing result in a whole series of hits. We can then use our biological knowledge to narrow these down. In other diseases we have found that we sometimes come up with a slightly surprising hit which can then be worked up. It is this kind of novel perspective that somehow gives a new insight into quite a big area.

Evans: Sometimes it is almost easier to go for a known set of compounds and do the biology, working back to the hits.

References

Erbel P, Schiering N, D'Arcy A et al 2006 Structural basis for the activation of flaviviral NS3 proteases from dengue and West Nile virus. Nat Struct Mol Biol 13:372–373

Yin Z, Patel SJ, Wang WL et al 2006a Peptide inhibitors of Dengue virus NS3 protease. Part 1: Warhead. Bioorg Med Chem Lett 16:36–39

Yin Z, Patel SJ, Wang WL et al 2006b Peptide inhibitors of dengue virus NS3 protease. Part 2: SAR study of tetrapeptide aldehyde inhibitors. Bioorg Med Chem Lett 16:40–43

Structural and functional analysis of dengue virus RNA

Diego E. Alvarez, Maria F. Lodeiro, Claudia V. Filomatori, Silvana Fucito, Juan A. Mondotte and Andrea V. Gamarnik[1]

Fundación Instituto Leloir, Patricias Argentinas 435, Buenos Aires, Argentina

Abstract. Sequences and structures present at the 5′ and 3′ UTRs of RNA viruses play crucial roles in the initiation and regulation of translation, RNA synthesis and viral assembly. In dengue virus, as well as in other mosquito-borne flaviviruses, the presence of complementary sequences at the ends of the genome mediate long-range RNA–RNA interactions. Dengue virus RNA displays two pairs of complementary sequences (CS and UAR) required for genome circularization and viral viability. In order to study the molecular mechanism by which these RNA–RNA interactions participate in the viral life cycle, we developed a dengue virus replicon system. RNA transfection of the replicon in mosquito and mammalian cells allows discrimination between RNA elements involved in translation and RNA synthesis. We found that mutations within CS or UAR at the 5′ or 3′ ends of the RNA that interfere with base pairing did not significantly affect translation of the input RNA but seriously compromised or abolished RNA synthesis. Furthermore, a systematic mutational analysis of UAR sequences indicated that, beside the role in RNA cyclization, specific nucleotides within UAR are also important for efficient RNA synthesis.

2006 New treatment strategies for dengue and other flaviviral diseases. Wiley, Chichester (Novartis Foundation Symposium 277) p 120–135

The genome-length RNA of dengue virus is infectious. Delivery of this RNA molecule into a susceptible cell triggers a complete round of viral replication. Once in the cytoplasm of the host cell, the viral genome participates in at least three different processes: it serves as mRNA to direct the synthesis of viral proteins, it acts as template for genome amplification, and it is packaged along with structural proteins during viral assembly. The molecular mechanisms controlling the utilization of the viral RNA in each step of the viral life cycle are still poorly understood. Because the 5′ and 3′ ends of the viral RNA are the places where translation initiation and synthesis of positive and negative strand RNA occur, we are interested

[1]This paper was presented at the symposium by Andrea Gamarnik, to whom correspondence should be addressed.

in investigating the function of RNA structures and sequences located at the ends of the viral RNA.

The genome of dengue and other flaviviruses is about 11 kb long and encodes one open reading frame flanked by 5′ and 3′ untranslated regions (UTRs) (Rice 2001). The 5′ UTR of dengue virus is around 100 nucleotides long and the sequence conservation is almost complete among different dengue virus serotypes (Markoff 2003). The predicted structure consists of a large stem loop (SLA) and a second short stem loop (SLB) containing at the 3′-terminal sequences the translation initiation codon (Fig. 1). It is likely that these sequences and structures influence translation initiation, which presumably takes place by a cap dependent scanning mechanism. Another major function of the 5′ UTR probably resides in the negative strand, which serves as a site for positive strand RNA synthesis. Deletions engineered into the 5′ UTR of dengue virus 4 were lethal (Cahour et al 1995), suggesting an important role of these RNA structures in viral replication.

The 3′ UTR of dengue virus is around 450 nucleotides, lacks a poly(A) tail, but contains a number of conserved RNA structures (Fig. 1) (Shurtleff et al 2001). The viral genome ends in a very conserved 3′ stem loop (3′ SL). Detailed analysis of the structure-function of the 3′ SL in West Nile virus, Kunjin virus, dengue virus and yellow fever virus revealed an absolute requirement of this RNA element for viral replication (Brinton et al 1986, Zeng et al 1998, Rauscher et al 1997, Proutski et al 1997, Men et al 1996, Yu & Markoff 2005, Tilgner et al 2005, Elghonemy et al 2005). Upstream of the 3′ SL there is another essential RNA element for viral replication, the conserved sequence CS1 (Men et al 1996). This element contains the CS sequence, which is complementary to a sequence present within the coding region of protein C at the 5′ end of the genome (Hahn et al 1987). 5′–3′ long-range RNA–RNA interactions through these complementary sequences have been proposed to be necessary for replication of different mosquito-borne

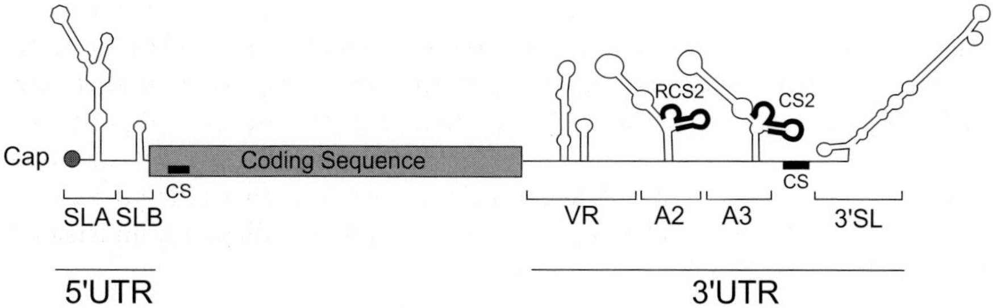

FIG. 1. Schematic representation of dengue virus genome. The predicted secondary structures of defined domains at the 5′ and 3′ UTR are indicated: stem loop A (SLA), stem loop B (SLB), variable region (VR), domain A2, domain A3, and the 3′ stem loop (3′SL). Also, the conserved sequences CS, CS2 and RCS2 are shown.

flaviviruses (Alvarez et al 2005b, 2005a, Lo et al 2003, Khromykh et al 2001, Corver et al 2003). In addition, 5′–3′ CS base pairing has also been reported to be important for *in vitro* activity of dengue and West Nile virus RNA polymerases (You et al 2001, Nomaguchi et al 2004). The mechanism by which the flavivirus replicase machinery initiates RNA synthesis specifically at the viral 3′ UTR is still not clearly understood. The RNA replication complex assembles on cellular membranes and involves the viral RNA dependent RNA polymerase-methyltransferase NS5, the helicase-protease NS3, the glycoprotein NS1, the hydrophobic proteins NS2A and NS4A, and presumably host factors (Westaway et al 1999, 1997, Mackenzie et al 1998).

In addition to the 3′ SL, other RNA structures and conserved motifs are present within dengue virus 3′ UTR. Folding algorithms predicts two almost identical structures designed A2 and A3 preceding the 3′ SL. Recent experiments using recombinant dengue virus 2 carrying deletions of domain A2 and/or A3 showed viral attenuation with defects in RNA synthesis, suggesting an important role of these RNA structures in viral replication (Alvarez et al 2005a). Within domains A3 and A2 there are highly conserved regions known as CS2 and repeated CS2 (RCS2), respectively (Shurtleff et al 2001). A recombinant virus with a deletion of 30 nucleotides between CS2 and RCS2 is currently under study as a dengue virus vaccine candidate (Durbin et al 2001). CS2 and RCS2 sequences can be found in Japanese encephalitis virus, West Nile virus, Murray Valley encephalitis virus, and dengue virus types 1 to 4 (for review see Markoff 2003), suggesting a conserved function of these elements in flavivirus replication. However, it is not clear the mechanisms by which these RNA structures participate in viral replication.

Here, we discuss the nature and requirements of long-range RNA–RNA interactions in the viral genome during dengue virus replication. Using genomic and subgenomic dengue virus RNAs together with biochemical tools, we analysed the role of secondary and tertiary structures of the viral RNA during translation and RNA synthesis.

Results and discussion

Flavivirus genomes possess inverted complementary sequences at the ends of the RNA, similar to that observed in the negative strand RNA viruses (bunya-, arena- and orthomyxoviruses) (Kohl et al 2004, Barr & Wertz 2004, Hsu et al 1987, Raju & Kolakofsky 1989, Mir & Panganiban 2005, Ghiringhelli et al 1991). These complementary sequences have been suggested to allow the ends of the genome to associate through base pairing, leading to circular conformations of the RNA (panhandle-like structures). Do flavivirus genomes acquire a circular conformation? Is the long-range RNA–RNA interaction required for dengue virus replication?

To investigate a possible association between the 5′ and 3′ ends of dengue virus genome, we analyzed the formation of RNA–RNA complexes using electrophoresis mobility shift assays (EMSA) with *in vitro* transcribed radiolabelled RNA molecules. We found that an RNA molecule corresponding to the first 160 nucleotides of dengue virus RNA (5′UTRC62 RNA) specifically binds to a second RNA molecule carrying the last 106 nucleotides of the viral genome (3′SL Probe, Fig. 2A). This RNA–RNA interaction shows high affinity (K_d about 8 nM) and is absolutely dependent on the presence of Mg^{2+}. In dengue and other mosquito-borne flaviviruses it was proposed that the complementary sequences 5′–3′ CS are a potential cyclization element. Folding predictions of the sequences representing the RNA–RNA complex formed by the ends of the dengue virus genome show two pairs of complementary regions (Fig. 2B). One of these regions is the 5′–3′ CS, the second is located at the 5′ end just upstream of the initiator AUG and at the 3′ end within the stem of the 3′ SL (named UAR, upstream AUG region, Fig. 2B). To investigate the RNA determinants for complex formation, we generated 5′ UTRC62 RNA molecules with substitutions generating mismatches in either 5′–3′ CS or 5′–3′ UAR. We tested the binding ability of these mutated 5′RNAs in EMSA using a 3′ SL wild-type probe. Mutations in 5′ CS or 5′ UAR greatly decreased the binding of the mutated 5′UTRC62 RNAs to the radiolabelled 3′ SL, confirming that both complementary sequences were necessary for RNA–RNA complex formation (Fig. 2C).

To investigate whether the RNA–RNA contacts observed between two RNA molecules representing the ends of dengue virus genome also occur in a single RNA molecule as long-range interactions, we analysed the conformation of individual molecules by atomic force microscopy (AFM). The *in vitro* transcribed full-length dengue virus RNA was deposited on mica, dried and visualized by tapping mode AFM. The single-stranded RNA molecules acquire compact structures, precluding visualization of intramolecular contacts (Fig. 3A). To overcome this problem, we hybridized the viral RNA with an antisense molecule of 3.3 kb (complementary to a region encompassing NS4B–NS5), which yielded an extended double stranded region flanked by the single stranded ends of the viral RNA. These molecules were observed in both circular and linear conformations in the absence of proteins (Fig. 3B and C). Statistical analysis of this and other model RNA molecules carrying specific mutations confirmed that viral RNA circularizes through direct RNA–RNA contacts involving CS and UAR sequences (Alvarez et al 2005b).

Previous reports have suggested that base pairing between 5′ and 3′ CS of flavivirus genomes is necessary for viral replication. Using Kunjin and West Nile virus replicons it has been shown that substitution mutations in either 5′ or 3′ CS that disrupted base pairing were lethal for RNA replication (Khromykh et al 2001, Lo et al 2003). However, when both CS sequences were mutagenized with respect

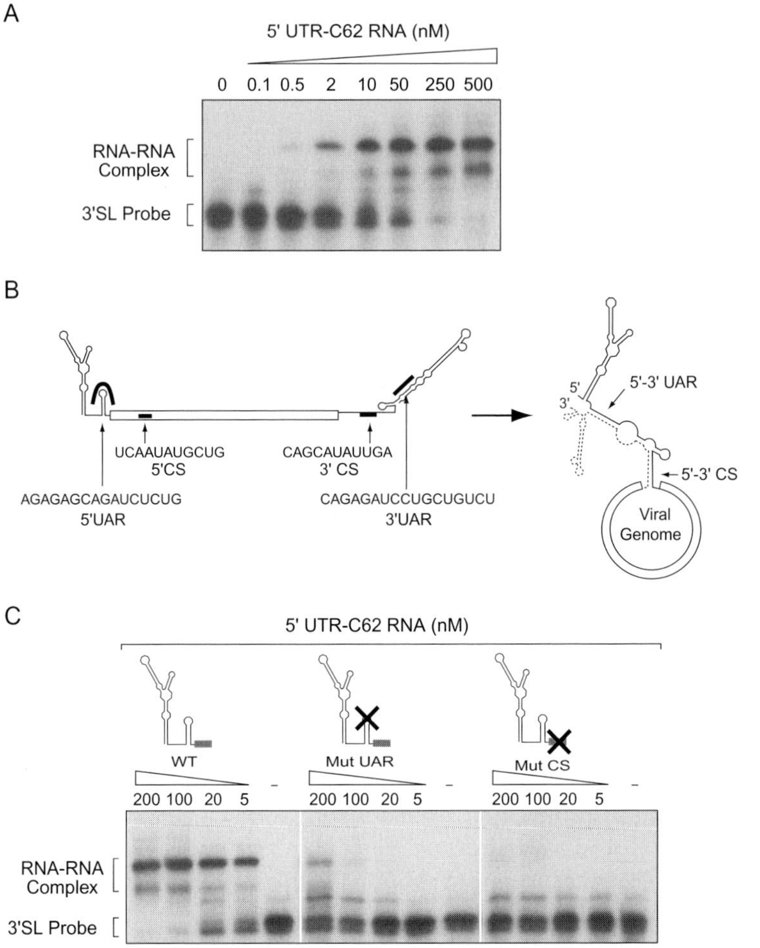

FIG. 2. RNA–RNA complex formation between the end sequences of dengue virus RNA. (A) Mobility shift assays shows RNA–RNA complex formation. Uniformly labelled 3′ SL RNA (1 nM, 30 000 cmp), corresponding to the last 106 nucleotides of dengue virus 2, was incubated with increasing concentrations of the 5′UTR–C62 RNA corresponding to the first 160 nucleotides of the viral genome, as indicated at the top of the gel. (B) Schematic representation of dengue virus genome showing the location and nucleotide sequence of 5′ CS, 5′ UAR, 3′ CS, and 3′ UAR. Folding prediction of the proposed circular conformation of the RNA is also shown. (C) RNA mobility shift analysis showing the effect of mutations in CS or UAR within the 5′ UTR–C62 RNA on binding to the 3′ SL RNA. Uniformly labelled 3′ SL RNA was incubated with increasing concentrations of wild-type and mutated 5′ UTR–C62 RNAs as indicated at the top of the gel.

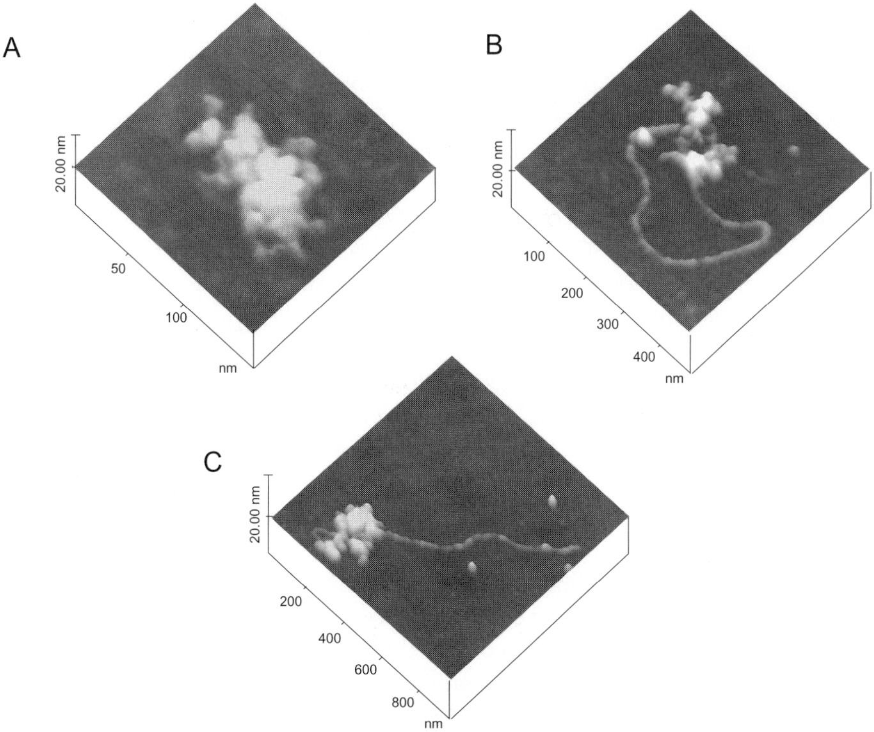

FIG. 3. Visualization of genome–length dengue virus RNA of 10.7 kb by tapping mode AFM. (A) Visualization of a representative single stranded dengue virus RNA molecule. (B, C) Images of circular and linear conformations of individual dengue virus RNA molecules, respectively. The 10.7 kb RNA molecule was hybridized with an antisense RNA of 3.3 kb resulting in a linear double stranded region with single stranded regions of 6970 and 451 nucleotides at the 5′ and 3′ ends, respectively.

to the wild-type sequence such that their capacity to base pair was maintained, RNA replication was restored. We performed similar experiments using recombinant dengue viruses to address the importance of 5′–3′ UAR complementarity during viral replication. Our data showed that specific substitutions within 5′ or 3′ UAR, in the context of the infectious dengue virus 2, yielded no viable viruses. Importantly, mutations at the 5′ and 3′ UAR that restored complementarity were sufficient to rescue viral replication. The replication of this recombinant virus displayed slow growth and small plaque phenotype when compared with the wild-type virus, suggesting that the nucleotide sequence/structure of 5′ and/or 3′ UAR is also important for efficient viral replication (Alvarez et al 2005b).

Our results indicate that long range RNA–RNA interactions result in circular conformation of the viral genome. In addition, the functional studies on dengue virus together with previous reports obtained with other flaviviruses strongly suggest that sequence complementarity is required for viral replication. However, many questions remain open: what is the role of the long-range 5′–3′ end interactions during dengue virus replication? Is the cyclization of the viral genome necessary for efficient translation, similar to that observed for cellular mRNAs? Is the complex between the 5′ and 3′ ends of the genome required for NS5 polymerase binding during RNA synthesis? Does the structure of the RNA involving both ends of the genome play a role in coordinating translation and RNA synthesis? Does the long range RNA–RNA interaction constitute a signal for RNA encapsidation?

To dissect the role of the cyclization sequences during the viral processes, we constructed a dengue virus replicon that allows discrimination between translation of the input RNA and RNA synthesis (Alvarez et al 2005a). Similar replicons have been recently developed for West Nile and yellow fever viruses (Lo et al 2003, Jones et al 2005). In the context of dengue virus 2, we replaced the viral structural proteins by the firefly luciferase coding sequence (Luc). The trans membrane domain (TM) corresponding to the C-terminal 24 amino acids of E protein was retained in order to maintain the topology of the viral protein NS1 inside of the ER compartment (Fig. 4A). The Luc was fused in-frame to the first 102 nucleotides of the capsid protein (C), which contain the 5′ CS sequence. To ensure proper release of the Luc from the viral polyprotein, we introduced the *cis*-acting FMDV 2A protease (Fig. 4A).

Dengue virus replicon was efficiently translated and amplified in transfected BHK and mosquito cells. After RNA transfection with lipofectamine, Luc activity increases as a function of time reaching the highest levels between 8 and 10 h, reflecting the translation of the input RNA. After 24 h of transfection, Luc activity increases exponentially as a result of replicon RNA amplification (Fig. 4A). A replicon with a mutation in the GDD active site of the RNA dependent RNA polymerase NS5 (MutNS5) showed the translation peak, but after 24 h the levels of Luc were indistinguishable from the background, showing the lack of RNA amplification. These results indicate that replicon RNA amplification by the viral replicase machinery can be monitored through the expression of Luc as a function of time in transfected cells.

We used the replicon system to generate specific substitutions disrupting 5′–3′ CS or 5′–3′ UAR complementarity. To this end, we incorporated substitutions within the 3′ CS sequence, generating 4 mismatches (the wild-type 3′ CS CAGCAUAUUGA was replaced by **U**A**U**CAU**U**U**G**GA, Mut.3′CS replicon). In addition, a second replicon was designed carrying point mutations at the 5′ CS that restored sequence complementarity with the mutated 3′CS (Rec.5′–3′CS replicon). RNAs corresponding to the wild-type, Mut.NS5, Mut.3′CS, and the double mutant Rec.5′–3′CS were *in vitro* transcribed and equal amounts of the RNA were trans-

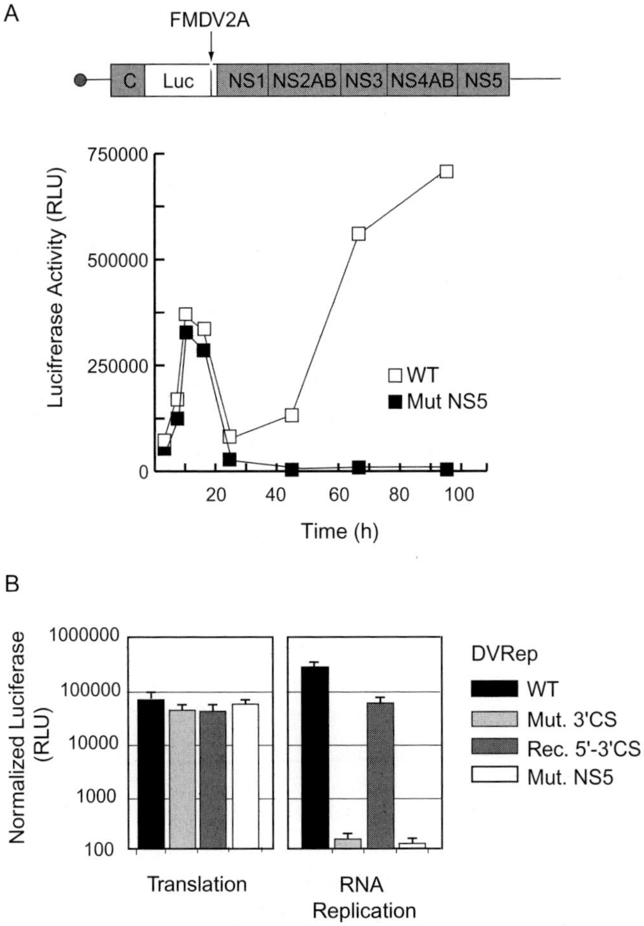

FIG. 4. Dengue virus replicon allows discrimination between translation and RNA replication in transfected cells. (A) At the top, schematic representation of dengue virus replicon. Boxes denoting coding sequences of capsid (C), Luciferase (Luc), and non–structural proteins (NS) are shown. Also, the position of FMDV2A protease is indicated by an arrow. At the bottom, replication of dengue virus replicon in BHK cells is shown. Time-course of luciferase activity was detected in cytoplasmic extracts prepared from BHK cells transfected with wild-type dengue virus replicon (WT) or replication-incompetent MutNS5 RNAs. (B) Translation and RNA replication of WT, 3′ CS mutant (3′CSMut), double mutant at the 3′ and 5′ CS (Rec.5′–3′CS), and replication-incompetent MutNS5 were analysed in transfected BHK cells. Normalized Luc levels are shown in logarithmic scale at 10 h after transfection to estimate translation of input RNA and at 3 days after transfection to evaluate RNA replication.

fected into BHK cells (Fig. 4B). Renilla Luc mRNA was cotransfected with all the replicons and used to standardize the transfection efficiency in each time point. After 10 h of transfection no significant differences in Luc activity were observed between cells transfected with the wild-type or mutated replicons, suggesting that translation of the input RNA was not dependent on 5′–3′ CS interactions. In contrast, RNA synthesis of the Mut 3′CS replicon was undetectable. Reconstitution of 5′–3′ base pairing in the Rec.5′–3′CS replicon also restored RNA synthesis (Fig. 4B). The level of RNA replication detected with this replicon was lower than the one observed with the wild-type RNA, suggesting that the highly conserved WT sequences within CS provide an advantage during RNA amplification.

To study the importance of 5′–3′ UAR complementarity during viral translation and RNA synthesis we introduced specific mutations in these regions in the replicon system. UAR sequences are located within very conserved stem loops of the viral 5′ and 3′ UTRs, therefore, it is difficult to introduce mutations in 5′ or 3′ UAR sequences that disrupt complementarity without altering secondary structures. We designed point substitutions in both sides of the stem or loop of SLB (5′ UAR) and/or in their complementary sequences within the stem of the 3′SL (3′ UAR) (Fig. 5A). We generated three groups of mutations (Mut 1, Mut 2 and Mut 3), each group was composed of three different replicons: (a) with mutations in 5′ UAR, (b) with mutations in 3′ UAR, and (c) with mutations a + b in the same replicon, restoring 5′–3′ UAR complementarity with sequences that differed from the wild-type sequences (Fig. 5A). The RNAs corresponding to the wild-type and the nine mutated replicons described above were transfected into BHK cells and Luc activity was measured as a function of time. The translation efficiency was determined measuring Luc activity after 10 h of transfection. The results showed that all mutant replicons were translated with similar efficiencies to that observed with the wild-type replicon (data not shown). To evaluate RNA synthesis, we compared the RNA amplification of each mutated replicon with the wild-type levels and expressed the results as a percentage of the wild-type (Fig. 5B). From this analysis we observed that: (i) single mutations disrupting 5′–3′ UAR base pairing decreased or abolished RNA synthesis; (ii) in the three groups of mutants the reconstitution of the 5′–3′ complementarity (5′3′Mut 1, 5′3′Mut 2 and 5′3′Mut 3) also increased the levels of RNA synthesis; (iii) not all base pairings within 5′–3′ UAR are equally important for RNA synthesis (compare the replication levels of 3′Mut-2 with 3′Mut-3, both with only one mismatch); and (iv) in some cases reconstitution of 5′–3′ complementarity is not sufficient to rescue RNA synthesis to wild-type levels, suggesting that specific nucleotides within the 5′ or 3′ UAR sequences can be critical. In this regard, it is important to mention that the sequence of 5′ and 3′ UAR are absolutely conserved in all dengue virus serotypes, suggesting that even though co-evolution of the two complementary sequences could occur there must be a growth advantage to preserve the wild-type nucleotide sequences.

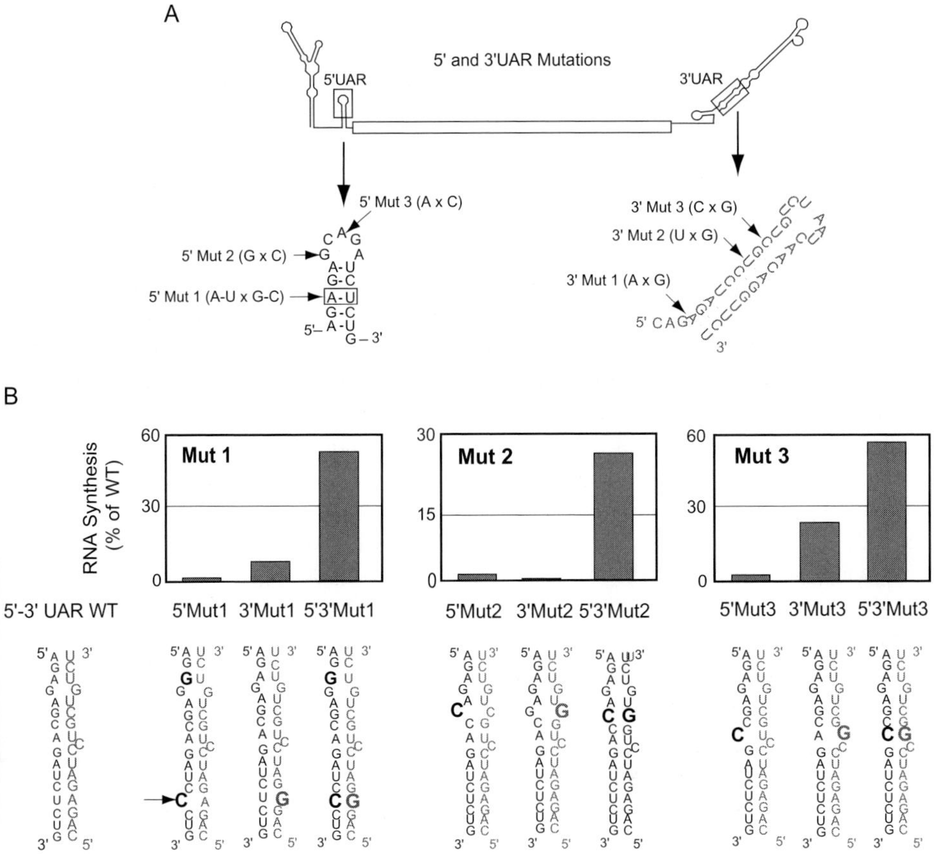

FIG. 5. Mutations within 5′ or 3′ UAR alter replicon RNA amplification (A) Schematic representation of dengue virus genome showing the predicted secondary structure of the RNA elements containing 5′ and 3′ UAR (shown in boxes). Nucleotide sequences of wild-type and the substitutions introduced at the 5′ UAR (5′Mut1, 5′Mut2, and 5′Mut3); and at the 3′ UAR (3′Mut1, 3′Mut2, and 3′Mut3) are shown. (B) RNA replication of mutant replicons disrupting and restoring 5′–3′ UAR complementarity in transfected BHK cells. Normalized luciferase levels determined 3 days after transfection are shown as a percentage of the wild-type replicon for the three groups of mutants within UAR sequences (Mut1, Mut2 and Mut3). Underneath of each plot, base pairing between sequences corresponding to 5′–3′UAR for the mutants are compared with the wild-type sequences.

Taken together, the results discussed here indicate that the complementary sequences 5′–3′ CS and 5′–3′ UAR are not necessary for efficient translation of the input RNA but they are essential during RNA synthesis. To better define the role of sequence complementarity during the process of RNA synthesis, we are currently investigating NS5 polymerase association to circular and linear conformations of the RNA and the impact of RNA conformation on enzyme activity. Disruption of 5′–3′ interactions did not alter significantly translation initiation of the input RNA, however, we cannot rule out a possible regulatory role of RNA-RNA interactions on translation after several rounds of protein synthesis had taken place in the infected cell. It is possible that dynamic RNA–RNA and RNA–protein interactions involving cellular and viral factors, induce conformational changes of the viral RNA rendering the viral genome more competent for translation, RNA synthesis, or encapsidation at different stages of viral infection. Understanding the role of secondary and tertiary structures of the viral RNA during the viral life cycle will help to clarify molecular details of dengue virus replication. At present neither specific antiviral therapy nor licensed vaccine exists to control dengue virus infections. Therefore, it is of crucial interest to investigate the biology of this virus at the molecular level as an essential step on designing novel antiviral strategies.

References

Alvarez DE, De Lella Ezcurra AL, Fucito S, Gamarnik AV 2005a Role of RNA structures present at the 3′UTR of dengue virus on translation, RNA synthesis, and viral replication. Virology 339:200–212

Alvarez DE, Lodeiro MF, Luduena SJ, Pietrasanta LI, Gamarnik AV 2005b Long-range RNA-RNA interactions circularize the dengue virus genome. J Virol 79:6631–6643

Barr JN, Wertz GW 2004 Bunyamwera bunyavirus RNA synthesis requires cooperation of 3′- and 5′-terminal sequences. J Virol 78:1129–1138

Brinton MA, Fernandez AV, Dispoto JH 1986 The 3′-nucleotides of flavivirus genomic RNA form a conserved secondary structure. Virology 153:113–121

Cahour A, Pletnev A, Vazielle-Falcoz M, Rosen L, Lai CJ 1995 Growth-restricted dengue virus mutants containing deletions in the 5′ noncoding region of the RNA genome. Virology 207:68–76

Corver J, Lenches E, Smith K et al 2003 Fine mapping of a cis-acting sequence element in yellow fever virus RNA that is required for RNA replication and cyclization. J Virol 77:2265–2270

Durbin AP, Karron RA, Sun W et al 2001 Attenuation and immunogenicity in humans of a live dengue virus type-4 vaccine candidate with a 30 nucleotide deletion in its 3′-untranslated region. Am J Trop Med Hyg 65:405–413

Elghonemy S, Davis WG, Brinton MA 2005 The majority of the nucleotides in the top loop of the genomic 3′ terminal stem loop structure are cis-acting in a West Nile virus infectious clone. Virology 331:238–246

Ghiringhelli PD, Rivera-Pomar RV, Lozano ME, Grau O, Romanowski V 1991 Molecular organization of Junin virus S RNA: complete nucleotide sequence, relationship with other members of the Arenaviridae and unusual secondary structures. J Gen Virol 72 (Pt 9): 2129–2141

Hahn CS, Hahn YS, Rice CM et al 1987 Conserved elements in the 3′ untranslated region of flavivirus RNAs and potential cyclization sequences. J Mol Biol 198:33–41

Hsu MT, Parvin JD, Gupta S, Krystal M, Palese P 1987 Genomic RNAs of influenza viruses are held in a circular conformation in virions and in infected cells by a terminal panhandle. Proc Natl Acad Sci USA 84:8140–8144

Jones CT, Patkar CG, Kuhn RJ 2005 Construction and applications of yellow fever virus replicons. Virology 331:247–259

Khromykh AA, Meka H, Guyatt KJ, Westaway EG 2001 Essential role of cyclization sequences in flavivirus RNA replication. J Virol 75:6719–6728

Kohl A, Dunn EF, Lowen AC, Elliott RM 2004 Complementarity, sequence and structural elements within the 3′ and 5′ non-coding regions of the Bunyamwera orthobunyavirus S segment determine promoter strength. J Gen Virol 85:3269–3278

Lo MK, Tilgner M, Bernard KA, Shi PY 2003 Functional analysis of mosquito-borne flavivirus conserved sequence elements within 3′ untranslated region of West Nile virus by use of a reporting replicon that differentiates between viral translation and RNA replication. J Virol 77:10004–10014

Mackenzie JM, Khromykh AA, Jones MK, Westaway EG 1998 Subcellular localization and some biochemical properties of the flavivirus Kunjin nonstructural proteins NS2A and NS4A. Virology 245:203–215

Markoff L 2003 5′ and 3′ NCRs in Flavivirus RNA. In: Chambers TJ, Monath TP (ed) Flaviviruses, Adv Virus Res, 60. Elsevier Academic Press, p 177–223

Men R, Bray M, Clark D, Chanock RM, Lai CJ 1996 Dengue type 4 virus mutants containing deletions in the 3′ noncoding region of the RNA genome: analysis of growth restriction in cell culture and altered viremia pattern and immunogenicity in rhesus monkeys. J Virol 70:3930–3937

Mir MA, Panganiban AT 2005 The hantavirus nucleocapsid protein recognizes specific features of the viral RNA panhandle and is altered in conformation upon RNA binding. J Virol 79:1824–1835

Nomaguchi M, Teramoto T, Yu L, Markoff L, Padmanabhan R 2004 Requirements for West Nile virus (−)- and (+)-strand subgenomic RNA synthesis in vitro by the viral RNA-dependent RNA polymerase expressed in Escherichia coli. J Biol Chem 279:12141–12151

Proutski V, Gould EA, Holmes EC 1997 Secondary structure of the 3′ untranslated region of flaviviruses: similarities and differences. Nucleic Acids Res 25:1194–1202

Raju R, Kolakofsky D 1989 The ends of La Crosse virus genome and antigenome RNAs within nucleocapsids are base paired. J Virol 63:122–128

Rauscher S, Flamm C, Mandl CW, Heinz FX, Stadler PF 1997 Secondary structure of the 3′-noncoding region of flavivirus genomes: comparative analysis of base pairing probabilities. RNA 3:779–791

Rice C 2001 Flaviviridae: the viruses and their replication. In: Fields Virology. In: (ed) 1. Lippincott-Raven, Philadelphia, p 991–1041

Shurtleff AC, Beasley DW, Chen JJ et al 2001 Genetic variation in the 3′ non-coding region of dengue viruses. Virology 281:75–87

Tilgner M, Deas TS, Shi PY 2005 The flavivirus-conserved penta-nucleotide in the 3′ stem-loop of the West Nile virus genome requires a specific sequence and structure for RNA synthesis, but not for viral translation. Virology 331:375–386

Westaway EG, Mackenzie JM, Kenney MT, Jones MK, Khromykh AA 1997 Ultrastructure of Kunjin virus-infected cells: colocalization of NS1 and NS3 with double-stranded RNA, and of NS2B with NS3, in virus-induced membrane structures. J Virol 71:6650–6661

Westaway EG, Khromykh AA, Mackenzie JM 1999 Nascent flavivirus RNA colocalized in situ with double-stranded RNA in stable replication complexes. Virology 258:108–117

You S, Falgout B, Markoff L, Padmanabhan R 2001 *In vitro* RNA synthesis from exogenous dengue viral RNA templates requires long range interactions between 5′- and 3′-terminal regions that influence RNA structure. J Biol Chem 276:15581–1591

Yu L, Markoff L 2005 The topology of bulges in the long stem of the flavivirus 3′ stem-loop is a major determinant of RNA replication competence. J Virol 79:2309–2324

Zeng L, Falgout B, Markoff L 1998 Identification of specific nucleotide sequences within the conserved 3′–SL in the dengue type 2 virus genome required for replication. J Virol 72:7510–7522

DISCUSSION

Harris: When you add NS5, does it increase the percentage of circular RNA molecules?

Gamarnik: That's one of the key questions. We need to do AFM in solution. The images of NS5 bound to the viral RNA shown here were obtained in air. However, we would like to see the RNA–protein interaction in solution. Using different buffer conditions, we can monitor the percentage of RNA cyclization in the presence of the viral protein. We would like to see the polymerase activity in real time. Some people have been able to visualize transcription in real time using AFM (*http://alice.berkeley.edu/~cjblab/transcription.html*).

Harris: Presumably you never see NS5 binding to the linear molecule?

Gamarnik: That is not the case. We indeed see linear molecules binding to NS5. However, it is difficult to discriminate whether the protein binds to the 5′ or the 3′ end of the RNA molecules. What we most commonly see is that the circles are decorated with polymerase.

Padmanabhan: When we did gel shift assays with NS3, we also used NS5 and found exactly the same thing: it binds to subgenomic RNA which has both ends, but not to the 5′ or 3′ separately. This is consistent with the requirement for RNA synthesis.

Jans: I'm curious as to how you did some of these AFM experiments. What sort of buffers did you use? Did you look at ionic strength? And what control did you do for NS5?

Gamarnik: We haven't used other proteins as controls, but we need to do this. It is necessary to include Mg^{2+} in the buffers. If we add other divalent cations, such as Ni^{2+}, we don't see RNA–RNA interactions. In addition, a divalent cation is necessary to immobilize the RNA on the mica. For this purpose people use Ni^{2+}, however, in our case the Ni^{2+} interferes with RNA–RNA complex formation and we do not observe circular forms of the RNA in this condition. Also if we include both Ni^{2+} and Mg^{2+} in the mix, we observe less RNA–RNA interaction.

Jans: Do you dry rapidly?

Gamarnik: We heat the samples to 80 °C, mix the sense and antisense molecules, dry them with nitrogen on a freshly cleaved mica, and then look at them right away.

Jans: To what extent do you think these 5′ and 3′ structures are dynamic?

Gamarnik: The RNA is very dynamic. The presence of viral and cellular proteins could also modulate RNA conformations.

Jans: Do you estimate the free energies of mutant RNAs?

Gamarnik: When we generate mutant RNAs, we try to maintain the free energy.

Jans: To get RNA–RNA complexes using gel shift assays, do you need heating?

Gamarnik: We can see the RNA–RNA complexes with or without heating. However, the standard assay includes heating and refolding. We use the same conditions to compare binding affinities of mutants and wild-type RNA.

Canard: Did you use capped and uncapped RNAs?

Gamarnik: That's a very good question. We are currently working on that.

Canard: The two anchor points for the polymerase. One is the 3′ end of the genome and the other is the 5′ cap.

Gamarnik: In order to study binding of the polymerase and the methyltransferase domains to the RNA, we are planning to use the full-length NS5 protein. In the results shown here, we used just the polymerase domain of NS5 without the methyltransferase. Perhaps if we use a capped RNA and full length NS5 we will be able to study the binding of the protein to both the cap structure and the RNA.

Rice: Have you looked at the influence of the capsid protein on any of your 5′/3′ interactions?

Gamarnik: We cloned and expressed a recombinant C protein, but haven't yet done anything with it.

Rice: Even in terms of your replicon assays?

Gamarnik: We have made one replicon that contains the first 34 amino acids of C fused to the reporter. Also, we have constructed monocistronic and bicistronic full-length genomes with reporters. So, hopefully we will be able to study the role of C on RNA replication.

Vasudevan: Have you looked at natural variations in this region?

Gamarnik: Yes. Dengue 1, 2, 3 and 4 are identical. All the isolates that we have analyzed showed 100% conservation of the complementary regions Sequence complementarity was also observed in other mosquito-borne flaviviruses.

Vasudevan: Earlier I was talking to Paul Young about the ease of making replicons or infectious clones with different virus strains. Not all strains seem to be the same in terms of maintaining them stably as a plasmid.

Gamarnik: They are not. It seems that we will have to test whether the reporter works in different strains. We have used DENV-2 (16681). Since the viral polyprotein is processed both in the cytoplasm and in the ER, the topology of the polyprotein fused to the reporter is critical to get replicating RNAs.

Harris: We have generated several DENV reporter replicons, and it seems to make a difference whether we place an HDVr ribozyme at the end versus cleaving with a restriction enzyme. In one case, the replicons generate a higher translation (first) peak and less replication (second peak); and in the other, the replicons produce a lower translation peak and greater replication.

Gamarnik: We tried putting a ribozyme at the end, but we did not see improvement in the replication efficiency of our replicon.

Harris: Interestingly, the West Nile virus replicons (constructed by P.-Y. Shi), by chance recapitulated the same two systems, and they found the same differences in translation versus replication depending on the cleavage at the end (HDVr versus restriction enzyme).

Vasudevan: When Ricco-Hesse and colleagues did the strain severity work, they also looked at the 5′/3′ region and the free energy of stem loop structures. There was some sort of correlation with severity. This is why it would be interesting to look at full-genomes (at least 500) to see if there is much statistical correlation between virus sequence and disease severity.

Gamarnik: Yes, there are some differences in the stem loop A structure. There are two stem loops in the 5′ UTR of all dengue viruses, stem loop A and stem loop B. In DENV-4 from the end of the stem loop B to the AUG there are five nucleotides, while in DENV-1, -2 and -3 the AUG is right after stem loop B. We speculate that 5′–3′UAR hybridization would decrease translation efficiency since this structure is right upstream of the initiator AUG. However, we did not see an effect on translation when these sequences were mutated. In these experiments, we monitored translation efficiency only of the input RNA. It will be interesting to analyse what happens with translation when the polymerase, the viral protein NS3, and the other components of the replication machinery are present. In these conditions, it is possible that RNA cyclization could have an effect on translation. To study this, we are setting up an *in vitro* system to see translation and RNA synthesis in the presence of the viral non-structural proteins. We are testing a *Xenopus* oocyte system, where we can inject both the viral RNA and viral proteins. We have previously developed a similar system to study poliovirus replication.

Rice: Do you get initiation of minus strand synthesis in the oocytes?

Gamarnik: So far we haven't. But we haven't worked much in this system yet.

Rice: Are you just looking for a double-stranded product?

Gamarnik: We are looking for viruses now. We are injecting the RNA and making whole cytoplasmic extract and searching for viral particles.

Rice: The polio experience.

Gamarnik: We had a very good experience with polio! Actually, with that system we could discriminate between synthesis of negative and positive strand RNA.

Gu: Because both the 5′ and 3′ UTR interact with cellular proteins do you think you can try to see RNA–RNA interactions in the presence of protein?

Gamarnik: Absolutely. We haven't tried this with cellular proteins. We know that the viral genome interacts with viral and cellular proteins. We examined first RNA–RNA interactions in the absence of proteins but now we need to move to a more complex system and add other factors. We recently performed preliminary experiments adding infected and uninfected cell extracts to the RNA–RNA complex, and we observed a mobility shift in native gels. In addition, we observed some differences between the infected and uninfected extracts, which doesn't tell us much at this point. We have done some purification and separation of fractions, but we did not identify any protein yet.

Canard: It is always tempting to get more complicated with proteins from the cell, but we can't be sure that this is telling us more: you will never know whether you have all the proteins from the cells.

Gamarnik: I agree. I think one has to choose a strategy; we started with a simple system using viral RNA and purified viral proteins. Next, the idea is to include more viral and cellular proteins to get closer to the natural environment.

Organization of flaviviral replicase proteins in virus-induced membranes: a role for NS1′ in Japanese encephalitis virus RNA synthesis

Vijaya Satchidanandam, Pradeep Devappa Uchil and Priti Kumar

Department of Microbiology and Cell Biology, Sir C.V. Raman Avenue, Indian Institute of Science, Bangalore 560012, India

Abstract. The organization of flaviviral replicase proteins within the membrane-bound replication complexes of West Nile (WNV), dengue (DENV) and Japanese encephalitis viruses (JEV) was probed by investigating the combined effect of detergents and trypsin on both viral replicase activity and profile of metabolically labelled viral proteins. While trypsin treatment of virus-induced membrane fractions degraded the vast majority of replicase proteins, viral RNA-dependent RNA polymerase (RdRp) activity remained completely unaffected. Solubilization of the membranes with deoxycholate (DOC) however rendered the replicase accessible to trypsin. Triton X-100 (TX100) treatment reduced RdRp activity by half in WNV but totally destroyed RdRp activity in JEV. TX100 also dissociated NS1′ in addition to NS1 from NS5 and NS3 in JEV. Antibodies to NS3 coprecipitated NS1′ along with NS5 only from DOC-solubilized but not from TX100-treated extracts, the former of which alone retained RdRp activity. Exogenous addition of recombinant NS1′ to TX100 treated JEV-induced membranes restored the defect in the release step of RNA synthesis. Our results suggest for the first time a direct role for JEV NS1′ in viral RNA synthesis *in vitro.*

2006 New treatment strategies for dengue and other flaviviral diseases. Wiley, Chichester (Novartis Foundation Symposium 277) p 136–148

The increasing threat to human health from mosquito-borne flaviviruses such as West Nile (WNV), dengue (DENV) and Japanese encephalitis (JEV) viruses has led to the categorization of many members of this family as 'emerging infectious agents'. While WNV, first reported in 1999 in the USA has since spread to most of the states (Hayes 2001, Jia et al 1999), the disease incidence due to DENV, as well as the childhood encephalitis rates due to JEV in Asia, have recorded an extraordinary rise (Rice & Lindenbach 2001). The quest for much-needed measures to curtail flaviviral spread would thus be tremendously aided by a comprehensive understanding of the factors that influence the process of viral replication.

The flaviviral genome of single-stranded (ss) positive sense RNA ~11 kb long comprises a single long open reading frame flanked by 5′ and 3′ untranslated regions (UTRs) and codes for a ~3400 amino acid polypeptide that is processed to give three structural and seven non-structural (NS) proteins. NS5 and NS3 function as the RNA-dependent RNA polymerase (RdRp) and helicase respectively, in viral RNA synthesis. A role in negative strand RNA synthesis has been ascribed (Lindenbach & Rice 1999) to the secreted NS1 protein. NS4a, an integral membrane protein, might serve as a protein bridge between NS1 with which it specifically interacts (Lindenbach & Rice 1999), and the flaviviral replication complex (RC), thus tethering the RC with its numerous proteins to the membrane (Westaway et al 1997). The demonstrated specific binding of elongation factor 1α and Mov34 to the 3′ UTRs of flaviviral genome (Blackwell & Brinton 1997, Ta & Vrati 2000) additionally implicates host protein(s) in flaviviral replication although a direct role remains to be unequivocally established.

Flaviviral RNA replication takes place in the host cell cytoplasm on membrane bound viral RC. Electron microscopic analysis carried out on Kunjin (KUN)- as well as DENV-infected cells (Mackenzie et al 1996, Westaway et al 1997) has displayed a dominant association of the replicating viral RNA with membranous vesicle packets. Moreover, our recent findings on the architecture of the RNA in the RC of flaviviruses WNV, JEV and DENV have offered conclusive biochemical evidence that at least two membranes with differing detergent solubilization properties encase the RF template present within the RC (Uchil & Satchidanandam 2003a). We also showed that extensive protease treatment of heavy membrane fractions from JEV-infected cells did not compromise the *in vitro* RdRp activity (Uchil & Satchidanandam 2003a) despite loss of most of the detectable replicase proteins NS5, NS3 and NS1. Thus, a bounding membrane protects the RC from protease action and only catalytic amounts of replicase proteins sufficed to manifest all the detectable RdRp activity. Herein we extend these observations for the three flaviviruses WNV, DENV and JEV with definitive biochemical proof for location of the proteins responsible for replicase activity behind a membrane barrier that also encloses the replicase. We also demonstrate preliminary evidence for a role for NS1′ of JEV specifically in the release step of RNA synthesis based on reconstitution of RdRp activity *in vitro.*

Materials and methods

Cells and viruses

Maintenance of cell lines and growth of viruses have been described previously (Uchil & Satchidanandam 2003a).

Preparation of flaviviral replication complexes and in vitro RdRp assay

PS cells infected with WNV, JEV or DENV at a multiplicity of infection (m.o.i.) of 10 were used as source of replication complex (RC) harvested by centrifugation at 800 × *g* at 18–22h post-infection (p.i.) and washed with ice-cold phosphate buffered saline (PBS). Preparation of subcellular fractions and assay for RdRp as well as partially denaturing gel electrophoresis and visualization of labelled products were carried out as reported earlier (Uchil & Satchidanandam 2003a).

Trypsin and detergent treatments of flavivirus-infected cell 16KP fractions

These methods have been described previously (Uchil & Satchidanandam 2003a).

Metabolic labelling of proteins

Subconfluent monolayers of PS cells were mock or flavivirus-infected at an m.o.i. of 10. At 16p.i., cells were incubated in methionine- and cysteine-deficient medium containing 10μg/ml AMD for 1h. Cells were then labelled with 50μCi of $[^{35}S]$methionine-cysteine (EXPRE^{35}S^{35}S, NEN, 1175.0Ci/mmol) per ml for 30min in the presence of 3μg/ml of AMD and 0.1% bovine serum albumin (BSA, Sigma). Protein samples were electrophoresed and labelled bands visualized as reported earlier (Uchil & Satchidanandam 2003b).

Western blot analysis

16KP fractions with or without prior treatments were resolved using SDS-10% PAGE, blotted onto a nitrocellulose membrane and blocked with 0.5% gelatin for 2h in Tris-buffered saline containing 0.1% Tween-20, pH 8.0. The membranes were incubated with polyclonal rabbit anti-JEV NS3 (1:4000), anti-JEV NS5 (1:2000), anti JEV envelope (1:2000) and anti-JEV NS1′ (1:2000) (a kind gift from Dr Gadkari, NIV, Pune, India), washed and developed using HRP-conjugated goat anti-rabbit IgG and diaminobenzidine (Sigma).

Radioimmunoprecipitation analysis

16KP fractions from $[^{35}S]$methionine-cysteine-labelled JEV-infected PS cells were treated with 1% TX100 or DOC and centrifuged at 16000 × *g* for 20min at 4°C to obtain TX100 and DOC extracts of 16KP fractions. The samples were then immunoprecipitated with protein A-sepharose beads (Amersham Pharmacia Biotech) bound to either pre-immune mouse serum or mouse anti-JEV NS3 serum.

The beads were washed appropriately in the TNMg buffer containing either 1% TX100 or DOC and 140 mM NaCl prior to electrophoresis and subsequent fluorography.

Expression, purification and in vitro reconstitution of NS1′

The Expand high fidelity reverse-transcriptase polymerase chain reaction (RT-PCR) system (Roche) was used to amplify the region corresponding to the NS1′ gene (2478-3713 nt in the JEV genome) using total RNA from JE-infected PS cell as template using appropriate primer pairs as described (Kumar et al 2003). The 1.26 kb PCR product initially cloned pFASTBAC Htb (GIBCO BRL) was released from using the *Nco* I and *Stu* I sites and cloned into the *Nco* I and Klenow polymerase blunted *Eco* RI sites of pRSET B *Escherichia coli* expression vector (Invitrogen). The expressed protein was purified using Ni-NTA agarose (Qiagen) chromatography according to manufacturer's instructions. GFP cloned at the same site and purified similarly was used as control in *in vitro* RdRp reconstitution assays. The 16KP fraction from JE-infected PS cells was treated with 1% TX100 twice, and subsequently solubilized in TnMg buffer containing 1% DOC. RdRp assay was then carried out in presence of varying concentrations of NS1′ or the control protein as stated.

Results

Flavivirus replicase proteins are present behind a membrane barrier and are required in catalytic amounts

As reported by us previously for JEV (Uchil & Satchidanandam, 2003b), trypsin treatment of heavy membrane fractions from WNV and DENV-infected PS cells did not compromise *in vitro* RdRp activity (Fig. 1, lanes 2–3 for WNV and 15–16 for DENV). This resistance to trypsin indicated the presence of a membrane barrier behind which not only the template RF as shown by us previously (Uchil & Satchidanandam 2003a) but also the protein components of the flaviviral replication machinery were located. Complete solubilization of this barrier with the ionic detergent DOC that was known to preserve RdRp activity is expected to render the replicase proteins NS5 and NS3 accessible to trypsin. In keeping with this premise, pretreatment with 1% DOC followed by trypsin digestion resulted in complete destruction of RdRP activity (Fig. 1, lanes 5–6, 11–12 and 18–19 for WNV, JEV and DENV, respectively). Analysis of the viral protein profile from metabolically labelled WNV, DENV and JEV-infected cells following trypsin digestion in the absence and presence of DOC showed that trypsin degraded most of the detectable viral replicase proteins even in intact membranes in the absence

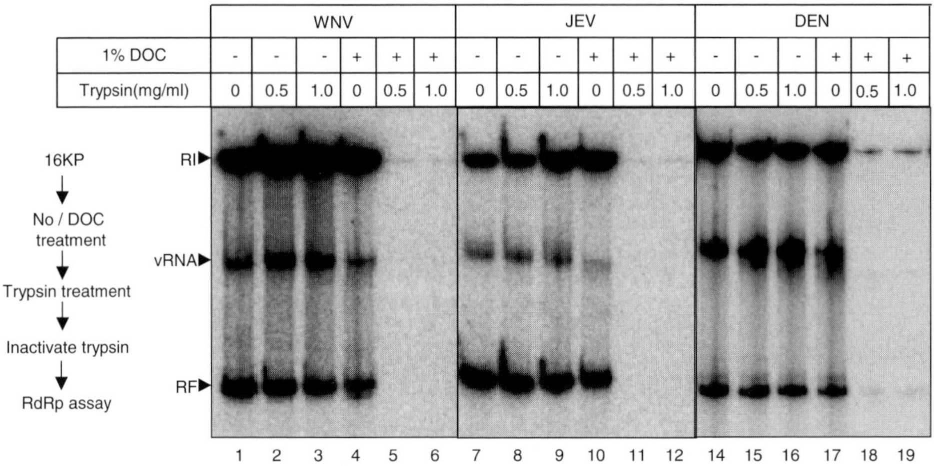

FIG. 1. Flaviviral replication complexes are present behind a membrane barrier. Heavy membrane (16KP) fractions from WNV (lanes 1–6), JEV (lanes 7–12) and DENV (lanes 14–19) infected cells were subjected to increasing concentrations of trypsin with (+) or without (–) prior treatment with 1% sodium deoxycholate (DOC) as depicted in the flow chart before carrying out the RdRp assays using [α^{32}P]GTP. The position and identity of the three viral RNA species RI, RF and vRNA following partially denaturing 7M urea 3% polyacrylamide gel electrophoresis (urea-PAGE) are denoted by the arrowheads.

of DOC without attendant loss of RdRp activity (Fig. 2). Thus trace amounts of NS5 and NS3 that are protected from trypsin digestion obviously suffice for manifesting the total detectable RdRp activity as previously demonstrated for JEV (Uchil & Satchidanandam 2003b).

Fractionation of proteins from TX100 treated membrane-bound flaviviral replication complexes

Our earlier studies had revealed that the RNA in the flaviviral RC was housed within virus-induced double-membrane structures, where the inner vesicles harboured the double-stranded RF template used by the flaviviral replicase (Uchil & Satchidanandam 2003a). When we carried out trypsin digestion of TX100 treated 16KP fractions, as seen for samples not exposed to detergent, most of the detectable viral replicase proteins were degraded by this treatment (Fig. 2). JEV RdRp activity was sensitive to TX100 (Fig. 3A, lane 6), in contrast to RdRp activity of WNV which was reduced but not completely lost in the presence of this detergent (Fig. 3A, compare lanes 1 and 2), similar to that seen with KUN RdRp (Chu & Westaway 1987). This residual activity partitioned almost equally between the

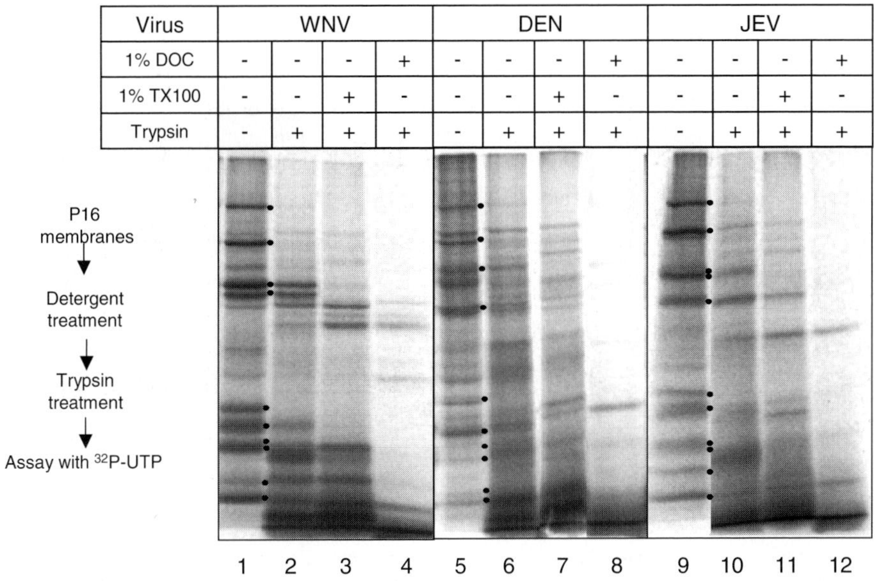

FIG. 2. Effect of *in vitro* trypsin treatment on metabolically labelled flaviviral proteins. 16KP fractions metabolically labelled with [^{35}S]methionine-cysteine from WNV (lanes 1–4), DENV (lanes 5–8) and JEV (lanes 9–12) infected cells were subjected to trypsin at 4°C for 20 min without (lanes 2, 6 and 10) or with prior treatment with 1% sodium deoxycholate (DOC; lanes 4, 8 and 12) or 1% triton X-100 (TX100; lanes 3, 7 and 11). The processed samples were electrophoresed on SDS-10% polyacrylamide gel followed by autoradiography. The dots indicate the locations of flavivirus proteins.

pellet and supernatant fractions following centrifugation at 16 000 × g for 10 min (Fig. 3A, lanes 3 and 4).

The differential effect of TX100 on the RdRp activity of JEV prompted us to probe the identity of protein(s) that might be released from their association with the flavivirus replicase by this detergent. Metabolically labelled flaviviral proteins from infected cells were analyzed following detergent extraction. As previously demonstrated (Uchil & Satchidanandam 2003a), these individual vesicles could be sedimented in an ultracentrifuge at 150 000 × g for 5 h. Comparison of the proteins present in the pellet and supernatant fractions following ultracentrifugation using the high-resolution Tris-tricine buffer revealed a uniform enrichment of NS5 and NS3 of JEV in the pellet and NS1 in the supernatant fraction (Fig. 3B). Interestingly NS1′ which is unique to JEV and has not been reported in the other two viruses was also quantitatively released by TX100 into the supernatant (Fig. 3B, lane 2).

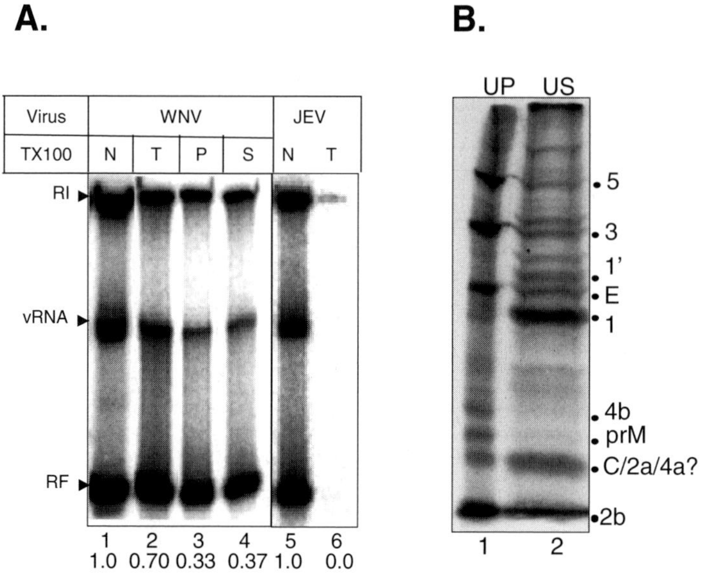

FIG. 3. Effect of TX100 on *in vitro* flaviviral RdRp activity. (A) 16KP fractions from WNV (lanes 1–4), or JEV (lanes 5–6) were treated (T; lanes 2, 6) or not treated (N; lanes 1, 5) with 1% TX100 and the supernatant fraction was sedimented at $16\,000 \times g$. *In vitro* RdRp assay was carried out and labelled RNA products were analyzed using urea-PAGE. (B) Metabolically labelled proteins from JEV-infected 16KP fractions extracted with 1% TX100 and sedimented at $16\,000 \times$ g to obtain pellet and supernatant fractions. Pellet (UP; pellet) and supernantant (US; supernatant) fractions obtained by ultra centrifugation at $150\,000 \times$ g of the latter were analysed on SDS-Tricine-10% PAGE system and the proteins visualized by autoradiography. The dots represent flavivirus-specific proteins absent in mock-infected cells with their putative identities mentioned alongside. The asterisks denote proteins of probable host origin.

TX100 but not DOC leads to dissociation of NS1′ protein of JEV from NS3 and NS5

The above observation pointed to a possible role for NS1′ in the TX100-induced loss of JEV RdRp activity. Having further confirmed the identity of the released proteins as NS1 and NS1′ by use of a high-resolution tris-tricine gel system (Fig. 4A) and by Western blotting to NS1 specific serum (Fig. 4B), we attempted to assess the interaction of NS1 and NS1′ with NS5 and NS3 proteins in the replicase complex by radioimmunoprecipitation of detergent extracts from JEV-infected cells using anti-NS3 serum. As seen in Fig. 4C (lanes 4 and 6), a comparable amount of NS5 was coprecipitated from both DOC and TX100-extracted membrane fractions. In corroboration with the near complete dissociation of NS1 from NS5 and NS3 following ultracentrifugation (Fig. 4, A and B), insignificant quantities of NS1 was associated with NS3 and NS5 (Fig. 4C, lanes 4 and 6). In keeping

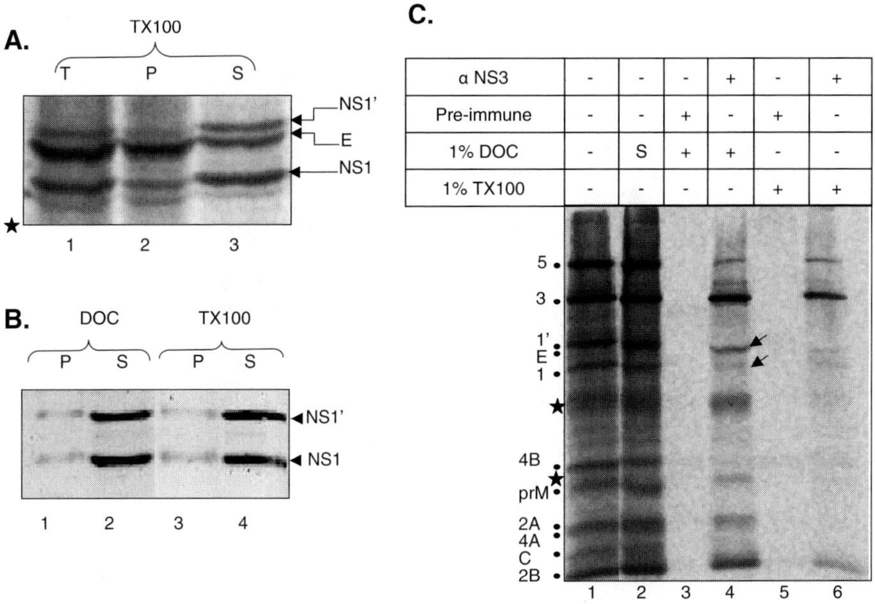

FIG. 4. NS1′ is co precipitated along with NS5 and NS3 by mouse anti-NS3 serum. (A) [^{35}S]methionine-cysteine labelled proteins from JEV-infected 16KP fractions were treated with 1% TX100 (T), fractionated into pellet (P) and supernatant (S) fractions following centrifugation at 16 000 × g. and analysed on SDS-Tricine-10% PAGE and visualized by autoradiography. (B) Western blotting of 1% deoxycholate or 1% TX100 fractionated proteins as in (A) from JEV-infected 16KP fractions using rabbit anti-NS1 antibodies. (C) Radioimmunoprecipitation (RIP) analysis of proteins released from JEV-infected 16KP fractions using anti-JEV NS3 antibodies. Labelled proteins released into supernatant fractions after DOC (S; lane 2) or TX100 treatment were immunoprecipitated using mouse pre-immune serum (lanes 3 and 5) or mouse anti-JEV NS3 antibodies (lanes 4 and 6). Arrows indicate the NS1′ and NS1 co precipitated with NS3 in DOC extracts of 16KP fractions. The dots represent flavivirus-specific proteins with their putative identities mentioned alongside. The asterisks represent proteins presumably of host origin.

with the ability of DOC-solubilized membranes to retain RdRp activity, multiple viral proteins were coprecipitated by anti-NS3 antibody from these extracts in contrast to TX100-extracted membranes (Fig. 4C, lanes 4 and 6). The most striking observation was however the sizeable quantities of NS1′ that coprecipitated only from DOC but not TX100 treated JEV-infected 16KP fractions, suggesting a vital role for this alternatively processed form of NS1 in JEV RNA synthesis. This protein brought down by anti-NS3 antibodies was confirmed to be NS1′ and not the similar sized envelope by Western blotting to antisera specific to the two proteins (data not shown).

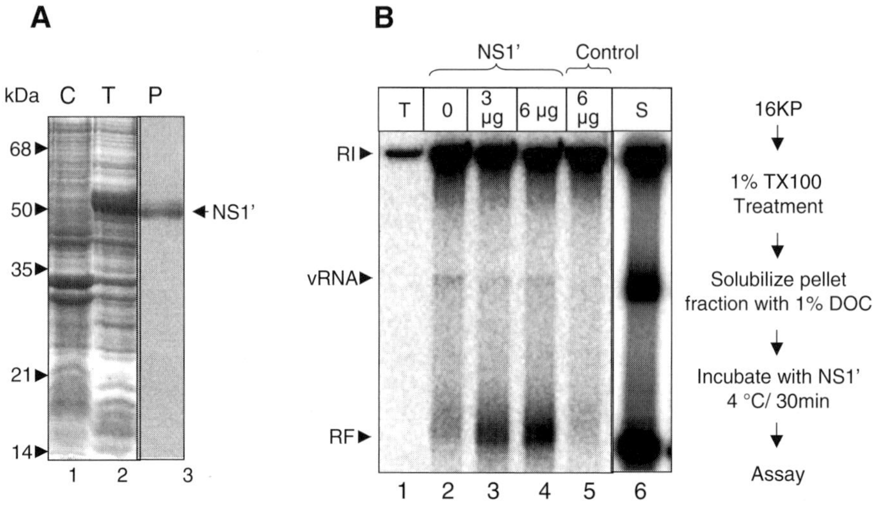

FIG. 5. Restoration of release function of JEV RdRp activity using recombinant NS1′ protein. (A) SDS-PAGE analysis of recombinant NS1′ of JEV expressed in *E. coli*. *E. coli* cells containing non-recombinant (C; lane 1) or recombinant pRSET B carrying the JEV NS1′ gene (T, lane 2) were induced with IPTG, proteins in the cell lysates were analysed using SDS-10% PAGE and visualized by Coomassie blue staining. The 55 kDa recombinant NS1′ protein was purified by Ni-NTA agarose affinity chromatography, (P; lane 3). (B) 16KP fractions were processed as depicted in the flowchart and RdRp assay with [α^{32}P]GTP carried out at the end of the incubation period with increasing amount of recombinant NS1′ (lanes 3–4) or 6 μg of control protein (lane 5). The labelled RNA products were analysed as before. Lanes 6 and 1 depict RdRp activity in 16KP fractions before (S) and after (T) TX100 extraction. Arrowheads denote the positions of RI, vRNA and RF.

Addition of exogenous NS1′ restores release function of TX100 treated JEV replicase

In order to directly ascertain the contribution of NS1′ to JEV RdRp activity, we depleted 16KP membranes obtained from JEV-infected cells of the endogenous NS1′ by two rounds of extraction using 1% TX100 after which no RdRp activity was observed (Fig. 5B, lane 1). The residual membranes were completely solubilized using 1% DOC following which incorporation of label into RI species alone was observed (Fig. 5B, lane 2) denoting restoration of elongation step of RNA synthesis by this detergent. Addition of recombinant NS1′ purified from *E. coli* expressing this protein (Fig. 5A) brought about a dramatic restoration of release function of RdRp activity (Fig. 5B, lanes 3 and 4) with significant amounts of label incorporated into the RF species indicating proficient generation of RF from replicative intermediates (RI) following release of vRNA. The lack of complete restoration of RdRp activity under these conditions however suggested that proteins other than NS1′ either of viral or host origin, required by the replicase complex may have also been depleted by TX100 treatment.

References

Blackwell JL, Brinton MA 1997 Translation elongation factor-1 alpha interacts with the 3′ stem-loop region of West Nile virus genomic RNA. J Virol 71:6433–6444

Chu PW, Westaway EG 1987 Characterization of Kunjin virus RNA–dependent RNA polymerase: reinitiation of synthesis in vitro. Virology 157:330–337

Hayes CG 2001 West Nile virus: Uganda, 1937, to New York City, 1999. Ann N Y Acad Sci 951:25–37

Jia XY, Briese T, Jordan I et al 1999 Genetic analysis of West Nile New York 1999 encephalitis virus. Lancet 354:1971–1972

Kumar P, Uchil PD, Sulochana P et al 2003 Screening for T cell-eliciting proteins of Japanese encephalitis virus in a healthy JE-endemic human cohort using recombinant baculovirus-infected insect cell preparations. Arch Virol 148:1569–1591

Lindenbach BD, Rice CM 1999 Genetic interaction of flavivirus nonstructural proteins NS1 and NS4A as a determinant of replicase function. J Virol 73:4611–4621

Mackenzie JM, Jones MK, Young PR 1996 Immunolocalization of the dengue virus nonstructural glycoprotein NS1 suggests a role in viral RNA replication. Virology 220:232–240

Rice CM, Lindenbach BD (eds) 2001 *Flaviviridae*. Lippincott Williams and Wilkins, Philadelphia, PA

Ta M, Vrati S 2000 Mov34 protein from mouse brain interacts with the 3′ noncoding region of Japanese encephalitis virus. J Virol 74:5108–5115

Uchil PD, Satchidanandam V 2003a Architecture of the flaviviral replication complex: protease, nuclease, and detergents reveal encasement within double-layered membrane compartments. J Biol Chem 278:24388–24398

Uchil PD, Satchidanandam V 2003b Characterization of RNA synthesis, replication mechanism, and in vitro RNA-dependent RNA polymerase activity of Japanese encephalitis virus. Virology 307:358–371

Westaway EG, Mackenzie JM, Kenney MT, Jones MK, Khromykh AA 1997 Ultrastructure of Kunjin virus-infected cells: colocalization of NS1 and NS3 with double-stranded RNA, and of NS2B with NS3, in virus-induced membrane structures. J Virol 71 6650–6661

DISCUSSION

Jans: What is your model for how things get into and out of the vesicles? Can we start with nucleotides—can they get in and out?

Satchidanandam: Yes, but they don't come out very easily. We have calculated the K_m for each of the ribonucleotides. We had to go to enormous lengths in order to extract the endogenous pool. We went to 2 M KCl washes, and only then were we able to get any increase in activity with exogenously added material. I don't think they go in and out very easily. But we must keep in mind that these membranes do not exist in an uninfected cell. We need to look at infected cells and get rid of the nucleus. If we do this about 12–14 h post-infection (the time point varies for the different viruses) for the postnuclear supernatant you collect which contains the mitochondria, you can get a visible pellet only if it is an infected cell. If you do the same thing in uninfected cells you hardly see a pellet. These are membranes that come up as a kind of stress response following viral infection. This has been referred to in the literature as endoplasmic reticulum (ER) stress, since several of the viral non-structural proteins are extremely hydrophobic and incorporate into the ER membrane/lumen following their synthesis on ER-attached ribosomes.

Jans: So you think that these nucleotides get trapped right at the beginning and everything is formed inside the membranes. But you can still add label and get it in?

Satchidanandam: The vRNA is coming out; we know that it is coming out and gets translated, but we don't know how this happens. In other viruses people have seen pores in these vesicles. They postulate that RNA is extruded out through these. We don't know what happens with flaviviruses.

Jans: You mentioned earlier that you have some of these structures in the nucleus.

Satchidanandam: Yes, these are closely in apposition to the inner nuclear membrane. Our EM pictures show this.

Jans: Why would you need some of these in the nucleus?

Satchidanandam: As far as we can tell the biochemical properties of the nuclear complex are almost identical to the cytoplasmic complexes.

Young: With regards to the *E. coli*-derived NS1, we have looked at the folding of this expressed protein with a panel of monoclonal antibodies. Apart from a very few, it doesn't react with them. It is clearly not folded correctly. One of the biggest problems is that the 12 cysteines don't form appropriate disulfide bridges. Of course, you are making NS1′, and this is probably the single explanation for the association under different detergent conditions. NS1 is loosely held to membranes by a hydrophobic patch, whereas NS1′ is a transmembrane protein. It has the N-terminal membrane-spanning domain of NS2A. This would explain the differential association. How would you explain any critical role for NS1′, given that there are only a small number of flaviviruses that can actually generate this?

Satchidanandam: It is only circumstantial. When we treat the membrane-bound replicase complex with detergents that inactivate the polymerase activity, we notice that we also lose NS1. The detergent that does not cause loss of activity, such as deoxycholate, despite complete solubilization, permits the association of NS1′ with NS3 and NS5. Under conditions in which we lose activity using Triton X100, despite not solubilizing the replicase complexes or the NS3 and NS5 residing within them, we find that the interaction of NS1′ with NS3 and NS5 is lost. Our inability to reconstitute complete activity by exogenous addition of NS1 may be because these proteins have to interact with each other and get into the nascent replication complex as it is forming and/or as they are folding.

Young: Even trying to regenerate the eukaryote-derived NS1 recombinant is going to be difficult.

Satchidanandam: We have not purified NS1 from an infected cell.

Padmanabhan: What is the earliest species you see with respect to the three replicative RNA species according to Westaway's system?

Satchidanandam: It is exactly the same as Westaway showed with Kunjin virus. The first species to pick up radioactive label is the RI. As it starts decaying you will see increasing amounts of label getting into both the vRNA (vRNA) and RF species

Screaton: Some of the double membrane structures you showed were quite tightly packed. Is this autophagy?

Satchidanandam: I don't know what goes on, but I wouldn't think so.

Screaton: It looks like it with these double membrane structures. Has anyone tried inhibiting autophagy?

Satchidanandam: No, but it would be worthwhile.

Harris: I was curious about the two host proteins: what was their molecular weight?

Satchidanandam: One was around 35 kDa, the other is smaller. The one at 35 kDa could be the Mov34 shown by Malancha Ta and Sudhanshu Vrati, which is a nuclear localized protein. Those were *in vitro* studies (Ta & Vrati 2000).

Harris: The other one looked like it was close to the size of PRM.

Satchidanandam: Yes. We don't know whether they are related.

Ng: Are the Triton-resistant vesicles the small membrane vesicles we commonly see which often have thread-like structures within them? Are you referring to the same thing?

Satchidanandam: Those are what are generated when we treat these membrane structures from the infected cells *in vitro.*

Ng: Normally in infected cells we see bags of smooth membrane vesicles, sometimes with thread-like structures within them.

Satchidanandam: Within the bigger bag there are smaller vesicles.

Ng: Is this thread-like structure the replication complex?

Satchidanandam: These membrane bags often have protrusions or extensions. If you look at the immunostained EMs, those thread-like structures are PDI positive, so they are ER-derived. What Westaway shows is that intracellular membranes, derived from Golgi as well as ER reorganize following viral infection. There is a lot of membrane reorganization. I am not sure what the threads you are talking about are.

Rice: Can you comment on the level of activity you are seeing when you make your cell-free extracts? What is actually going on in the cell? You made the comment that the majority of the protein can be digested away and this does not affect the *in vitro* activity. Can we extrapolate the *in vitro* results to what is taking place in the cell?

Satchidanandam: We labelled *in vivo* with ^{32}P, but then the specific activity is almost impossible to compute since we do not know what proportion of the viral NS3 and NS5 proteins within an infected cell actually belong inside active replicase complexes. All I can say from looking at these pictures is that when we isolate these membranes which carry the replication complexes (these are post-nuclear supernatants collected at 15 000 rpm), they display the same amount of viral RNA synthesizing activity before and after treatment with trypsin.

Rice: It is an important conceptual issue for the field. If that is the case in an infected cell, it raises the issue of the role of the non-structural replicase protein. Is it simply made in excess for no good reason? Or is it doing things worth thinking about as potential target activities?

Satchidanandam: When we fractionate the cell extract, in the cytosol after we pelleted all the membranes, we found enormous quantities of NS5. We don't see NS3 in the cytosol, just NS5. We have always wondered how these proteins do their job. They might be doing different things in different compartments.

Rice: You have done a lot of the biochemistry on this replication complex. Are we going to get to a point where either by initiating the system by a cell free translation system or by reconstituting purified protein, we ever get to a system where we have a cell-free replication assay?

Satchidanandam: We have done the following experiments. We have taken insect cells and expressed each of these three proteins, NS1, NS3 and NS5. Then we coinfected Sf9 and Sf21 cells with the three baculoviruses together, or only NS3 and NS5. They did not incorporate label into RNA when we provided viral 3′ and 5′ end templates, but this could be because we need the membrane-bound proteins to tether NS1, and then bring in the 3′ end. In the absence of the membrane-bound proteins these three NS proteins won't get into any cell membrane on their own. Thus we didn't see any activity in either the soluble fraction or the membrane fraction of these coinfected insect cells.

Rice: It is certainly possible that coupled translation and production of these proteins is needed in order to get the replication complex. It may be a bit of a biochemical challenge to do this.

Reference

Ta M, Vrati S 2000 Mov34 protein from mouse brain interacts with the 3′ noncoding region of Japanese encephalitis virus. J Virol 74:5108–5115

CRM1-dependent nuclear export of dengue virus type 2 NS5

Melinda J. Pryor, Stephen M. Rawlinson, Peter J. Wright* and David A. Jans[1]

*Department of Biochemistry and Molecular Biology, and *Department of Microbiology Monash University, Victoria 3800, Australia*

Abstract. The dengue virus multidomain RNA polymerase NS5 has been observed in the nucleus in mammalian infected cell systems. We previously showed that NS5 nuclear localization is mediated by two nuclear targeting signals within the NS5 interdomain region that are recognized by distinct members of the importin superfamily of intracellular transporters. Intriguingly, we have recently found that NS5 also possesses the ability to be exported from the nucleus by the importin family member CRM1 (exportin 1) both in Vero cells transfected to express NS5, and in dengue virus type 2 infected Vero cells, based on use of the CRM1-specific inhibitor leptomycin B (LMB). LMB treatment of Vero cells resulted in increased nuclear accumulation in both systems, and interestingly in the latter, resulted in an alteration in the kinetics of virus production. Our results imply that subcellular trafficking of NS5 at particular times in the infectious cycle may be central to the kinetics of virus production; perturbing this trafficking may represent a viable approach to develop new antiviral therapeutics.

2006 New treatment strategies for dengue and other flaviviral diseases. Wiley, Chichester (Novartis Foundation Symposium 277) p 149–163

The dengue virus non-structural protein NS5 is the largest protein encoded by the viral genome and the most highly conserved of the flavivirus proteins. It is a multifunctional protein comprising an N-terminal methyltransferase (MTase) (Egloff et al 2002) and a C-terminal RNA-dependent RNA polymerase (RdRp) (Bartholomeusz & Wright 1993, Tan et al 1996, Ackermann & Padmanabhan 2001) domains, separated by an interdomain region containing functional nuclear targeting signals (Forwood et al 1999, Brooks et al 2002) (see Fig. 1). Similar to other animal RNA viruses, flavivirus RNA replication occurs in the cytoplasm of host cells in membrane-induced perinuclear vesicles (Westaway et al 1999). A hypophosphorylated form of dengue virus type 2 (DENV-2) NS5 has been shown to

[1]This paper was presented at the symposium by David A. Jans, to whom correspondence should be addressed.

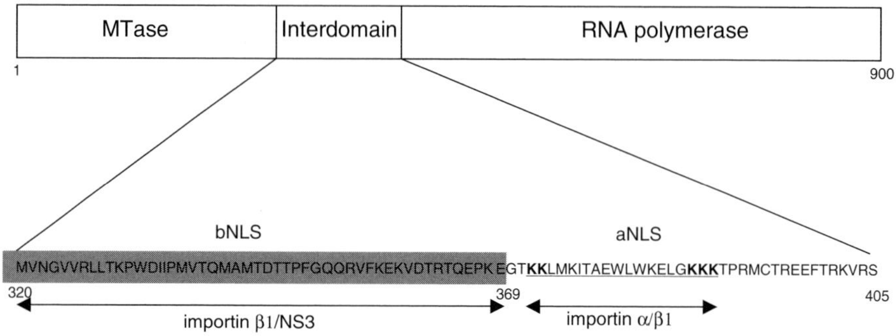

FIG. 1. Schematic diagram of the NS5 protein highlighting the interdomain region (residues 320–405). The single letter amino acid code is used with the bNLS highlighted and aNLS underlined. The basic residues in the aNLS are shown in bold. The binding sites for importin β, NS3 and importin αβ are indicated.

interact with NS3 in the cytoplasm (Kapoor et al 1995), constituting part of the proposed replication complex that also includes the other non-structural proteins NS1, NS2A and NS4A (Mackenzie et al 1998). The 5′ end of the DENV-2 RNA genome has been shown to contain a type 1 cap (Cleaves & Dubin 1979). Capping of cellular mRNA is normally performed in the cell nucleus, but viruses like dengue that replicate in the cytoplasm must encode their own cytoplasmic capping enzymes. NS5 possess the nucleoside-2′-*O*-methyltransferase activity (Egloff et al 2002), one of the four enzymatic activities required for the capping process.

All the known functions of DENV-2 NS5 are believed to occur in the cytosol of infected cells, but a hyperphosphorylated form of NS5, that is unable to interact with NS3, has been reported in the nucleus of DENV-2 infected mammalian cells late in infection (Kapoor et al 1995). These results suggest that phosphorylation is involved in modulating NS5 subcellular localization; the NS5/NS5A proteins belonging to all three genera of the *Flaviviridae* are known to be phosphorylated (Reed et al 1998). The role of DENV-2 NS5 within the nucleus has yet to be established, but it has been speculated to be involved in the regulation of cellular genes expressed in response to viral infection (Kapoor et al 1995), and promoting a shift from RNA replication to the packaging phase of infection (Butcher 2003).

Nuclear import of proteins >45 kDa is generally dependent on transport receptors called importins that bind to proteins and transport them through the nuclear envelope localized nuclear pore complexes (NPCs) (Paschal 2002). There are two classes of well characterized classical nuclear localization sequences (NLSs) that confer recognition by importins; these include monopartite NLSs containing a single cluster of basic amino acids, and bipartite NLSs comprising two clusters of basic residues separated by a variable spacer region (Jans et al 2000). Additional non-basic NLSs have also been described. NLS-containing proteins are conventionally recognized in the cytoplasm by a heterodimer composed of importin α and β1 (Jans et al 2000). Importin α binds directly to the NLS (Conti et al 1998) and importin β1 is responsible for the translocation of the NLS-containing protein-importin complex into the nucleus by interaction with the protein constituents of the NPC (Gorlich & Kutay 1999). It has been shown that nuclear import can also be mediated by importin β1 or importin β homologues exclusively, whereby importin β1/the importin β homologue is able to bind the NLS directly without the need for an adaptor protein such as importin α, (Jans et al 2000). Dissociation of the NLS-containing protein–importin complex is facilitated by binding to importin β1/the importin β homologue by the nuclear guanine nucleotide binding protein Ran in activated GTP-bound form (Pemberton & Paschal 2005). Analogously, nuclear export of proteins is mediated by transport receptors of the importin β superfamily, called exportins, in association with Ran-GTP (Pemberton & Paschal 2005). The best studied exportin, CRM1, binds to leucine-rich nuclear export sequences (NESs) (Fornerod et al 1997), including that of HIV-Rev (Fischer et al 1995). Leptomycin B (LMB), a specific inhibitor of CRM1, has been shown to block this nuclear export pathway specifically by preventing NES-CRM1 interaction (Fukuda et al 1997).

The interdomain linker region of DENV-2 NS5 between the methyltransferase and polymerase domains has been shown to contain two functional NLSs (Fig. 1). NS5 residues 369–405 were initially demonstrated to contain a functional NLS recognized by the importin αβ heterodimer (Forwood et al 1999). Site-directed mutagenesis established that the two basic regions $K^{371}K^{372}$ and $K^{388}K^{389}$ separated by a 14 amino acid linker were critical for nuclear localization, referred to as the aNLS (Brooks et al 2002). Using the yeast two-hybrid system, residues 320–368 adjacent to the previously characterized NLS were found to confer interaction with importin β1 (Johansson et al 2001). This region was subsequently shown to be able to target β-galactosidase to the nucleus of mammalian cells in an *in vitro* reconstituted system, albeit at slower rates than that observed for aNLS (Brooks et al 2002). This region is referred to as the bNLS, although the critical residues of the NLS have yet to be identified.

The overall objective of this work was to investigate further the subcellular localization of DENV-2 NS5 in virus-infected and transfected mammalian cells. Confocal laser scanning microscopy (CLSM) of immunostained DENV-2-infected Vero cells was used to monitor the localization of NS5 during infection. Surprisingly, NS5 was detected in the nucleus early in infection, with this localization continuing throughout the course of infection. Using the CRM1-specific inhibitor LMB, we were able to establish that NS5 possesses the ability to be exported from the nucleus, and that disruption of nuclear export in virus-infected cells altered the kinetics of virus production. This latter implies that the subcellular trafficking of NS5 during infection may be important to the kinetics of virus production. Antiviral drugs designed to disrupt the trafficking of NS5 may represent a viable strategy to control the progression of dengue virus infection.

Methods

Cells, virus and viral titration

DENV-2 New Guinea C (NGC) strain was propagated on Vero cells at a multiplicity of infection (MOI) of approximately 10 in Eagle's medium M199 containing 2% fetal calf serum (FCS). Where indicated, 5 ng/ml LMB was added to the culture medium at 0 h post infection (p.i.). At specific times p.i. the culture media was collected for viral titration and the coverslips were fixed for indirect immunofluorescence.

Virus was titrated by plaque assay on *Aedes albopictus* C6/36 cells at 28 °C as previously (Gualano et al 1998). Viral titres were calculated as plaque forming units (p.f.u.)/ml.

Cells and transfection

Vero cells were grown in Dulbecco minimal medium (DMEM) containing 10% FCS. DNA transfection was performed using Lipofectamine 2000 (Invitrogen) according to the manufacturer's specifications. Where indicated, LMB (2.8 ng/ml) was added to the culture medium 5 h prior to live cell imaging 20–24 h post transfection (p.t.) using a BioRad MC 500 CLSM with a 40 × water immersion lens and Kalman filter mode. Confocal images were analysed using the ImageJ v1.28 software as previously (Harley et al 2003). The fluorescence intensity within the nucleus and cytoplasm was measured and background fluorescence subtracted in order to calculate the nuclear to cytoplasmic ratio (Fn/c).

Plasmid construction

pEPI-GFP-NS5 encodes the sequence of GFP (green fluorescent protein) in frame N-terminal to the sequence of full length DENV-2 NS5. The plasmid was constructed as follows: a full-length DENV-2 NS5 cDNA fragment flanked by *att*B1/*att*B2 recombination sites was generated by PCR (Platinum Taq High fidelity, Invitrogen) using primers 22.5 (5′-ggggacaagtttgtacaaaaaagcaggctctGGAAC TGGCAACATAGGAG DENV-2 cDNA in upper case) and 22.6 (5′ gggga ccactttgtacaagaaagctgggTCACCACAGGACTCCTGCCTCTTCC complementary to DENV-2 cDNA in upper case). The NS5 PCR product was firstly introduced into the Gateway™ pDONR207 vector using the BP clonase enzyme, and subsequently into the pEPI-GFP-DESTC expression vector (Ghildyal et al 2005) using the LR clonase enzyme following the manufacture's instructions. The integrity of the NS5 insert was established through DNA sequencing.

Indirect immunofluorescence and CLSM

Cells were fixed in methanol/acetone (1 : 1) for 2 min at room temperature. The intracellular distribution of NS5 was analysed by immunostaining using an anti-NS5 antibody raised in rabbits against residues 397-772 of DENV-2 NS5 (gift from Dr Keng Teo), and a goat anti-rabbit Alexa Fluor® 488 green fluorescent dye-conjugate (Molecular probes) as the secondary antibody. The cells were mounted in glycerol–2% propylgallate and viewed by CLSM using a 60 × oil immersion lens as above.

Results and discussion

Subcellular localization of the NS5 protein during DENV infection in Vero cells

To determine the subcellular location of the NS5 protein during dengue virus infection, Vero cells infected with DENV-2 NGC at an MOI of approximately 10 were examined at specific stages p.i., by indirect immunofluorescence using a polyclonal anti-NS5 and an Alexa Fluor® 488 green fluorescent dye-conjugated secondary antibody (Fig. 2A). Prior to 12 h p.i. (data not shown), NS5 was unable to be detected in either the cytoplasm or the nucleus of infected Vero cells, supporting the long latent period of DENV-2 infections described by Cleaves et al (1981). At 12 h p.i., a small percentage of cells showed staining for NS5 in both the cytoplasm and the nucleus of infected cells although the staining was predominantly nuclear. After 20 h p.i., the number of cells staining NS5 significantly increased, with the predominantly nuclear localization of NS5 and the diffuse cytoplasmic staining continuing throughout the virus infection. The observed localization of NS5 in the nucleus early in infection is surprising in view of the fact that NS5 has a major role in cytoplasmic replication of the virus, although it is known that only small quantities of NS5 are required for replicative activity, as shown for several flaviviruses (Chu & Westaway 1992, Grun & Brinton 1987, Uchil & Satchidanandam 2003). NS5 has also been detected in the nucleus of yellow fever virus infected cells (Buckley et al 1992), but not in the case of Kunjin and Japanese encephalitis virus (Edward & Takegami 1993, Westaway et al 1997). Several other flavivirus proteins have been detected in the nucleus during infection; including the capsid protein of dengue (Bulich & Aaskov 1992, Makino et al 1989, Wang et al 2002),

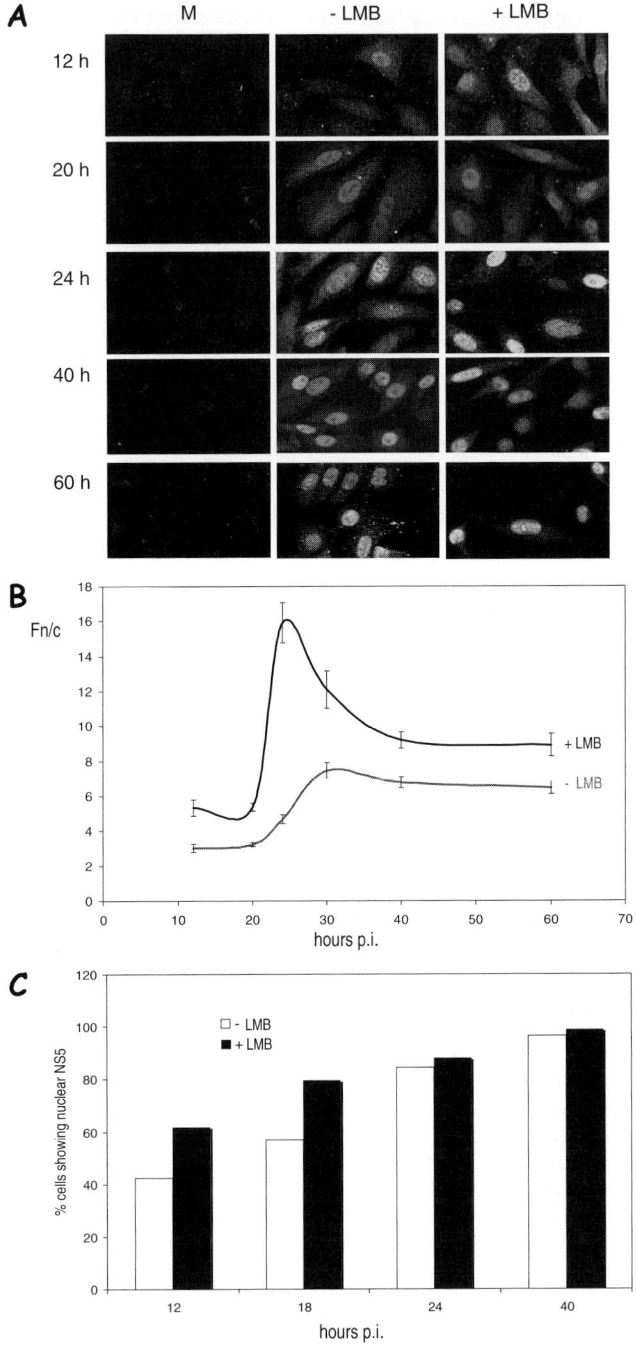
A
M
- LMB
+ LMB
12 h
20 h
24 h
40 h
60 h
B
Fn/c
+ LMB
- LMB
hours p.i.
C
% cells showing nuclear NS5
- LMB
+ LMB
12
18
24
40
hours p.i.

and Japanese encephalitis (Mori et al 2005) and both the capsid and NS4B proteins of Kunjin (Westaway et al 1997).

LMB treatment increases nuclear localization of DENV-2 NS5 in infected cells

The presence of DENV-2 NS5 in the nucleus of infected cells very early in infection prompted us to investigate if NS5 may shuttle between the nucleus and cytoplasm during infection. To assess the possibility that NS5 may possess a CRM1 dependent nuclear export pathway, Vero cells infected with DENV-2 were treated with LMB at 0 h p.i., and NS5 nuclear accumulation compared to that of non-treated samples at specific stages during infection (Fig. 2A). Image analysis of digitized CLSM images (see Materials and Methods) was performed to determine the nuclear to cytoplasmic ratio (Fn/c) for the infected cells treated without or with LMB (Fig. 2B). All samples had a mean Fn/c of greater than 2, indicative of nuclear accumulation, although the Fn/c was significantly higher at each time point in the case of the LMB-treated cells. Further, nuclear accumulation peaked more rapidly (at *c.* 24 h) in the presence of LMB, compared to *c.* 30 h in its absence. These results suggest that DENV-2 NS5 possesses the ability to be exported from the nucleus by CRM1, which presumably represents a mechanism to control NS5 subcellular localization during infection. This would appear to be of great importance since all of the known functions of DENV-2 NS5 occur in the cytoplasm of infected cells.

LMB treatment of DENV-2 infected cells was also found to increase the percentage of cells showing NS5 nuclear localization during the early stages of infection (Fig. 2C). At 12 and 18 h p.i., the number of cells showing nuclear staining of NS5 was significantly higher than in untreated cells. This difference was insignificant at times later than 18 h p.i, when the majority of cells treated without or with LMB show NS5 nuclear localization. These results show that blocking the CRM1 nuclear export pathway can alter the kinetics of NS5 nuclear accumulation during infection.

FIG. 2. Nuclear accumulation of NS5 in DENV-2-infected Vero cells is increased by treatment with the CRM1 inhibitor LMB. (A) CLSM images of the subcellular localization of NS5 in DENV-2 infected Vero cells, treated without and with LMB (5 ng/ml), as visualized by indirect immunofluorescence. Mock (left panels) or DENV-2-infected Vero cells were fixed at the indicated times p.i. and stained using polyclonal anti-NS5 antiserum and a goat anti-rabbit Alexa Fluor® 488 green fluorescent dye-conjugate. (B) Quantitative analysis of the levels of NS5 nuclear accumulation in DENV-2 infected Vero cells ±06 LMB. Mean Fn/c values (n = ≥ 70) ± SEM were calculated from CLSM images such as those in (A) using the Image J public domain software as previously (Harley et al 2003). (C) The percentage of DENV-2-infected Vero cells (±LMB) showing NS5 nuclear localization. NS5 nuclear localization was determined and expressed as a percentage of the total cells ($n \geq 110$, except LMB treated virus at 60 h p.i., where $n = 37$).

LMB treatment increases NS5 nuclear accumulation in transfected cells

To study the subcellular localization of DENV-2 NS5 independently of the other DENV-2 proteins, plasmid pEPI-GFP-NS5, encoding GFP fused N-terminally in frame with the full-length DENV-2 NS5, was transfected into Vero cells treated with and without LMB treatment for 5 h prior to live cell imaging by CLSM at 20–24 h post transfection (p.t.) (Fig. 3A). In the absence of LMB, GPF-NS5 was

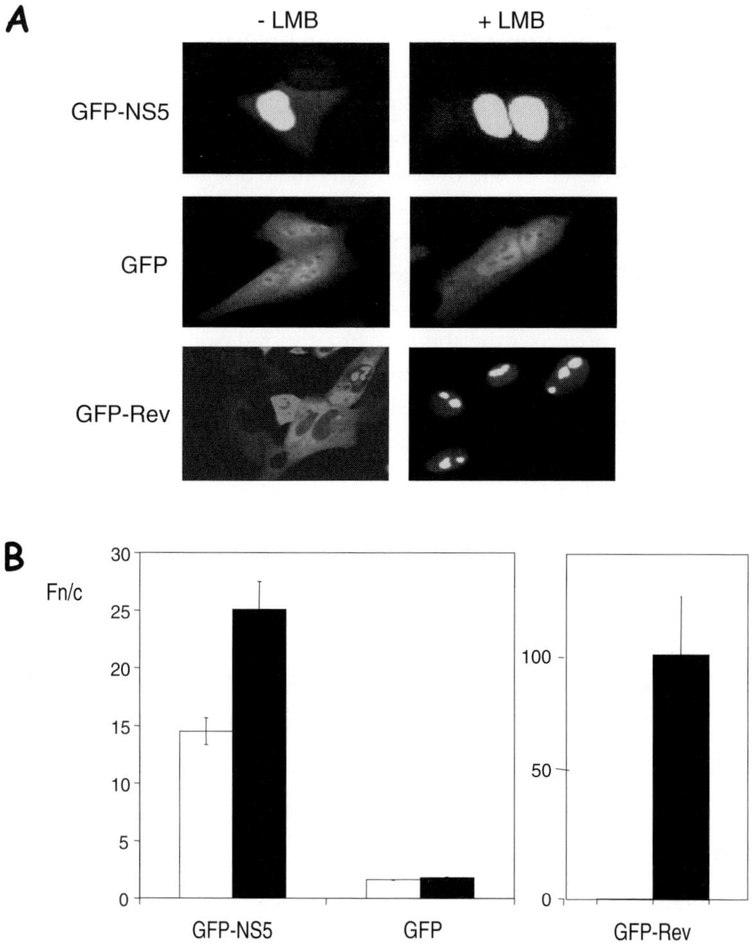

FIG. 3. Nuclear accumulation of GFP-NS5 fusion protein in transfected Vero cells is increased by treatment with the CRM1 inhibitor LMB. (A) CLSM images of the subcellular localization of GFP-NS5 in transfected Vero cells treated without and with LMB. Cells were treated with LMB (2.8 ng/ml) 5 h prior to live cell imaging by CLSM 20–24 h p.t. (B) Quantitative analysis of the levels of GFP-NS5 nuclear accumulation in Vero cells ± LMB. Mean Fn/c values ($n = \geq 80$) ± SEM were calculated from CLSM images such as those in (A) using the Image J public domain software.

found predominantly in the nucleus, in contrast to the GFP control that showed a diffuse localization throughout the nucleus and the cytoplasm. That GFP-NS5 was nuclear in the absence of the other DENV-2 proteins supported the previous observation that NS5 contains a functional NLS (Brooks et al 2002). Analogous to the results in DENV-2 infected cells, LMB effected a significant ($P < 0.0001$) increase in the Fn/c compared to untreated samples (Fig. 3B). These results confirm that NS5 possess a CRM1-dependent pathway, and thereby is capable of shuttling between the nucleus and cytoplasm in the absence of the other DENV-2 proteins. A plasmid expressing the CRM1-recognised HIV-1 Rev protein fused to GFP (GFP-Rev) was also transfected into Vero cells as a control. In the presence of LMB, GFP-Rev is localized in the nucleus and nucleolus of cells, a dramatic change from the largely cytoplasmic staining in untreated cells (Fig. 3A, 3B).

CRM1-recognized NESs are typically short peptide sequences with regularly spaced leucine or hydrophobic amino acids. Examination of the DENV-2 NS5 protein sequence has revealed several candidate NESs; these sequences are currently being evaluated by site-directed mutagenic approaches.

Effect of LMB on the DENV-2 NGC virus production

The observation that LMB treatment of DENV-2-infected Vero cells altered the kinetics of NS5 nuclear localization prompted us to test if LMB treatment affects virus production. Viral titres were determined from culture supernatants harvested at various time points in DENV-2-infected Vero cells treated in the absence without or with LMB (Fig. 4). In the presence of LMB, a marked increase in virus production was detected for all the time points excluding 72 h p.i. These results support our earlier findings that blocking the CRM1 nuclear export pathway altered the kinetics of NS5 nuclear accumulation early in infection (Fig. 2C). Taken together, these results suggest that LMB treatment of DENV-2-infected Vero cells results in increased accumulation of NS5 within the nucleus, which correlates with the acceleration of the rate of virus production.

Summary

Here we report for the first time that DENV-2 NS5 RNA polymerase possesses the ability to be exported from the nucleus by the importin family member CRM1. Previously, we demonstrated that NS5 possesses two functional NLSs, recognized by importin αβ and β1, which mediate its nuclear import. Together, these results imply that NS5 has the ability to shuttle between nuclear and cytoplasmic compartments through distinct NLS- and NES-dependent nuclear transport pathways. From as early as 12 h p.i., we observed DENV-2 NS5 predominantly localized in the nucleus, suggesting a specific role for NS5 in the nuclear compartment,

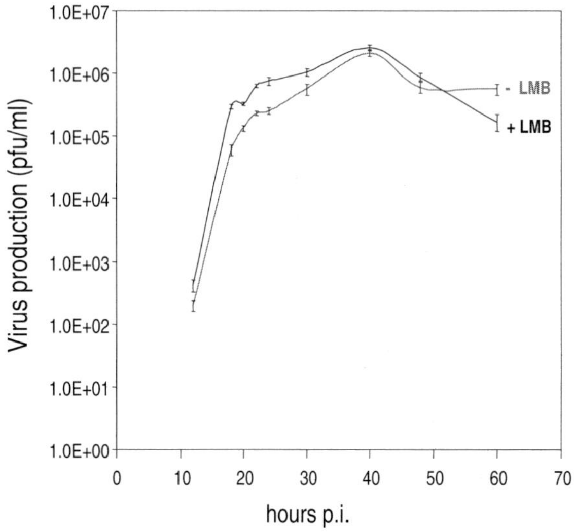

FIG. 4. The kinetics of DENV-2 NGC virus production are altered by LMB treatment. Vero cells were infected with DENV-2 NGC at a MOI of 10, ± LMB (5 ng/ml) at 0 h p.i. Virus titration of the culture supernatant at the indicated times post infection were performed on C6/36 cells. The mean titre (±SD) is graphed against time (hour p.i.) for each sample.

additional to its cytoplasmic role in viral replication. This role remains unclear, but in a similar fashion to other cytoplasmic RNA viruses, DENV-2 NS5 may be involved in assisting replication, inhibiting antiviral responses and/or interfering with host cell functions (Hiscox 2003).

We also show that both the nuclear import and export of NS5 is independent of other DENV-2 proteins, including the capsid protein that has also been identified in the nucleus of dengue virus infected cells (Bulich & Aaskov 1992, Makino et al 1989). The control over NS5's distribution in the nucleus and cytoplasm through its competing NLS and NES sequences remains to be determined, but may involve protein phosphorylation. The NS5/NS5A proteins belonging to all three genera of the *Flaviviridae* are known to be phosphorylated (Reed et al 1998); in the case of dengue, the nuclear form of NS5 is hyperphosphorylated by comparison with the cytoplasmic form (Kapoor et al 1995). It thus does not seem inconceivable that phosphorylation may represent a switch between the competing import and export mechanisms.

Blocking the CRM1 export pathway in DENV-2-infected cells was observed here to increase NS5 nuclear accumulation in parallel with the apparent acceleration of virus production. Although these data imply that CRM1-dependent nuclear export is important to the kinetics of virus production, we have not formally shown that this is due to effects on NS5 specifically; clearly LMB will inhibit nuclear export of a number of host factors which could impact on infection kinetics. Demonstrating that NS5 nuclear export is a key factor in dengue infection will require identification of the sequences responsible and reverse genetic approaches. In conclusion, the development of antiviral drugs designed to disrupt the trafficking of NS5, specifically by blocking NS5 interaction with importin αβ, importin β, CRM1 or interfering with the phosphorylation of NS5, may represent a viable strategy to control the progression of dengue virus infection. The results here with respect to the alteration of the kinetics dengue virus replication by LMB treatment represent an initial proof-of-principle in this regard.

References

Ackermann M, Padmanabhan R 2001 De novo synthesis of RNA by the dengue virus RNA-dependent RNA polymerase exhibits temperature dependence at the initiation but not elongation phase. J Biol Chem 276:39926–39937

Bartholomeusz AI, Wright PJ 1993 Synthesis of dengue virus RNA in vitro: initiation and the involvement of proteins NS3 and NS5. Arch Virol 128:111–121

Brooks AJ, Johansson M, John AV et al 2002 The interdomain region of dengue NS5 protein that binds to the viral helicase NS3 contains independently functional importin beta 1 and importin alpha/beta-recognized nuclear localization signals. J Biol Chem 277:36399–36407

Buckley A, Gaidamovich S, Turchinskaya A, Gould EA 1992 Monoclonal antibodies identify the NS5 yellow fever virus non-structural protein in the nuclei of infected cells. J Gen Virol 73:1125–1130

Butcher R 2003 Mutagenesis of dengue virus type-2 NS5 using a genomic-length cDNA clone. PhD thesis, Monash University, Clayton, Victoria, Australia

Bulich R, Aaskov JG 1992 Nuclear localization of dengue 2 virus core protein detected with monoclonal antibodies. J Gen Virol 73:2999–3003

Chu PW, Westaway EG 1992 Molecular and ultrastructural analysis of heavy membrane fractions associated with the replication of Kunjin virus RNA. Arch Virol 125:177–191

Cleaves GR, Dubin DT 1979 Methylation status of intracellular dengue type 2 40 S RNA. Virology 96:159–165

Cleaves GR, Ryan TE, Schlesinger RW 1981 Identification and characterization of type 2 dengue virus replicative intermediate and replicative form RNAs. Virology 111:73–83

Conti E, Uy M, Leighton L, Blobel G, Kuriyan J 1998 Crystallographic analysis of the recognition of a nuclear localization signal by the nuclear import factor karyopherin alpha. Cell 94:193–204

Edward Z, Takegami T 1993 Localization and functions of Japanese encephalitis virus nonstructural proteins NS3 and NS5 for viral RNA synthesis in the infected cells. Microbiol Immunol 37:239–243

Egloff MP, Benarroch D, Selisko B, Romette JL, Canard B 2002 An RNA cap (nucleoside-2′-O-)-methyltransferase in the flavivirus RNA polymerase NS5: crystal structure and functional characterization. Embo J 21:2757–2768

Fischer U, Huber J, Boelens WC, Mattaj IW, Luhrmann R 1995 The HIV-1 Rev activation domain is a nuclear export signal that accesses an export pathway used by specific cellular RNAs. Cell 82:475–483

Fornerod M, Ohno M, Yoshida M, Mattaj IW 1997 CRM1 is an export receptor for leucine-rich nuclear export signals. Cell 90:1051–1060
Forwood JK, Brooks A, Briggs LJ et al 1999 The 37-amino-acid interdomain of dengue virus NS5 protein contains a functional NLS and inhibitory CK2 site. Biochem Biophys Res Commun 257:731–737
Fukuda M, Asano S, Nakamura T et al 1997 CRM1 is responsible for intracellular transport mediated by the nuclear export signal. Nature 390:308–311
Ghildyal R, Li D, Peroulis I et al 2005 Interaction between the respiratory syncytial virus G glycoprotein cytoplasmic domain and the matrix protein. J Gen Virol 86:1879–1884
Gorlich D, Kutay U 1999 Transport between the cell nucleus and the cytoplasm. Annu Rev Cell Dev Biol 15:607–660
Grun JB, Brinton MA 1987 Dissociation of NS5 from cell fractions containing West Nile virus-specific polymerase activity. J Virol 61:3641–3644
Gualano RC, Pryor MJ, Cauchi MR, Wright PJ, Davidson AD 1998 Identification of a major determinant of mouse neurovirulence of dengue virus type 2 using stably cloned genomic-length cDNA. J Gen Virol 79:437–446
Harley VR, Layfield S, Mitchell CL et al 2003 Defective importin beta recognition and nuclear import of the sex-determining factor SRY are associated with XY sex-reversing mutations. Proc Natl Acad Sci USA 100:7045–7050
Hiscox JA 2003 The interaction of animal cytoplasmic RNA viruses with the nucleus to facilitate replication. Virus Res 95:13–22
Jans DA, Xiao CY, Lam MH 2000 Nuclear targeting signal recognition: a key control point in nuclear transport? Bioessays 22:532–544
Johansson M, Brooks AJ, Jans DA, Vasudevan SG 2001 A small region of the dengue virus-encoded RNA-dependent RNA polymerase, NS5, confers interaction with both the nuclear transport receptor importin-beta and the viral helicase, NS3. J Gen Virol 82:735–745
Kapoor M, Zhang L, Ramachandra M et al 1995 Association between NS3 and NS5 proteins of dengue virus type 2 in the putative RNA replicase is linked to differential phosphorylation of NS5. J Biol Chem 270:19100–19106
Mackenzie JM, Khromykh AA, Jones MK, Westaway EG 1998 Subcellular localization and some biochemical properties of the flavivirus Kunjin nonstructural proteins NS2A and NS4A. Virology 245:203–215
Makino Y, Tadano M, Anzai T et al 1989 Detection of dengue 4 virus core protein in the nucleus. II. Antibody against dengue 4 core protein produced by a recombinant baculovirus reacts with the antigen in the nucleus. J Gen Virol 70:1417–1425
Mori Y, Okabayashi T, Yamashita T et al 2005 Nuclear localization of Japanese encephalitis virus core protein enhances viral replication. J Virol 79:3448–3458
Paschal BM 2002 Translocation through the nuclear pore complex. Trends Biochem Sci 27:593–596
Pemberton LF, Paschal BM 2005 Mechanisms of receptor-mediated nuclear import and nuclear export. Traffic 6:187–198
Reed KE, Gorbalenya AE, Rice CM 1998 The NS5A/NS5 proteins of viruses from three genera of the family flaviviridae are phosphorylated by associated serine/threonine kinases. J Virol 72:6199–6206
Tan BH, Fu J, Sugrue RJ et al 1996 Recombinant dengue type 1 virus NS5 protein expressed in *Escherichia coli* exhibits RNA-dependent RNA polymerase activity. Virology 216:317–325
Uchil PD, Satchidanandam V 2003 Characterization of RNA synthesis, replication mechanism, and in vitro RNA-dependent RNA polymerase activity of Japanese encephalitis virus. Virology 307:358–371
Wang SH, Syu WJ, Huang KJ et al 2002 Intracellular localization and determination of a nuclear localization signal of the core protein of dengue virus. J Gen Virol 83:3093–3102

Westaway EG, Khromykh AA, Kenney MT, Mackenzie JM, Jones MK 1997 Proteins C and NS4B of the flavivirus Kunjin translocate independently into the nucleus. Virology 234:31–41

Westaway EG, Khromykh AA, Mackenzie JM 1999 Nascent flavivirus RNA colocalized in situ with double-stranded RNA in stable replication complexes. Virology 258:108–117

DISCUSSION

Gamarnik: Has anyone looked at the phosphorylation sites of NS5? Have they been mutated?

Jans: Not as far as I know. It is clearly phosphorylated, so inhibiting host kinases may prevent getting it in or out.

Satchidanandam: In your leptomycin B experiments you saw an increase in the viral titre. At what point post-infection did you add the leptomycin?

Jans: We added it immediately, at 0 h. Leptomycin is a labile compound that is active for about 6 h. We like this because we can put it in at different times during infection. We added it straight away because we thought it had to be in the cytoplasm for its replication function, but we saw the reverse of what one might have thought. If we force it into the nucleus more we seem to accelerate the kinetics of infection rather than the reverse.

Satchidanandam: So you added the leptomycin B very early, and you said that the drug is known to be active for 6 h. You are looking at virus production perhaps 24 h later. Does this mean that during the first 6 h following infection, all the NS5 would have been inside the nucleus, and later on there will be a lot of NS5 in the cytosol?

Jans: I don't think we can say that all the NS5 will be in the nucleus. We could probably say twice as much.

Satchidanandam: It does seem that nuclear polymerase activity is going to help in pushing up viral titres.

Jans: Let's just say that having NS5 in the nucleus clearly isn't detrimental to the course of infection. If anything, it enhances infection.

Satchidanandam: Perhaps what you should be looking at is the enzyme activity in the nucleus in the presence and absence of leptomycin.

Canard: Do you have any indication that RNA is present in NS5?

Jans: No. There are various approaches we would like to try to attempt to understand the role of NS5 in the nucleus. The way we would like to go is by looking at protein–protein interactions rather than using Affymetrix chips or arrays. We are currently doing yeast two-hybrid screening, and we have some interesting candidates. The other technique we are using is a simple pull-down.

Harris: When you talk about a signal for virus production, you are presumably referring to signal to form new vesicles and start the secretion process. You have the RNA being synthesized in the cytoplasm. What sort of signals are we talking about here?

Jans: An effect on transcription or something else in the nucleus that then is going to have an effect in the cytoplasm. It is a continuum rather than a switch. It must be an 'amount' thing. The signal has to accumulate sufficiently in the cytoplasm.

Satchidanandam: When you treat infected cells with leptomycin B, can you look to see whether it delays the lysis of the cell or the degradation of the viral proteins or RNA?

Jans: We could do those experiments better, but our impression is that there isn't much difference at all. We haven't shown that the reason leptomycin B is making such a difference to virus production is that NS5 isn't getting exported. Leptomycin will inhibit lots of processes in the cell. But it is an intriguing observation.

Stephenson: You hypothesized some good reasons for NS5 being in the nucleus. But why is there so much of it? You'd have thought that most of it should be in the cytoplasm, not the nucleus. Many years ago some experiments we carried out with the nucleocapsid protein from measles virus had a similar problem. There, the amount that is in the nucleus is proportionate to the dysfunction of the infection. With a non-productive infection just about everything is in the nucleus, but with a productive infection there is much more in the cytoplasm. The hypothesis there is that the nucleus is just acting as a repository for nuclear protein, which wasn't being used in the replication complex. Is this a potential interpretation for your results?

Jans: I don't think so because of the effect on kinetics. If we push it more into the nucleus, we get more virus production. I don't think it is in that sense non-productive. I should also add that some of the mutated infectious clone viruses that we have made, which don't accumulate NS5 in the nucleus, seem to show this slow infectious profile as well. They produce less virus and produce it later.

Rice: I guess one concern is that when you are inhibiting import into the nucleus, it might be doing other things to these cells that could affect the kinetics of virus production. How can you dissect out these effects from NS5 import?

Jans: The way to do this is reverse genetics. If we can make an infectious clone with the mutations at the right point, we can establish whether nuclear import or export is critical for infection.

Rice: I guess the issue is going to be ruling out the hinge region which contains nuclear localization signals. It is a multifunctional part of the protein, so if you change the basic character of it you are not only affecting nuclear import of protein but also perhaps some other function that is deleterious for virus replication independent of that. A bias in the positive-stand virus field, probably based on working with long RNAs and finding that RNA degradation was a problem, was that these viruses are vulnerable until they make minus-strand RNA. There is a rush to get into the cell and make that first minus-strand RNA. On the other hand we can establish whether nuclear import or export is critical for infection we have data suggesting that non-structural replicase proteins may be made in vast excess.

In the case of the alphaviruses the polymerase is the last non-structural protein and is the critical enzymatic component to make RNA. It is the last thing that is made. In alphaviruses polymerase production is not equimolar. It is tightly regulated so it is degraded if it is made in excess. This should probably change our thinking a little bit. Perhaps the decision to make that first minus strand is a very important one for the virus, and it is not the first thing on its mind when it gets into a cell. One possibility for nuclear sequestration of NS5 would be to prevent it from initiating replication.

Schul: One technique now used in GFP localization studies is photobleaching. Nothing needs to be added; the nuclear fluorescence is bleached and then you see how fast it comes back.

Jans: We certainly are able to do those experiments. This would tell us about shuttling. We didn't think that transport kinetics *per se* represented a burning question here.

Kuhn: What is the reason for having two NLS sequences in NS5? Why are they so well conserved?

Jans: That's a good question, and the answer is that we don't know. Either one can work independently. When they are together, I think there is a regulatory phenomenon going on and phosphorylation is likely to be the key to this. The question is also, why do you need two different nuclear import pathways? One might be regulated and the other not.

Canard: If I understood correctly, your hypothesis is that NS3 is holding in NS5 out of the nucleus. So there shouldn't be any NS3 inside the nucleus with NS5.

Jans: I think it is unlikely, but we haven't tested this.

Canard: For me, it is important for capping. NS3 first dephosphorylates the 5′ triphosphate RNA, and the second step would be to have the RNA inside the nucleus for a putative guanylyl transferase reaction. It is one hypothesis.

Satchidanandam: Has anyone looked at the capping status of RNA in the packaged virion? You pointed out that these viruses don't seem to be in a hurry to make the negative strand, but are first concerned with making their protein. One explanation could be that this would be to facilitate the membrane tethering, which precedes RNA production. I don't know whether you need to cap what will be packaged.

Rice: I think that is where the cap structure was analysed, but I don't know whether there was an analysis of what fraction might be uncapped.

Lim: Do you know whether the nuclear NS5 is still full length?

Jans: If we do Westerns we see full length protein in the nucleus.

Rice: Has anyone looked at replication of dengue in enucleated cells?

Satchidanandam: In my assays it is just like Japanese encephalitis virus. We see activity, but I don't have the antibodies to dengue proteins to localize NS3 or NS5 in the nucleus.

Rice: It would be interesting to enucleate cells with cytochalasin and see whether they support normal levels of viral replication.

T cell responses and dengue haemorrhagic fever

Gavin Screaton and Juthathip Mongkolsapaya

Department of Immunology, Hammersmith Hospital, Imperial College, Du Cane Road, London W12 0NN, UK

Abstract. The enhancement of severe disease upon secondary infection makes dengue almost unique among infectious pathogens and presents a serious challenge to vaccine design. Several key observations have been made which shed light onto this phenomenon particularly that antibodies can enhance Fc receptor-dependent uptake of virus into macrophages thereby increasing virus replication. Furthermore there seems to be a relationship between the peak virus load and disease severity. However, a second key feature of dengue is that the life-threatening symptoms do not correlate with the period of high viraemia; instead they occur at a time when the virus load is in steep decline. The coincidence of severe disease manifestations with defervescence and virus control suggests that the symptoms may be a consequence of the immune response to the virus rather than virus induced cytopathology. One of the key elements in the immune response to viruses are T cells which can both secrete a host of inflammatory cytokines and also be directly cytotoxic to infected cells. There are a number of experimental models of T cell-induced immunopathology including in responses to viruses. Particularly interesting in this respect are models of RSV-induced immunopathology, which have direct relevance to vaccine design as a formalin-inactivated vaccine to RSV actually enhanced disease in children when they became naturally infected with RSV, an echo of the disease enhancement seen in dengue. We will present an analysis of $CD8^+$ T cell responses to a number of novel T cell epitopes during dengue infection and also analyse the function and cytokine secretion of these cells. We suggest that an exaggerated and partially misdirected T cell response seen in secondary dengue infection may be part of the complex series of events leading to dengue haemorrhagic fever and shock.

2006 New treatment strategies for dengue and other flaviviral diseases. Wiley, Chichester (Novartis Foundation Symposium 277) p 164–176

Infection by dengue virus has become a major public health threat in tropical and sub tropical countries. The virus belongs to the family flaviviridae and circulates as four major serotypes. The virus is still evolving and the serotypes differ in sequence by around 30%. Dengue is transmitted to humans following a bite from an infected mosquito, usually *Aedes aegypti*.

Following the bite there is an incubation period of 3–8 days which is followed by a symptomatic phase, although school based serological surveys suggest that

50% or more of infections are asymptomatic (Endy et al 2002). Symptoms, which range from a mild undifferentiated fever to more severe fever, headache, muscle, joint and bone pains accompanied later on in the illness by a maculopapular rash. This more severe clinical syndrome is classified as dengue fever and may be accompanied by thrombocytopaenia leucopaenia and petechial haemorrhage (World Health Organization 2002).

More severe manifestations of the disease are classified as dengue haemorrhagic fever (DHF). This is a potentially life threatening condition with up to 20% mortality without expert medical care. DHF is characterised by plasma leakage and thrombocytopaenia and can be divided into four clinical grades with increasing severity of bleeding, vascular leakage, hypovolaemia and shock.

One highly characteristic and interesting feature of DHF is that the severe symptoms of bleeding and circulatory collapse seem to occur coincidentally with the time of defervescence, the point at which the fever subsides. Before this and for the 2–7 days of fever these patients will resemble cases of dengue fever.

The mainstay of treatment for DHF is careful management of fluid status with replacement with isotonic saline, colloidal plasma expanders or blood where required. With careful monitoring and attention to fluid balance the mortality can be reduced to substantially below 1%.

Despite this dengue infection still remains a major public health issue in a number of tropical and subtropical countries. About 2.5 billion people are at risk of dengue infection and there are estimated to be 50–100 million infections annually. The disease occurs in an epidemic fashion and therefore puts a huge strain on health care services; in Thailand for instance there are around 100 000 cases of DHF annually.

The incidence of dengue infection has increased rapidly since the Second World War. DHF which was rare or absent in many parts of the world has also become much more frequent. Many explanations of this huge expansion of disease have been put forward including a drive toward urbanisation and travel allowing rapid dissemination and cocirculation of viral strains (Mackenzie et al 2004).

Careful epidemiological analyses often in island populations have yielded very valuable information about the pathogenesis of severe dengue infection or DHF. Some of the best of such evidence has come from Cuba (Guzman et al 2000). In 1977 there was an epidemic of dengue fever cased by the DENV-1 serotype of virus. The majority of the population was dengue naïve and interestingly during this outbreak there were few cases of severe disease. Infection with the DENV-1 serotype reduced in subsequent years until in 1981 there was a second epidemic caused by the DENV-2 serotype. On this occasion the spectrum of disease was markedly different; there were nearly 350 000 cases of dengue fever, over 10 000 cases of DHF and around 150 deaths.

This and a number of similar studies have shown that the risk of developing DHF is much higher in people who have been previously exposed to one serotype of dengue and who are subsequently exposed to a secondary or sequential infection with a virus of different serotype (Sangkawibha et al 1984). In this case it therefore appears that pre-existing immunity or memory for the previous infection can actually potentiate disease. A further epidemic of DENV-2 in Cuba in 1997 demonstrates that the risk from previous infection (DENV-1 1977) can last for 20 years or more.

Antibody dependent enhancement has been suggested as a mechanism of disease potentiation (Halstead & O'Rourke 1977). This was proposed by Scott Halstead in 1977 and was based on similar observations on the related flavivirus tick-borne encephalitis (Phillpotts et al 1985). It is proposed that following a primary dengue infection an antibody response is mounted to envelope proteins on the infecting virus. Since different viral serotypes will differ by around 30% in sequence, antibodies directed to one serotype may not confer complete neutralising protection to another serotype. Following primary infection as the titre of these antibodies falls it is proposed that there becomes a point where the antibodies may bind to virus and rather than neutralize infection completely actually target uptake of opsonized virus by Fc-receptor bearing cells, principally macrophages thereby driving higher viral replication.

This phenomenon can be demonstrated *in vitro* and there is some evidence that it can operate *in vivo* (Halstead 1979). Indeed, antibody-dependent enhancement has been invoked to explain a small peak in severe dengue infection in children below the age of one year which can occur during a primary exposure and is proposed to be driven by the fall in titre of maternally acquired anti-dengue antibodies which occurs with age (Halstead et al 2002).

Recently, viral loads have been measured throughout the course of infection using either virus culture or reverse transcriptase PCR. There appears to be a correlation between higher viral titres and more severe disease (Vaughn et al 2000). Interestingly however, the virus load falls precipitously at the time that the fever remits. At this point, when symptoms peak, virus titre is low or undetectable. The lack of correlation between virus load and severe symptoms has led some to speculate that the sequelae of severe dengue infection may be driven more by the immune response to the virus than virus load *per se*.

Several groups including our own have measured the levels of inflammatory cytokines during dengue infection. Many of these are raised and in some the peak in their levels coincides with the onset of severe symptoms. High levels of tumour necrosis factor (TNF)α, interferon (IFN)γ, interleukin (IL)10, IL6, etc. have been recorded and many of these can be the products of activated T cells (Rothman & Ennis 1999). These findings have led some to suggest that it is T cells that cause the damage and vascular leak characteristic of dengue either by causing direct cytotoxic tissue damage or by producing a variety of inflammatory cytokines.

Like B cells, T cells belong to the acquired immune system in that the antigen receptor is not germline encoded as with molecules characteristic of the innate immune system but instead constructed following a series of genomic recombinations which occur in the thymus. The T cell receptor is a non-covalently linked heterodimer, the major form consisting of a pairing of α and β chains whilst a minor form consists of a pairing of γ and δ chains (Janeway et al 2001). T cell receptors recognise short antigenic peptides which are bound in the antigen binding groove of MHC molecules. MHC-I presents 8–11mer peptides which are derived predominantly from intracellularly produced proteins to cytotoxic T cells bearing the CD8 co-receptor. MHC-II presents slightly longer, predominantly endocytosed exogenous antigen to helper T cells bearing the CD4 co-receptor.

To study T cell responses in detail it is necessary to define the sequence of the antigenic peptides from dengue presented to T cells. Since MHC is highly polymorphic there will likely be a large number of dengue peptides which can bind to different MHC molecules. Several T cell epitopes for dengue have been previously described but these are mostly restricted by Western MHC types which make their utility to examine T cell responses in endemic areas such as SE Asia limited.

To overcome this problem, we undertook a systematic search for dengue T cell epitopes using an overlapping peptide approach (Altfeld et al 2000). MHC molecules bind linear peptides between 8–11 amino acids in length and the peptide binding groove can accommodate, although with less efficiency, short peptide extensions at either end. Therefore, a nested set of peptides 15–20 amino acids in length and overlapping by 10–11 amino acid will contain all possible 10–11 linear amino acid sequences in effect covering all possible antigens and can be used for antigen discovery.

The presence of responding T cells can then be assessed by culturing peripheral blood mononuclear cells from dengue immune individuals with pools of these overlapping peptides. If responding T cells are present then they will be activated by peptide and this activation can then be read by assaying cytokine production from the activated T cells. We used the IFNγ Elispot assay for this purpose where a sandwich ELISA for IFNγ allows the enumeration of the responding T cells.

Using this assay we were able to discover a number of T cell epitopes for both $CD8^+$ and $CD4^+$ positive T cells. We were particularly interested in those epitopes which were restricted by commonly expressed MHC-I alleles in SE Asia such as A11, A24 and A33. An immunodominant epitope from the NS3 protein GTSG-SPIIDKK which was restricted by HLA-A11 was found and studied in detail (Mongkolsapaya et al 2003). T cell responses to this epitope were examined in a cohort of children admitted to hospital in Khon Kaen in North East Thailand.

Since dengue virus seroytypes show about 30% amino acid difference we searched sequence databases to ascertain whether differences were found in this A11 epitope and found six variants of the epitope (Table 1) which were expressed by the variant viruses. Responses to these variant epitopes were analysed in the patients' samples using both Elispot and MHC tetramer analysis using fluorescence-activated cell sorting (FACS) (McMichael & O'Callaghan 1998).

All of the patients in this study were suffering from secondary dengue infections and they showed diverse T cell responses to the variant peptide epitopes derived from the different dengue serotypes. During the acute illness dengue-specific T cells were highly activated and almost all also showed signs of cell proliferation and apoptosis. There was also a correlation between the magnitude of the T cell response and the severity of the dengue illness (Fig. 1).

When we looked at the fine specificity of the T cell response we found paradoxically that many of the T cells had a relatively low affinity for the currently infecting viral serotype and showed higher affinity for serotypes which we presume had been encountered before (Fig. 2).

This phenomenon whereby a response to a variant of a previously encountered epitope is constructed mainly from memory T or B cells rather than being generated *de novo* by fresh priming is termed original antigenic sin (Fazekas de St & Webster 1966b,a, McMichael 1998). Original antigenic sin has the advantage that a cross-reactive response can be rapidly recalled from memory but has the disadvantage that the response may not be optimal and may contain some lower affinity T cells. In some experimental circumstances such as murine infection with variants of the lymphocytic choriomeningitis virus original antigenic sin has been shown to be detrimental leading to slower clearance of virus in a secondary as opposed to primary infection (Klenerman & Zinkernagel 1998).

In summary, it seems likely that the increased severity of a secondary or sequential dengue infection is immunologically driven and multifactorial. Antibody

TABLE 1 A list of GTS variants obtained from dengue sequences published in Genbank

Variant	*Sequence*	*Number of sequences found/total*
Den 1.1	GTSGSPI**VNRE**	5/5
Den 2.1	GTSGSPIIDKK	29/39
Den 2.2	GTSGSPI**V**DKK	6/39
Den 2.3	GTSGSPI**V**D**R**K	3/39
Den 2.4	GTSGSPI**A**DKK	1/39
Den 3.1	GTSGSPII**NRE**	2/2
Den 4.1	GTSGSPII**NR**K	2/2

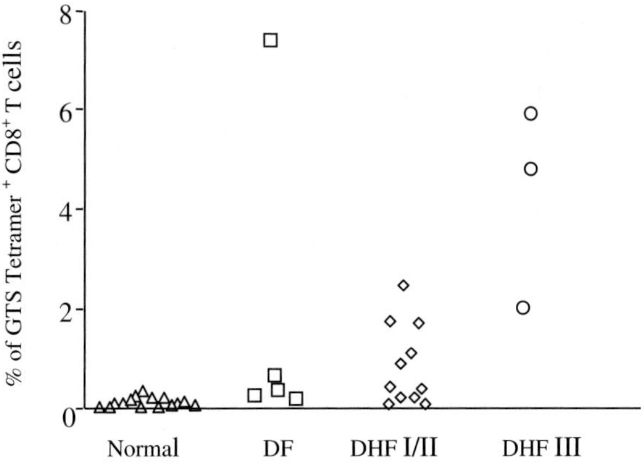

FIG. 1. Frequency of dengue-specific CD8$^+$ T cells at convalescent day 4 from patients with different disease severities; dengue fever (DF), dengue haemorrhagic fever grades I and II (DHF I/II) and DHF grade III compared with normal healthy dengue-immune individuals (normal).

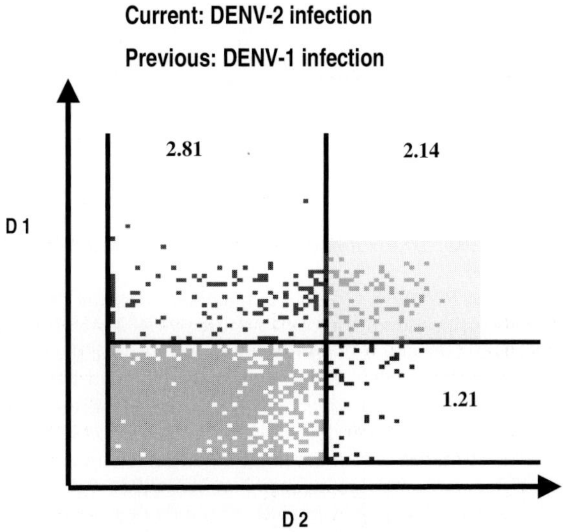

FIG. 2. Secondary dengue infection showing the original antigenic sin phenomenon. PBMCs from DENV-2-infected patients were simultaneously stained with the GTS tetramers from DENV-1 (D1) and DENV-2 (D2) and analysed by Flow cytometry. A substantial number of T cells are of higher affinity for the previously encountered DENV-1 as opposed to the currently infecting virus DENV-2.

dependent enhancement may drive virus internalisation into macrophages and enhance viral replication. At the same time original antigenic sin may initially delay an effective high affinity T cell response to the virus allowing further viral replication whilst a high affinity T cell response is generated. This will lead to a dangerous situation where there is finally the collision of a large number of T cells with a high antigen load. This will lead to massive T cell activation which will cause both direct T cell mediated cytotoxicity and the release of a variety of inflammatory cytokines. These in turn will lead to tissue damage and the syndrome of vascular leak which occurs coincidentally with viral clearance in DHF.

References

Altfeld MA, Trocha A, Eldridge RL et al 2000 Identification of dominant optimal HLA-B60- and HLA-B61-restricted cytotoxic T-lymphocyte (CTL) epitopes: rapid characterization of CTL responses by enzyme-linked immunospot assay. J Virol 74:8541–8549

Endy TP, Chunsuttiwat S, Nisalak A et al 2002 Epidemiology of inapparent and symptomatic acute dengue virus infection: a prospective study of primary school children in Kamphaeng Phet, Thailand. Am J Epidemiol 156:40–51

Fazekas de St G, Webster RG 1966a Disquisitions of original antigenic sin. I. Evidence in man. J Exp Med 124:331–345

Fazekas de St G, Webster RG 1966b Disquisitions on original antigenic sin. II. Proof in lower creatures. J Exp Med 124:347–361

Guzman MG, Kouri G, Valdes L et al 2000 Epidemiologic studies on Dengue in Santiago de Cuba, 1997. Am J Epidemiol 152:793–799; discussion 804

Halstead SB 1979 In vivo enhancement of dengue virus infection in rhesus monkeys by passively transferred antibody. J Infect Dis 140:527–533

Halstead SB, O'Rourke EJ 1977 Antibody-enhanced dengue virus infection in primate leukocytes. Nature 265:739–741

Halstead SB, Lan NT, Myint TT et al 2002 Dengue hemorrhagic fever in infants: research opportunities ignored. Emerg Infect Dis 8:1474–1479

Janeway C, Travers P, Walport M, Shlomachik M 2001 Antigen recognition by B-cell and T cell receptors. Immunobiology: the immune system in health and disease. Garland Publishing, New York 3-1-3-19

Klenerman P, Zinkernagel RM 1998 Original antigenic sin impairs cytotoxic T lymphocyte responses to viruses bearing variant epitopes. Nature 394:482–485

Mackenzie JS, Gubler DJ, Petersen LR 2004 Emerging flaviviruses: the spread and resurgence of Japanese encephalitis, West Nile and dengue viruses. Nat Med 10:S98–109

McMichael AJ 1998 The original sin of killer T cells. Nature 394:421–422

McMichael AJ, O'Callaghan CA 1998 A new look at T cells. J Exp Med 187:1367–1371

Mongkolsapaya J, Dejnirattisai W, Xu XN et al 2003 Original antigenic sin and apoptosis in the pathogenesis of dengue hemorrhagic fever. Nat Med 9:921–927

Phillpotts RJ, Stephenson JR, Porterfield JS 1985 Antibody-dependent enhancement of tick-borne encephalitis virus infectivity. J Gen Virol 66 (Pt 8):1831–1837

Rothman AL, Ennis FA 1999 Immunopathogenesis of dengue hemorrhagic fever. Virology 257:1–6

Sangkawibha N, Rojanasuphot S, Ahandrik S et al 1984 Risk factors in dengue shock syndrome: a prospective epidemiologic study in Rayong, Thailand. I. The 1980 outbreak. Am J Epidemiol 120:653–669

Vaughn DW, Green S, Kalayanarooj S et al 2000 Dengue viremia titer, antibody response pattern, and virus serotype correlate with disease severity. J Infect Dis 181:2–9
World Health Organization 2002 Dengue and dengue haemorrhagic fever. Fact sheet no. 117

DISCUSSION

Harris: Regarding the apoptosis versus the reactivity, you have the apoptosis peaking at day 0 but the reactivity of the Elispot really starting at days 1–3.

Screaton: That's a good point. What we are seeing in the periphery is a lot of exhausted T cells which are dumping cytokines. The Elispot assay is a synthesis assay: the T cell needs to make cytokines. A T cell that has shredded its nucleus won't be able to make the RNA to make the cytokine. We are looking at nearly dead cells circulating. One of the remarkable observations here is that 80% of cells circulating are dying apoptoptically, which is very high. We have never seen anything like this with other viruses. These cells are very efficiently phagocytosed.

Harris: Does this mean that the actual cytokine production is happening earlier?

Screaton: Yes, in the tissue probably.

Farrar: We have long agonized over why these T cells seem to come up later during the infection. Could this be an artefact of the Elispot system? Those cells are also being taken from a patient at a time of viraemia, even if it is not peak viraemia. The interferon responses seem to be downgraded on array studies early during infection. Could it be that the T cells you are looking at are not responding within an Elispot assay, which is dependent on the making of interferon? It could be that there are lots of T cells there and you are just not seeing them as part of the Elispot assay.

Screaton: You are right. If you look at the frequency of T cells enumerated by the tetramer, they are definitely present. There are T cells there which have the potential to respond, but they don't make IFNγ. Either the secretion of INFγ is downregulated, or they are dying and can't synthesize the mRNA.

Farrar: We spent months agonizing over why we see this peak. We wondered whether pathogenesis and protection are two completely different, separate events. It may just be, though, that if we have another assay other than an Elispot, we might see a different picture. IFN may not be the best thing to measure on these cells early in infection.

Screaton: It is also possible that the T cells we are seeing by tetramer are the ones that don't do anything anyway. Not all T cells produce IFNγ.

Hombach: Have you had the chance to make this sort of analysis on a relatively severe primary infection?

Screaton: No, but we would love to. Out of the first 100 we looked at, three were primaries.

Halstead: We need some leadership from the T cell community. In 1966 I spent three months in Thailand looking at autopsies from Bangkok Children's Hospital. In those days almost every child that died of DHF was autopsied. I went back because the previous year we had made the association between a secondary-type antibody response and a severe form of DHF. The problem was that infants less than one had a primary antibody response. I knew this could not be DHF: how could you have secondary infection in older children and primary infection in younger children? The only problem was, that after having looked at 100 autopsies I could not throw out those infants less than one year. They are in. In 1970, I reported that 5%, year in, year out, of all DHF is in infants younger than one (Halstead et al 1970). On a per hundred thousand basis, the disease is more severe in infants during primary infections than in children with secondary dengue infections. Data are available from all over southeast Asia (Halstead et al 2002) You implied that it is very rare, but 5% is not rare. You are going to have to study infants younger than one. The whole concept that a secondary type T cell response is integral to the outcome of severe disease is simply not right.

Screaton: First, this sort of severe disease in infants less than one is rare. In Thailand, it is between less than 1 and 1.5%.

Halstead: Have you read my paper?

Screaton: Yes, which is what prompted me to look at the frequency in Thailand. Having more severe disease in infants is common in many other diseases, such as RSV, which in adults isn't too bad but which is severe in infants.

Halstead: This is an age-specific phenomenon that occurs in the latter six months of the first year of life and disappears after 12 months. There is no DHF between 12 and 24 months. It is not a phenomenon of infancy; it is a phenomenon related specifically to a period of development in infants born to dengue-immune mothers.

Screaton: I don't think anyone in this room will dispute the fact that you can get severe disease in a primary infection. This is true. In a way, this is just semantics. Is less than 2% rare or common? I can say for sure that it is less than 2%.

Halstead: I disagree. It is 5%.

Rice: There could be multiple predisposing factors that give rise to similar pathogenesis. It could be through original antigenic sin.

Screaton: One of the difficulties here is primacy. I am not saying that the only thing in this disease that kills children is T cells; I would be naïve to do so.

Halstead: I'd be happy to agree that it is T cells. The concept that it is has to be 'antigenic sin' is wrong. A primary T cell response is adequate to generate all the phenomena that exist in this disease.

Screaton: I think we agree. Primary infection exists, kids get ill with primary infection, and there is no original antigenic sin in the primary infection by definition.

Rice: It would still be interesting to look at the frequency and the phenotype of those T cells.

Screaton: We would love to look at this.

Farrar: Studying infants under the age of 1 is a crucial bit of dengue research. Everyone would agree with this. They are difficult studies to do, because in order to study the immunology we still require big volumes of blood. We have recruited over 100 infants under the age of one in the last couple of years, but it is difficult work.

Screaton: What percentage of severe disease are you finding?

Farrar: We don't see 5%. I would put it closer to 1–2%.

Holmes: Do you get more 'sin' in tertiary infections than secondary infections? If so, have you tried to tease this out in populations?

Screaton: It is difficult enough to define secondary, let alone tertiary. Lots will be tertiary. I would predict that as you have more infections your T cell pool will become smaller.

Halstead: We have data on cohort studies with a few well documented tertiary infections. For many years we thought that all the risk was in secondary infections.

Satchidanandam: I want to make a point on the IFNγ that you think isn't good in T cells. Diane Griffin showed in measles that IFNγ can clear measles virus in culture. We have found in Japanese encephalitis (JE) patients who have recovered after encephalitis that the level of IFNγ from T cells correlates not only with low severity during encephalitis but also with lack of sequelae and damage to the CNS. I don't know whether IFNγ could be protective.

Screaton: IFNγ is a good antiviral cytokine. In JE, where this is a virus-induced problem, IFNγ is probably good. What I am saying is that with IFNγ there is a fine line between good and too much. There are several immunopathology models where too much IFNγ can be a problem.

Young: I'd like to tease out some of the discussion about the temporal response, and when the T cells come into play. When do you see the T cells in the secondary infection?

Screaton: They start to appear before defervescence and they peak at two weeks. One of the difficulties with these studies is that we are looking at peripheral blood.

Young: What are their targets? One would have assumed that the virus has gone by then.

Screaton: If one uses the analogy of having the tap on and the plug out, I suspect the tap is full on during infection and the plug is out in terms of apoptosis. Once the virus is cleared the antigen goes down and a plug is put in the sink, so the water comes up.

Zinkernagel: Scott Halstead, you asked for a T cell immunologist to describe what is going on. In general terms there is one serious problem in T cells: we cannot discriminate between T cells that are positive in tetramers or IFNγ or whatever readout, and the T cells that really matter in terms of effector function in immunopathology. It is a bit like comparing ELISA antibodies versus neutralizing antibodies. We don't have reasonable measures to distinguish them. Secondly, in dengue, as far as I have tried to find out, no one really knows what the relative distribution of infected cells is during the course of infection. Is it only on endothelial cells? Is it on dendritic cells, macrophages, or neurons? At the end of the day, immunopathology is targeted. If it is in endothelial cells it is fine. Then perforin, or some of these granule-positive effector T cells, plus tetramer, will give you a reasonable readout of whether the ideas we have are real or not. At the moment it is very difficult, and I would be extremely surprised if it were only the T cells. It must be distribution and kinetics of both T cells and the virus. The kinetics of the virus and its distribution is one of the most notable gap in our discussions of the pathogenesis of the disease. If I had money, I would put a lot of it into biopsies and autopsies in an interesting area of the epidemiology: we need these data.

Screaton: One of the problems with the post mortem data is that the children die when they have cleared the virus.

Zinkernagel: Do you think that storms of IFN have an influence not only on the integrity of certain cells but also on coagulation and bleeding tendency? In particular, what is known about megakaryocytes during the whole course of infection? Are they a target for dengue or not?

Halstead: Early in the infection there is almost total pan-cytopaenic wipe-out of the bone marrow. At the time patients go into shock, megakaryoctes show maturation arrest (Bierman & Nelson 1965). Radioisotope studies have shown that the thrombocytopaenia is not just due to reduced production. There is actually destruction of platelets (Mitrakul et al 1977).

Flamand: What is the level of heterogeneity of the immunodominant epitopes among the four dengue serotypes? Can you predict whether there might be some sequential events of infection that might make the patient more at risk of developing a severe form of disease, with regard to the identity of the first infecting serotype?

Screaton: There are very substantial differences in the sequences of the whole virus among the different serotypes. This is sufficient to make it likely that there will be variants in a given 9-mer or 10-mer peptide. We don't know nearly enough about these epitopes. I have just described one which is relatively immunodominant, but we don't know enough to be able to predict sequential infections. However, there are some data suggesting that some sequences are good and others are bad. These are disputed.

Malasit: I agree that we need more pathology study. Pathology study should be ideally conducted using material obtained during the critical phase of the disease, when the virus has been cleared from the circulation, coinciding with the period when shock and leakage occur. In Thailand, the chance of studying such autopsy material (from DHF patients who die early in the disease) is very poor. The reasons are many. First, the mortality incidence in the country is low and death rarely occurs during the acute phase due to early admissions and adequate treatments. Second, most autopsy studies in the country were conducted in patients who died late in the course of disease, usually during the second week of admission; and causes of death are not directly related to the immuno-pathogenesis of DHF, but rather to the associated complications.

I would like to comment on the immuno-pathogenesis processes and pose some questions. Active, if not excessive, T cell responses—not dissimilar to the graft-versus-host (GVH) reaction, acute cytokine/chemokine production—severe thrombocytopenia, and complement activation are known to occur and participate in the pathogenesis of shock, leakage and haemorrhagic diathesis in DHF. A similar model would be in bacterial sepsis where LPS activates comparable processes resulting in vascular leakage, shock and haemorrhage. But unlike GVH or sepsis in which mortality is high and recovery is slow, patients with DHF recover from the acute episode rapidly (if adequate and appropriate fluid has been given), usually without sequelae. The mechanisms of shock and leakage in DHF are therefore unique, even though on the surface, they share common features with GVH and/or sepsis. Very little is known about where leakage occurs in DHF, but lung and gut are two most likely sites. As regard to thrombocytopenia, one of the unanswered questions is whether the megakaryocytes are infected by the dengue virus and/or the mechanisms leading to rapid consumption of platelets.

Zinkernagel: The short version of the answer is that we don't know. The longer answer is that dengue is not the only haemorrhagic fever caused by viruses. There are south American haemorrhagic fevers caused by Arena viruses. There must be common parameters. I am sure that interferon α and β, or something like this, is one of the major factors acting as a common denominator. Antibody enhancement or change in relative distribution of infected cells is another likely factor. Complement-dependent types of lyses (although complement isn't currently fashionable) is probably another. But I don't think I currently have the courage to voice a rational explanation of the various things you listed. At the moment we only look at the T cell side or the antibody side. We probably need both, plus a knowledge of the geography within the system.

Fairlie: In septic shock it is known that complement is a driver. It starts with an antibody/antigen interaction, and then the complement cascade begins and there is release of the anaphylotoxins C3a, C4a and C5a. These are known to bind to G protein-coupled receptors (GPCRs) on macrophages, T cells and neutrophils. This results in the release of TNFα, IL1β, IL6 and so on. It seems to me that there is a direct link here between complement-mediated septic shock and dengue-induced haemorrhagic shock. I suspect that the pathology is similar. We have to consider all these things together and I don't think we can isolate T cells as the cause of haemorrhagic shock.

References

Bierman HR, Nelson ER 1965 Hematodepressive virus diseases of Thailand. Ann Intern Med 62:867–884

Halstead SB, Nimmannitya S, Cohen SN 1970 Observations related to pathogenesis of dengue hemorrhagic fever. IV. Relation of disease severity to antibody response and virus recovered. Yale J Biol Med 42:311–328

Halstead SB, Lan NT, Myint TT et al 2002 Dengue hemorrhagic fever in infants: research opportunities ignored. Emerg Infect Dis 12:1474–1479

Mitrakul C, Poshyachinda M, Futrakul P, Sangkawibha N, Ahandrik S 1977 Hemostatic and platelet kinetic studies in dengue hemorrhagic fever. Am J Trop Med Hyg 26:975–984

The evolutionary biology of dengue virus

Edward C. Holmes

Department of Biology, The Pennsylvania State University, University Park, PA 16802, USA

Abstract. Studies of the evolution of dengue virus (DENV) have blossomed during the last 20 years, in part due to the increasing availability of viral gene sequence data. Herein I review key aspects of the evolutionary biology of DENV focusing on extent and structure of genetic diversity in DENV, the time and place of DENV origin, the major mechanisms of DENV evolution, and the evolution of DENV virulence. A central conclusion is that despite the high mutation rates common to RNA viruses in general, there are important constraints against adaptive evolution in DENV in particular. These have implications for the escape from immunological recognition and hence for vaccine design. Finally, I note that a fuller understanding of DENV evolution will require more extensive study of the sylvatic cycle, particularly in Africa, which has been largely ignored to date.

2006 New treatment strategies for dengue and other flaviviral diseases. Wiley, Chichester (Novartis Foundation Symposium 277) p 177–192

The genetic diversity of DENV

Many studies of dengue virus (DENV) evolution have assessed the extent and structure of genetic diversity at various phylogenetic levels. The most basic observation here is that the virus exists as four phylogenetically distinct serotypes that diverge at ~30% across their polyprotein, with the relative proportion of amino acid variation among genes in the order NS2A, C, NS1, NS2B, E, NS4A, NS5, M, NS3, NS4B, from the most to the least variable. Although this is equivalent to the level of divergence seen among some different species of flavivirus, there is no strong evidence for major biological differences among serotypes; all cause the same spectrum of disease, have the same virion structure and genome organization, and infect the same hosts and vectors.

There is also abundant genetic variation within each serotype in the form of phylogenetically distinct clusters of sequences dubbed 'subtypes' or 'genotypes' (Rico-Hesse 1990). At the time of writing, three subtypes can be identified in DENV-1, 6 in DENV-2 (one of which is only found in non-human primates), four in DENV-3 and four in DENV-4, with another exclusive to non-human primates

(Fig. 1). It is probable that more extensive sampling will identify additional subtypes, and that other subtypes will disappear as 'gaps' in the phylogenetic tree are filled in.

A more notable observation is that subtypes often have differing geographical distributions, with some more widespread than others, indicating that both population subdivision and gene flow are important in structuring genetic diversity. This is best described in DENV-2 where two subtypes are apparently restricted to Southeast Asia and another to the Americas (Twiddy et al 2002). In contrast, a 'Cosmopolitan' subtype has been sampled from a wider range of localities, although it is unknown whether this is due to intrinsic differences in fitness. Other conclusions that can be drawn from the phylogenetic analysis of DENV diversity are (i) that subtypes frequently co-circulate within the same locality, (ii) that there is a possible distinction between 'endemic' subtypes that have circulated within particular localities for extended periods of time and 'epidemic' subtypes that seem to spread rapidly through populations, (iii) that Southeast Asia harbours the greatest amount of DENV genetic diversity, suggesting that it acts as a 'source' population, generating strains that then ignite epidemics elsewhere, and (iv) that there is a relatively high rate of clade (including subtype) extinction, with periodic fluctuations in genetic diversity (Klungthong et al 2004, Sittisombut et al 1997, Wittke et al 2002). Determining the biological processes that underlie clade extinction is a key topic for future study.

It has also been proposed that some subtypes differ in virulence, in this case manifest as their capacity to cause dengue haemorrhagic fever (DHF)/ dengue septic shock (DSS) (Rico-Hesse et al 1997). Although this issue is discussed below, it is important to note here that the genetic basis of these fitness differences are usually unknown. As a case in point, suggestions that variation in the RNA secondary structure of the 3′ UTR were in part responsible for the different disease associations among American and Asian DENV-2 viruses (Leitmeyer et al 1999) have not yet been borne out (Shurtleff et al 2001).

The origins of DENV

As with most infectious diseases, inferring the origin of DENV from historical records alone is a difficult exercise. In particular, the clinical symptoms of DENV are often not sufficiently diagnostic to exclude other infectious agents. This withstanding, historical records reveal three key dates in the evolutionary history of DENV; (i) the earliest proposed description of DENV infection from a Chinese medical encyclopaedia dated to 992 AD, (ii) the first description of a large-scale DENV epidemics at the end of the 18th century, and (iii) the first

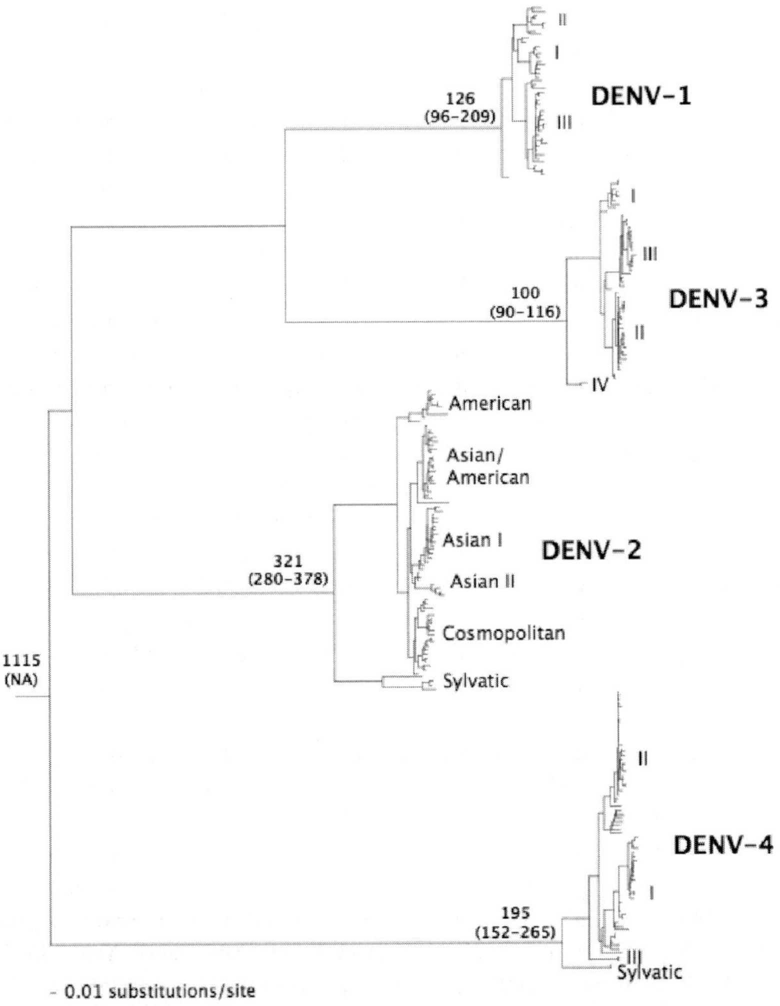

FIG. 1. The evolutionary history of DENV inferred through phylogenetic analysis. The phylogeny was reconstructed using a representative sample of 250 E gene sequences from all serotypes and subtypes of DENV under a maximum likelihood method and employing the HKY85+I+Γ_4 model of nucleotide substitution. The subtype designation is shown next to the relevant groupings and estimates of the age of the most recent common ancestor (with confidence intervals in parenthesis) are shown for key nodes. All age estimates were taken from Twiddy et al (2003). The tree is mid-point rooted for purposes of clarity only and all horizontal branch lengths are drawn to a scale of the number of nucleotide substitutions per site.

description of DHF/DSS epidemics following World War II (reviewed in Gubler 1997).

Luckily, viral gene sequence data offer an alternative means to infer the evolutionary history of DENV. Estimates for the age of DENV based on viral molecular clocks are relatively consistent (Twiddy et al 2003, Weaver & Barrett 2004, Zanotto et al 1996). For example, estimates for the date of the origin of DENV itself places this event within the last few thousand years, satisfactory adjacent to the earliest reports of dengue-like illness. Similarly, most molecular clock estimates for age of the genetic diversity within each serotype are similar, ~100–300 years ago (Fig. 1). These dates most likely correspond to either the time of cross-species transmission from non-human primates to humans, or to the acquisition of *Aedes aegypti* as the principle vector for DENV transmission in humans (Moncayo et al 2004). However, one puzzle remains: why the genetic diversity within the four serotypes appears at approximately the same time? The most likely explanation is that the spill-over of DENV strains from non-human primates to humans has occurred throughout history but that only in the last few centuries have human populations been sufficiently large to ensure the development of sustained transmission networks (Zanotto et al 1996).

Could these molecular clock dates be grossly inaccurate? The biggest source of error will be if substitution rates vary dramatically between humans and other species, for example because rates of amino acid change increase as the virus adapts to new hosts. However, given that similar levels of genetic diversity have been observed in both humans and mosquito populations (Craig et al 2003, Lin et al 2004) as well as the relatively low rate of amino acid substitution in DENV (see below), any sporadic elevation in the rate of amino acid replacement is unlikely to greatly affect dating estimates.

There is also uncertainly as to where DENV originated. The most popular idea is that DENV has an African origin (Gaunt et al 2001) because many of the mosquito-borne flaviviruses related to DENV are found in Africa, implying that this clade as a whole originated here, and because *A. aegypti* is also believed to have originated in Africa. However, both these inferences are open to question in that the phylogeny of the flaviviruses is by no means certain, particularly as it is undoubtedly the case that many of the viruses most closely related to DENV have yet to be sampled, and because it is likely that *A. aegypti* has only recently been adopted as a DENV vector (Moncayo et al 2004). Indeed, because the four DENV serotypes are also present in both humans and monkeys in Asia (Wang et al 2002) it is also theoretically possible that DENV has an Asian origin. To resolve the question of the origin of DENV it will be necessary to undertake a wider sample of viruses in Africa, from where relatively little DENV has been sampled to date.

Mechanisms of evolution in DENV

Mutation

Like all RNA viruses, the fuel of evolutionary change in DENV is provided by high mutation rates, a by-product of an intrinsically error-prone RNA polymerase. Although mutation rates have not been measured directly in DENV, estimates of its substitution rate, at $\sim 1 \times 10^{-3}$ substitutions per site, per year, are of the same order of magnitude as seen in other RNA viruses.

When coupled to large population sizes and rapid replication rates, high mutation rates mean that populations of DENV viruses show abundant genetic variation, both within and among hosts. However, this does not necessarily mean that this virus forms a quasispecies (Lin et al 2004, Wang et al 2002). In particular, although abundant genetic variation is necessary for quasispecies formation, it is a long way from being sufficient. The reality is that for viral populations to form quasispecies the rate of mutation must be sufficiently high, and the strength of genetic drift sufficiently weak, that there is a 'mutational coupling' among all the variants in the population. This in turn means that the fitness of individual genomes also depends on that of neighbours in sequence space so that natural selection acts on the whole viral population. Hence, it is a particular form of mutation–selection balance that defines the quasispecies, rather than the presence of genetic diversity. As there is (as yet) no evidence that natural selection acts on DENV populations as a whole, it also follows that there is no strong evidence that this virus forms quasispecies.

Natural selection

Understanding the process of natural selection is perhaps the most important aspect of evolutionary studies of DENV. The most notable observation here is that positive (or Darwinian) selection is a relatively rare process, with only sporadic occurrences across the genome and among serotypes (Twiddy et al 2002). Accordingly, most of the mutational changes fixed in DENV populations are neutral and subject to genetic drift rather than because they enhance fitness. There is one important caveat here: current computational methods are limited in that they usually cannot detect selection acting on mutations that have occurred once on a single lineage. It is therefore certain that more amino acid residues have been fixed by selection during DENV evolution than can be identified at present. Experimental assessments of fitness are therefore likely to provide a more precise description of the selection pressures facing individual mutations.

Despite this, it is clear that the dominant selection pressure acting on DENV is purifying selection; most amino acid changes are deleterious, reduce fitness and are hence removed from the viral population. The most compelling evidence comes from studies of the relative rates of non-synonymous (d_N) to synonymous (d_S) substitutions per site, with $d_S > d_N$ indicative of purifying selection. Not only is $d_S >> d_N$, showing that purifying selection is extremely strong, but d_N/d_S values differ greatly within and among hosts. Specifically, d_N/d_S within hosts approximates 1.0, the value expected if all mutations were equally likely and little selection of any kind has been imposed (Holmes 2003). In contrast, far lower d_N/d_S values are observed within host populations. From this large difference it can be estimated that roughly 90% of the amino acid variants that arise within hosts are deleterious. Such strong purifying selection appears to be true of arthropod-borne RNA viruses as a whole (Woelk & Holmes 2003), most likely a function of strong antagonistic pleiotropy; mutations that are beneficial or neutral in one environment (for example the mosquito vector) are deleterious in another (the human host). It is also probable that a significant proportion of the synonymous mutations that arise in DENV are deleterious (or slightly so), probably a function of constraints imposed by RNA secondary structure which seems to be common in RNA viruses (Simmonds et al 2004).

The lack of strong positive selection also raises important questions regarding the relative roles of antibody and T cell escape in DENV evolution. The lack of strong selection in the envelope (E) glycoprotein argues that escape from antibody recognition is not a key aspect of DENV evolution (and it is noteworthy that the E protein is not the most variable in the DENV genome). The role of T cell escape in shaping patterns of genetic diversity is harder to assess, particularly as T cell epitopes are relatively abundant across the DENV genome (Mathew et al 1998, Zeng et al 1996), and the process has been documented in NS3 (Mongkolsapaya et al 2003, Simmons et al 2005). Because CTL epitopes often fall within structural genes, T cell-mediated selection is expected to be a complex process as mutations that confer CTL escape might have other detrimental affects on fitness.

Recombination

The most debated mechanism of evolution in DENV is recombination (Monath et al 2005). This controversy has two principle causes, that some of proposed recombination events are undoubtedly due to laboratory error (da Silva & Messer 2004, Goncalvez et al 2002) and that there have been claims that occurrence of recombination undermines the use of live attenuated DENV vaccines (Seligman & Gould 2004).

To some, recombination needs to be established experimentally, using genetic makers. In contrast, all evidence for recombination in DENV has come from comparative sequence analysis, via the identification of incongruent phylogenetic trees. However, this is a legitimate and powerful way to detect recombination and has been used in many other viral systems with little controversy. Indeed, if there is a criticism of the phylogenetic approach it is that it is overly conservative and can only detect recombination among sequences that are relatively divergent. Under these criteria the evidence for recombination in DENV is strong, as it has been observed in a number of studies not all of which can be explained by laboratory error (Holmes & Twiddy 2003, Craig et al 2003). Second, there is no compelling biological reason why DENV virus should not recombine, unlike the negative-sense RNA viruses. Indeed, recombination has been demonstrated in other members of the *Flaviviridae* (Becher et al 2001, Colina et al 2004, Kalinina et al 2002), and given the huge numbers of humans and mosquitoes infected with DENV it is predictable that the occasional mixed infection resulting in recombination will occur.

Despite this, it must also be acknowledged that recombination in DENV is a relatively uncommon process and does not have a major effect on population level evolution. Indeed, most DENV recombinants, like most DENV point mutations, will undoubtedly be deleterious, reduce fitness and therefore be cleared by purifying selection (and recombination among closely related strains is not expected to have a major impact on fitness). Hence, most recombinants will only exist transiently and are most likely to detected with intra-host polymorphism data (Craig et al 2003). The strength of purifying selection against DENV recombinants also means that this process is not a major concern for vaccine design. While it is theoretically possible that recombination could occur between a wild and vaccine DENV strain, the vast majority will be removed by purifying selection. The key parameter in this context is therefore the relative rate of *successful* recombination (r) to that of nucleotide substitution per site (k). All studies indicate that $k >> r$ in DENV, so that recurrent mutation must be a more important evolutionary force than recombination.

The evolution of DENV virulence

The evolution of virulence is one of the most interesting, yet difficult, issues in evolutionary biology. The question of virulence evolution in DENV has been spurred by proposals that a 'high virulence' Asian strain of DENV-2 has out-competed a lower virulence 'American' strain of this virus and hence will continue to become more frequent in the future (Cologna et al 2005). Although possible in theory, there are a number of reasons why it is dangerous to predict such long-term trends. First, the division between Asian and American strains of DENV-2 is an overly simplistic representation of the data. In particular, it is clear that there is abundant genetic diversity within the Asian group (Twiddy et al 2002). Moreover, these Asian subtypes can differ at key amino acid sites, most notably E390, a major virulence determinant in experimental studies of mice. Consequently, a far wider array of Asian strains need to be studied to understand the intricate nature of virulence in DENV. Second, it is evident that viral strain is not the only factor that determines the clinical outcome of a DENV infection. In particular, most DENV infections do not result in DHF or DSS, even if infected with Asian DENV-2 strains, and there is evidence that both the genetic make-up of the host, manifest as differing T cell responses (Mongkolsapaya et al 2003), and the influence of co-circulating serotypes, through antibody-dependant enhancement (Thein et al 1997), determine the aetiology of severe dengue disease.

A more detailed phylogenetic analysis of the Asian DENV-2 strains also suggests that the evolution of virulence in DENV is a highly complex trait. Specifically, (i) if particular strains of Asian DENV were consistently more virulent than others then there should be a general clustering by disease type on phylogenetic trees which is not observed, and (ii) if virulence were selectively advantageous to DENV, because DHF/DSS are associated with higher viral loads (Vaughn et al 2000) which in turn elevates the rate of transmission, the high virulence strains of Asian DENV-2 should be increasing in frequency, and there is no evidence that this is the case.

A far more complex problem with predicting the evolution of virulence from experimental studies is that the host immune response is ignored. In particular, it is possible that complex immunological interactions among serotypes (Endy et al 2004) mean that strains associated with a higher virulence will not necessarily be successful in the long-term because of cross-protective immunity conferred by previous DENV infections in the population. Conversely, viral strains that are individually associated with low virulence may sometimes cause serious disease if they are subject to antibody-dependent enhancement. Understanding these complex immunodynamics should be a priority for future research. Indeed, it is evident that for a fuller understanding of the evolution of virulence in DENV it will be necessary to undertake more long-term prospective studies, in which genetic data are collected from the majority of DENV patients who experience asymptomatic infections.

Future research

From an evolutionary perspective, perhaps the biggest gap in our knowledge of DENV concerns the sylvatic cycle, particularly in African populations and including both sylvatic hosts and vectors. Only a handful of sylvatic DENV isolates have been sequenced, none of these constitute whole genome sequences, and most were collected many years ago. Given the low probabilities of sampling DENV from wild primate populations, the most profitable way this research can proceed is by undertaking extensive surveys of sylvatic vectors, using DNA collected from the blood meal as a means of identifying the primate host.

Obtaining more information on sylvatic DENV will enable us to address a number of key questions. First, it will provide essential information on the extent of genetic diversity in potential reservoir populations, including the description of viral strains that could eventually emerge in humans. Indeed, it is possible that more than the known four serotypes of DENV exist in non-human populations and, as mentioned above, it is unclear why these serotypes of DENV emerged in humans on roughly the same time-scale. Second, revealing the extent of genetic diversity of sylvatic DENV is central to understanding the adaptive basis (if any) of how sylvatic DENV acquired its human transmission cycle. Experimental evidence suggests that sylvatic DENV would have had to have accumulated adaptive changes in order to replicate efficiently in *Aedes* vectors (Moncayo et al 2004) and phylogenetic studies have been used to identify these mutations (Wang et al 2000). The goal for the future must be to marry the survey of DENV in wild populations with the phylogenetic analysis of complete genome data and experimental studies of viral fitness. Only with such a combined approach will we come to a full understanding of the evolutionary biology of this important emerging pathogen.

References

Becher P, Orlich M, Thiel HJ 2001 RNA recombination between persisting pestivirus and a vaccine strain: generation of cytopathogenic virus and induction of lethal disease. J Virol 75:6256–6264

Colina R, Casane D, Vasquez S et al 2004 Evidence of intratypic recombination in natural populations of hepatitis C virus. J Gen Virol 85:31–37

Cologna R, Armstrong PM, Rico-Hesse R 2005 Selection for virulent dengue viruses occurs in humans and mosquitoes. J Virol 79:853–859

Craig S, Thu HM, Lowry K, Wang X-F, Holmes EC, Aaskov J 2003 Diverse dengue type 2 virus populations contain recombinant and both parental viruses in a single mosquito host. J Virol 77:4463–4467

de Silva A, Messer W 2004 Arguments for live flavivirus vaccines. Lancet 364:500

Endy TP, Nisalak A, Chunsuttitwat S et al 2004 Relationship of preexisting dengue virus (DV) neutralizing antibody levels to viremia and severity of disease in a prospective cohort study of DV infection in Thailand. J Infect Dis 189:990–1000

Gaunt MW, Sall AA, de Lamballerie X, Falconar AKI, Dzhivanian TI, Gould EA 2001 Phylogenetic relationships of flaviviruses correlate with their epidemiology, disease association and biogeography. J Gen Virol 82:1867–1876

Goncalvez AP, Escalante AA, Pujol FH et al 2002 Diversity and evolution of the envelope gene of dengue virus type 1. Virology 303:110–119

Gubler DJ 1997 Dengue and dengue hemorrhagic fever: its history and resurgence as a global public health problem. In: Gubler DJ and Kuno G (eds) Dengue and dengue hemorrhagic fever. CAB International, New York, p 1–22

Holmes EC 2003 Patterns of intra- and inter-host nonsynonymous variation reveal strong purifying selection in dengue virus. J Virol 77:11296–11298

Holmes EC, Twiddy SS 2003 The origin, emergence and evolutionary genetics of dengue virus. Infect Genet Evol 3:19–28

Kalinina O, Norder H, Mukomolov S, Magnius LO 2002 A natural intergenotypic recombinant of hepatitis C virus identified in St. Petersburg. J Virol 76:4034–4043

Klungthong S, Zhang C, Mammen Jr MP, Ubol S, Holmes EC 2004 The molecular epidemiology of dengue virus serotype 4 in Bangkok, Thailand. Virology 329:168–179

Leitmeyer KC, Vaughn DW, Watts DM et al 1999 Dengue virus structural differences that correlate with pathogenesis. J Virol 73:4738–4747

Lin SR, Hsieh SC, Yueh YY et al 2004 Study of sequence variation of dengue type 3 virus in naturally infected mosquitoes and human hosts: implications for transmission and evolution. J Virol 78:12717–12721

Mathew A, Kurane I, Green S et al 1998 Predominance of HLA-restricted CTL responses to serotype crossreactive epitopes on nonstructural proteins after natural dengue virus infections. J Virol 72:3999–4004

Monath TP, Kanesa-Thasan N, Guirakhoo F et al 2005 Recombination and flavivirus vaccines: a commentary. Vaccine 23:2956–2958

Moncayo AC, Fernandez Z, Ortiz D 2004 Dengue emergence and adaptation to peridomestic mosquitoes. Emerg Infect Dis 10:1790–1796

Mongkolsapaya J, Dejnirattisai W, Xu X-N et al 2003 Original antigenic sin and apoptosis in the pathogenesis of dengue hemorrhagic fever. Nat Med 9:921–927

Rico-Hesse R 1990 Molecular evolution and distribution of dengue viruses type-1 and type 2 in nature. Virology 174:479–493

Rico-Hesse R, Harrison LM, Salas RA et al 1997 Origins of dengue type 2 viruses associated with increased pathogenicity in the Americas. Virology 230:244–251

Seligman SJ, Gould EA 2004 Live flavivirus vaccines: reasons for caution. Lancet 363:2073–2075

Shurtleff AC, Beasley DWC, Chen JJY et al 2001 Genetic variation in the 3′ non-coding region of dengue viruses. Virology 281:75–87

Simmonds P, Tuplin A, Evans DJ 2004 Detection of genome-scale ordered RNA structure (GORS) in genomes of positive-stranded RNA viruses: Implications for virus evolution and host persistence. RNA 10:1337–1351

Simmons CP, Dong T, Chau NV et al 2005 Early T-cell responses to dengue virus epitopes in Vietnamese adults with secondary dengue virus infections. J Virol 79:5665–5675

Sittisombut N, Sistayanarain A, Cardosa MJ et al 1997 Possible occurrence of a genetic bottleneck in dengue serotype 2 viruses between the 1980 and 1987 epidemic seasons in Bangkok, Thailand. Am J Trop Hyg Med 57:100–108

Thein S, Aung MM, Shwe TN et al 1997 Risk factors in dengue shock syndrome. Am J Trop Med Hyg 56:566–572

Twiddy SS, Woelk CH, Holmes EC 2002 Phylogenetic evidence for adaptive evolution of dengue viruses in nature. J Gen Virol 83:1679–1689

Twiddy SS, Holmes EC, Rambaut A 2003 Inferring the rate and time-scale of dengue virus evolution. Mol Biol Evol 20:122–129

Vaughn DW, Green S, Kalayanarooj S et al 2000 Dengue viremia titer, antibody response pattern, and virus serotype correlate with disease severity. J Infect Dis 181:2–9

Wang E, Ni H, Xu R et al 2000 Evolutionary relationships of endemic/epidemic and sylvatic dengue viruses. J Virol 74:3227–3234

Wang W-K, Sung T-L, Lee C-N, Lin T-Y, King C-C 2002 Dengue type 3 virus in plasma is a population of closely related genomes: quasispecies. J Virol 76:4662–4665

Weaver SC, Barrett ADT 2004 Transmission cycles, host range, evolution and emergence of arboviral disease. Nat Rev Micro 2:789–801

Wittke V, Robb TE, Thu HM et al 2002 Extinction and rapid emergence of strains of dengue 3 virus during an interepidemic period. Virology 301:148–156

Woelk CH, Holmes EC 2002 Reduced positive selection in vector-borne RNA viruses. Mol Biol Evol 19:2333–2336
Zanotto PM de A, Gould EA, Gao GF, Harvey PH, Holmes EC 1996 Population dynamics of flaviviruses revealed by molecular phylogenies. Proc Natl Acad Sci USA 93:548–553
Zeng L, Kurane I, Okamoto Y, Ennis FA, Brinton MA 1996 Identification of amino acids involved in recognition by dengue virus NS3-specific, HLA-DR15-restricted cytotoxic $CD4^+$ T-cell clones. J Virol 70:3108–3117

DISCUSSION

Evans: To a novice looking at your data for DENV-1, -2 and -3, I think you are missing the other half of the equation: is it just mosquito competency? There's a possibility that this hasn't been studied. You are looking at patterns over 7–17 years which are common with other parasites.

Holmes: We have looked at overall abundance of mosquitoes.

Evans: I am talking about one that might be better for transmitting DENV-4 versus -1, -2 and -3.

Holmes: If that were the case, why would dengue oscillate in frequency the way it does?

Evans: Because there are populations of different mosquitoes.

Holmes: I find that hard to believe. The pattern we see looks to me to be due to immune interactions. We have looked at mosquito abundance and it doesn't explain the pattern. If what you are suggesting is the case, then you'd expect DENV-4 to be in phase with the others, and it isn't.

Young: There might be competition between the different serotypes in the mosquitoes.

Gubler: We find considerable variation in the ability of mosquitoes to become infected. However, highly susceptible or competent strains are generally competent to all four viruses. You don't find one that is highly competent for DENV-1 and a poor vector for DENV-4, for example. One thing that has puzzled me for 30 years is that in 1976/77, there was an epidemic of DENV-3 in Jakarta, Indonesia. Dengue 3 was the predominant virus during the epidemic, but as soon as it started, DENV-1, -2 and -3 increased and spread all over Jakarta. All serotypes were fairly localized before the epidemic, but shortly after the epidemic began, transmission of all three viruses increased. DENV-4, which was relatively rare at the time, did not. This was a very rare virus at the time. Essentially the same thing happened in Puerto Rico, but with different viruses. Most of the epidemics in Puerto Rico in the past 25 years have been caused by DENV-4. In Indonesia most of the epidemics have been caused by DENV-3, and in Thailand most of them have been dengue 2. What does this mean? How do you explain three of the viruses or all four of the viruses increasing simultaneously, even though there is one predominant virus?

Holmes: One thing about dengue which I find quite shocking is how little is known about the sylvatic cycle and vectors. There are very few people who sample viruses in vectors. We should be looking at vector populations more fully. I think there are complicated interactions between these serotypes. It will be difficult to tease out.

Young: Getting back to the mosquito association question, it is not so much competence but competition. Do people know about the efficiency of different serotype competition within mosquitoes?

Gubler: You can have dual infections in the same mosquito. It is possible to have two serotypes in the same insect. I don't know whether there is more competition between DENV-4 and the other serotypes.

Holmes: DENV-4 caused the outbreaks in Puerto Rico, so it works in mosquitoes perfectly well (see Bennett et al 2003). In competition with other serotypes it might not, though. This is hard to measure: you need to do this in the context of the immunological landscape you are in, rather than just in an *in vitro* system.

Gubler: The Puerto Rico experience is unique in the sense that the American genotype of DENV-2 was there and caused a number of earlier epidemics. DENV-1 was introduced in 1977 and caused a major epidemic in 1978, and then it caused another small epidemic in 1981. DENV-4 was introduced into Puerto Rico in the middle of the 1981 DENV-1 epidemic: DENV-1 disappeared and 4 became the predominant virus. Over the next 20 years there were four major epidemics in Puerto Rico, one of which was DENV-2 and three of which were DENV-4. The Asian genotype of DENV-2 was introduced in 1984—we picked up the patient who brought it into Puerto Rico from Haiti. This virus didn't cause an epidemic in Puerto Rico for 10 years. The dynamics of these viruses are complex, and there is no good herd immunity explanation for this.

Holmes: DENV-1, -2 and -3 are strongly in phase, so they are not competing, but DENV-4 is very strange. The dengue serotypes are genetically as divergent as different flaviviruses.

Harris: You mentioned the stop mutation and then there were seven others. What were these?

Holmes: They are very rare changes in dengue virus. None of the other seven normally vary with dengue. This suggests to me that they are random deleterious mutations that are picked up by chance (see Aaskov et al 2006).

Harris: I remember doing a bunch of experiments involving coinfections of the same cells with viruses. We found a strong heterologous exclusion: if a cell is already infected with one dengue virus another can't get in. This was time dependent: it would carry on for the first eight hours or so.

Holmes: So it is less likely later?

Harris: No, it is more likely later. This is something seen *in vitro* with closely related viruses. There is an 8–10 h window where we can't get another virus in to already-infected cells.

Halstead: In humans, simultaneous infections do occur.

Harris: I'm talking here about a single cell. A person can be infected by two virus serotypes without the same cell having both. But without two serotypes in the same cell you won't get complementation.

Holmes: I can't think of any other mechanism for it than complementation. If that is the case then you have to have multiple infections in the same cell. This is an inference based on the pattern seen.

Harris: I think that can happen within other time frames in the cell, but you could also imagine complementation happening between cells.

Padmanabhan: Is there any evidence for recombination?

Holmes: No. I've checked this.

Rice: A stop codon in this position would be unprecedented for flavivirus RNAs that can replicate in *trans.*

Holmes: We've also found this stop codon mutation in Singapore samples. Many people who do dengue sequencing ignore stop codons. But if you see a stop codon consensus, this means that it is at high frequency in the population you are sampling. This tells us that something interesting is happening. I think it must be a common thing.

Canard: Perhaps a recombination event helps by-passing the stop codon.

Holmes: I checked those sequences really carefully and I can't see any recombination in them.

Canard: You reported a certain unexpected Y16 frequency.

Holmes: It's a different data set. That was DENV-2. If you look at clones, you also see recombinants, but there are no obvious recombinants in the dengue 4 data set.

Ng: Can you test the complementation in the cell?

Holmes: We could, yes. It is definitely a stop codon. Many people have argued that it is actually doing something—that in some way the virus benefits from having this around. Personally I don't believe that. The idea would be that the functional virus has some benefit from having a virus with a truncated E gene co-infect the same cell. Indeed, the capsid and the membrane proteins are probably functional, so it could be a way of controlling gene expression. I think this is unlikely. Or it gives you mutation in the population that you can allow to infect different cells. However, I think it is probably just a parasite.

Farrar: Couldn't it be acting as a decoy to the immune system? It will be a huge number of random mutations. The parasite could be a decoy because it will be quite immunogenic. It might protect the functional virus.

Canard: The gene products upstream of here are playing an important role which is not for replication of the virus but might be counteracting the host defence mechanisms.

Satchidanandam: I was thinking of the earlier discussion of original antigenic sin. When you attribute this sudden extinction to immune modulation, would you not, in light of these data, attribute it simply to viral fitness? When a flavivirus comes from a mosquito during a bite, it crosses to the human host from an invertebrate system with hardly any adaptive ability against a vertebrate immune system. If you had just original antigenic sin it will only stimulate pre-existing immune cells, then the extinction could simply be attributed to fitness.

Holmes: That is a good point. We have two clades, one of which goes from 1980–1994, and one which goes from 1990–2002. They coexist for four years and then one goes down and the other takes over (These results are described in Zhang et al 2005). These are exactly the same population, in the same hospital. This can't be a random bottle-neck effect, because this would affect both equally. The second serotype must be fitter than the first, but it has taken a number of years for this process to take place. It is not a dramatic loss of fitness. The decline of the first corresponds to the increase in DENV-4. This is an association, not a correlation. The other thing to note is that there are mutations in the E protein, which make the surviving clade of DENV-1 more dissimilar to DENV-4 E than that one that went extinct. They also differ in NS3 and other genes. It is a slow decline.

Satchidanandam: Does a DENV-4 infection always precede extinctions?

Holmes: We don't have many independent events. In the two we have this does fit. It is dangerous to think about viruses in a serotype as being homogeneous. There are genetic differences among them. Not all DENV-1 strains are the same; they differ in key sites in the E gene, NS3 and others.

Gubler: Didn't you speculate that in the Caribbean that these kinds of changes were associated with genetic drift.

Holmes: The model I had for this is that when these clades appeared, they were only in a DENV-1 environment. They drifted apart by chance. Then, when DENV-4 comes into the population, the DENV-1 strains which had picked up mutations by drift that make them more or less similar to DENV-4, then decline or survive. Their fitness is realised later on as the serotypes change in frequency.

Halstead: We are only studying an observable part of the iceberg. Unfortunately we are not studying everything that we would like to know. It is not so hard to study dengue when a brand new virus is introduced into a susceptible population, but in Thailand where all four viruses co-circulate, it is impossible to understand what is going on by just studying patients in a hospital. In a study in southeast Thailand, I showed years ago that DENV-1 was the virus causing the largest amount of infection in the population, but DENV-2 was the predominant cause of hospitalized disease. 90% of all dengue infections are silent. You have made a plea for more study of the sylvatic cycle; I'd like to make a plea for more study of the denominator. The American DENV-2 genotype that was discovered in Peru was associated with the secondary dengue infection which did not result in DHF, and was essentially picked up because of an ongoing study looking at mild disease. The American DENV-2 genotype made it all the way to Peru before the Asian DENV-2 genotype did; this correlates completely with your interpretation. But since we don't have mechanisms in place to study silent infections, no one knew how broadly spread the American DENV-2 genotype virus was.

Holmes: You are right: there is a huge sampling bias in the case studies. However, if you take viruses from people in hospitals they must have gone through people who are also not getting disease as well. The lineage in hospital must have gone through people who aren't ill, so I don't think the sequencing is biased. The epidemiology is, but not the sequencing.

Halstead: When did the ancestral dengue virus disappear? Has it disappeared?

Holmes: All I would say is that there must be a huge diversity of viruses like dengue in non-human primate populations. The sylvatic viruses we have sampled were from Al Rudnick, collected in the 1970s. There is a pool of sylvatic viruses that jumped four times into humans successfully. I suspect that viruses have jumped continually from primates to humans, causing localized outbreaks which burnt out in almost all cases. When the population became dense enough with urbanization, these lineages survived. We are seeing lineages that have survived, not the ones that crossed over. The ancestral dengue viruses are probably in a jungle somewhere.

Harris: We don't see isolates that are directly correlated with DHF, for instance. This is because the role of the host response is huge. It could also be that there is a huge selection pressure for the virus to be highly infectious and good at replicating. This is also the selection pressure we apply for isolation. Is it that the epitopes that are triggering some immunopathogenic response are not the ones we are isolating? Now that we have large-scale sequencing capacity it would be interesting to see whether there is diversity that correlates with severity, and clones that are not reaching consensus or the isolation.

Holmes: We have done this with the influenza virus, where we infected horses and sequenced them every few days. In one animal, the virus at day 2 shows lots of variability, but at day 3 there is a completely different set of mutations. There is a huge amount of diversity, and it turns over extremely quickly. This may have huge effects on key aspects of dengue biology.

Herrling: There must be a significant amount of coinfection between dengue and HIV. How do they influence each other? Do immune-deficient people get haemorrhagic fever?

Farrar: That is a great question. The HIV rates in Vietnam are still low, and they are a bit higher in Thailand. This is something we should look at.

Gubler: This hasn't been studied.

Herrling: It would tell you something about the T cells.

Farrar: Anecdotally, we see less coinfection with dengue and HIV than we would expect by chance.

Gubler: In this discussion it is important for us to remember that these different strains are not monolithic. They are changing as they move through different populations. Scott Halstead mentioned that the American genotype virus changed a lot over the years. It caused major epidemics in the early years: the 1969 epidemic, one of the worst in Puerto Rican history, was American genotype of DENV-2. It was the same genotype that caused the Taihiti, Fiji and New Caledonia epidemics in the early 1970s, both of which were explosive with severe disease. These viruses change quite dramatically genetically and phenotypically as they move through populations.

References

Aaskov J, Buzacott K, Thu HM, Lowry K, Holmes EC 2006 Long-term transmission of defective RNA viruses in humans and Aedes mosquitoes. Science 311:236–238

Bennett SN, Holmes EC, Chirivella M et al 2003 Selection-driven evolution of emergent dengue virus. Mol Biol Evol 20:1650–1658

Zhang C, Mammen MP Jr, Chinnawirotpisan P et al 2005 Clade replacements in dengue virus serotypes 1 and 3 are associated with changing serotype prevalence. J Virol 79:15123–15130

Developing vaccines against flavivirus diseases: past success, present hopes and future challenges

John R. Stephenson

Department of Infectious and Tropical Diseases, London School of Hygiene and Tropical Medicine, Keppel Street, London WC1E 7HT, UK

Abstract. Vaccination still remains the most cost-effective way of protecting large populations against infectious disease. Safe and effective vaccines are available against most pathogenic flaviviruses and in recent years substantial progress has been made in developing vaccines against dengue. Dengue vaccines based on conventional and recombinant DNA technologies are being assessed and initial results are encouraging. Many other experimental vaccines have been developed, but despite the intensity of effort, concerns about the safety of new vaccines appear to be hindering their development. With the global threat from dengue increasing, might it now be the time to consider a less risk-averse approach?

2006 New treatment strategies for dengue and other flaviviral diseases. Wiley, Chichester (Novartis Foundation Symposium 277) p 193–205

Commercially available vaccines

As they are agents of human and animal diseases with high morbidity and mortality, considerable efforts have been made to develop vaccines against several flaviviruses (Stephenson 1988, Venugopal & Gould 1994). Theiler's seminal discovery in the 1930s that repeated passage in mice or tissue culture reduced the pathogenicity of yellow fever (YF) virus, produced the 17D vaccine virus which has proved to be both safe and efficacious over several decades (Freestone et al 1977). Similar successes have been achieved with conventional killed vaccines, especially against tick-borne encephalitis (TBE) (Heinz et al 1980, Popov et al 1985, Klockmann et al 1989) and Japanese encephalitis (JE) (Monath 2002), and a veterinary vaccine against louping ill (Brotherston & Boyce 1969) is also available. Three TBE vaccines are available from Baxter (previously Immuno), Chiron (previously Behring) and the Chumakov Institute for Poliomyelitis and Viral Encephalitides in Moscow. The former two vaccines are derived from Western isolates of TBE virus grown in primary avian fibroblasts, and the Russian vaccine is derived from a far-eastern isolate grown in porcine kidney cells. Inactivated vaccines against JE are available from several manufacturers. However they are derived from rodent brains and products made from animal CNS preparations may not be acceptable for human use in the near future due to fears of transmitting spongiform encephalopathies such as BSE. Although all these products are well tolerated and efficacious they are frequently too expensive for widespread use in endemic areas. A live attenuated vaccine against JE has been developed in China (Monath 2002) and used extensively in that country. Its safety and efficacy has been independently assessed, but concerns over quality control during manufacture have limited its use outside of China. The recent invasion of North America by West Nile virus (WNV) has stimulated a substantial effort in producing vaccines against this largely ignored pathogen. Even though it was only six years ago that WNV was detected in North America there are already two veterinary vaccines on the market. A conventional inactivated cell culture-grown vaccine is available from Fort Dodge Animal Health (Wyeth) (Grosenbaugh et al 2003) and a recombinant vaccine based on canarypox virus, incorporating the preM&E genes of the NY1999 isolate of WNV (Recombitek–ALVAC) (Ng et al 2004) have been recently licensed for use in horses in the USA.

However, despite many years of research and substantial funding, a licensed vaccine against dengue is not currently available.

Improving vaccination protocols

Conventional immunization strategies for inactivated virus vaccines usually involve the administration of at least two doses, several weeks apart. However, studies using a murine model of TBE have demonstrated that much more rapid vaccination protocols are possible (Stephenson et al 1995). In this experimental animal model, which mimics human disease in both pathology and route of administration, vaccination schedules lasting only 2–3 days were able to give complete protection against lethal encephalitis. Moreover, this protection could be elicited against virus challenge at soon as 5 days, or up to 100 days after vaccination was commenced. The mechanism of this rapid protection is unknown, although stimulation of innate immunity could play a role. Alternatively, as the virus infection takes 9–11 days to kill the animals the rapid vaccination protocols could give the host adaptive immune system a sufficient 'head start' in the race against virus replication.

Immune stimulators have been investigated by many workers to improve vaccine efficiency, but few studies have utilized the stimulatory effect of growth hormones on the immune system. As human growth hormones are also active in mice we were able to study the effect of this hormone on TBE vaccine (Stephenson et al 1991). Mixing the hormone with a commercial vaccine gave complete protection against lethal disease with a single dose, whereas this was only achieved with two doses of vaccine administered alone.

Dengue vaccines under development

Although conventional vaccines have played a major role in combating many flavivirus diseases such as YF, JE and TBE, developing vaccines against dengue remains a significant challenge. Following the success of using classical virological techniques to develop a live attenuated YF vaccine, similar methods have been used to produce a tetravalent dengue vaccine (reviewed in Halstead & Deen 2002). Two tetravalent vaccines using viruses attenuated by serial passage in non-human cells have been developed and are undergoing clinical trials. Both these vaccines have completed safety testing through phase 2 trials, and phase 3 trials are either underway or planned. One vaccine developed at Mahidol University in Bangkok and licensed by Aventis Pasteur produced 80–90% seroconversion rates in young children to all four serotypes after administration of two doses. However, the levels of immunity induced to some of the serotypes may not be sufficient to provide lasting protection against disease. Another vaccine, produced by the Walter Reed Army Institute for Research in the USA and licensed by GSK, produced similar seroconversion rates in adult volunteers, but again with lower levels of immunity to some serotypes. These initial results are encouraging, but as the molecular basis of attenuation is not understood concerns have been raised that interference in replication between the serotypes and/or interference in immune stimulation may lead to an imbalanced immune response resulting in incomplete protection. Incomplete or low levels of protection pose more problems for dengue vaccine administration than for many other vaccines. It is now generally accepted that pre-existing immunity to one dengue virus serotype is a risk factor for DHF and DSS when an individual is infected with another serotype at a later time (Halstead 1982, Rothman 2003). However, there is not a general consensus on the mechanism of this immune enhancement of disease or the relative importance of antibodies, T cells or circulating cytokines (summarized in Stephenson 2005). High levels of vaccine-induced antibody to one serotype could enhance disease severity if the vaccinee is exposed to serotypes against which they have low levels of immunity. In addition, concerns have been raised (as with any attenuated vaccine virus with an RNA genome) that reversion to virulence through mutation, or through recombination between the vaccine components or with wild virus could occur (Seligman & Gould 2004).

Several attempts to overcome these concerns have been made and the vaccines that have advanced furthest through a formal evaluation process have employed genetically modified infectious virus clones (reviewed in Jacobs & Young 2003, Stephenson 2005). The best characterised of these is ChimeriVax-Dengue, licensed to Aventis-Pasteur, which uses the 17D YF vaccine virus as its genetic backbone and replaces the YF virion envelope protein genes with those from dengue viruses. A tetravalent vaccine containing four chimaeric viruses, each with the preM and E genes from one of the four dengue virus serotypes, has undergone extensive tests in experimental animals and human volunteers. Initial results have been very promising and neurovirulence tests in mice suggest that these chimaeric viruses may be even safer than the 17D parent. However immunization of primates indicated that levels of neutralizing antibodies to some serotypes were much lower than others (Guirakhoo et al 2002) and individual animals showed marked differences in their response to each of the four serotypes. Similar chimaeric vaccines have been developed against JE, WNV and St. Louis encephalitis (SLE).

Other groups have used cell-adapted DENV-2 viruses or dengue viruses with genetically engineered attenuating mutations as genetic backbones. The naturally attenuated DENV-2 virus developed at Mahidol University has been used as a backbone to produce chimaeric viruses in which their preM and E genes have been replaced with genes from each of the other three serotypes. This technology has also been used to develop a WNV vaccine. The group at NIAID has developed attenuated isolates from all four serotypes by deleting a stem-loop structure from the 3′ non-coding region of the parent virus. The attenuated dengue 4 virus has been used as a backbone to make chimaeric viruses by exchanging its preM and E genes for those of each of the other dengue serotypes. A WNV vaccine has also been developed using this attenuated dengue four virus. Furthermore, experimental chimaeric TBE vaccines using both the naturally attenuated DENV-2 virus and the genetically modified DENV-4 virus have been studied. A more radical approach has been adopted by the commercial company, Hawaii Biotech. This vaccine is based on the ectodomain of the E protein, combined with the NS1 protein. Proof-of-principle has been demonstrated with all these experimental vaccines and clinical trials are in progress with many of them.

New approaches to vaccine design

Vaccines based of the envelope protein

In addition to the whole genome vaccines described above there have been numerous attempts to use other strategies to produce vaccines containing the envelope proteins of several flaviviruses, especially dengue. Studies involving recombinant DNA viruses (reviewed in Stephenson 2001), and purified peptides (summarized in Volpina et al 2005) all report some measure of success in laboratory studies.

Experimental genetic vaccines using plasmids containing the preM and E genes of many flaviviruses have demonstrated protection in animals (summarized in Timofeev et al 2004). Those against JE, WNV, Murray Valley encephalitis (MVE), dengue, TBE and louping ill all show protection in experimental murine models. Moreover, the JE vaccine has been shown to protect against disease in pigs, and that against WNV to protect against disease in horses. Studies on non-human primates are limited, but work on TBE and dengue has been reported. In all the work described above a significant immune response has only been demonstrated with products which contain both the preM and E genes, as co-expression of preM protein is essential to ensure that when the E protein passes through the acid environment of the Golgi it does not unfold.

Vaccines based on non-structural proteins

Concerns about immune enhancement of disease and the relatively poor induction of CMI responses by the E protein have stimulated interest in exploring other virus proteins as potential vaccine candidates. The most studied of these is the NS1 protein as antibodies can be found in experimentally infected animals, vaccinees and patients with clinical disease or in convalescence. Moreover, it produces a good CMI response and was described in the early literature as the 'complement fixing antigen'. Our own work with the NS1 protein expressed by a defective recombinant adenovirus has shown it to produce high levels of antibody, proinflammatory cytokines and efficient protection in an experimental murine model, not only against the homologous virus but also against several other antigenically related human pathogens (Jacobs et al 1994). Genetic vaccination with the NS1 gene and purified peptides derived from a number of flaviviruses have been shown by several workers to elicit protection (summarised in Timofeev et al 2004, Volpina et al 2005).

Recently however some doubt has been expressed about the desirability of NS1 as a vaccine candidate as high levels of NS1 in the blood have been associated with disease severity (Alcon et al 2002). Furthermore cross-reactivity between NS1 antibodies and proteins on the walls of cells lining blood vessels have been reported (Lin et al 2003). Nevertheless, the latter phenomenon is unlikely to be clinically significant, as the presence of clinical symptoms does not correlate with the persistence of NS1 antibodies in the blood. The hypothetical toxicity of NS1 is also uncertain, as high levels may be the result of severe disease and not its cause. Moreover toxic effects have never been observed in experimental animals vaccinated with purified protein or with recombinant viruses expressing NS1 protein.

Although NS1 is a better inducer of CMI than the E protein, NS3 is significantly more effective than either and several workers have reported that NS3 is the most potent stimulator of CMI in dengue patients and in convalescents (Lobigs et al 1994). However the recent seminal work by Screaton and his colleagues describing the phenomenon of 'original antigenic sin' involved NS3-specific T cells. In these studies on cells taken from dengue patients, excessive T cell activation, coupled with rapid cell death was observed with domination of the cellular immune response by cells with a low affinity for the infecting virus. Thus the T cell epitopes on NS3 appear to have significant antigenic similarity which could give rise to an ineffective immune response to a secondary infection (Mongkolsapaya et al 2003).

Barriers to vaccine delivery

Many concerns (e.g. reversion to virulence and recombination with other viruses) about the implementation of new vaccines, especially those against dengue, can be overcome by careful design, good quality pre-clinical research and careful monitoring during clinical trials. However, recent experience in many industrialized countries has shown that unpredictable societal issues can be a bigger barrier to successful vaccination strategies (Amanna & Slifka 2005). Political atrocities and influential local religious leaders have had a deleterious effect on polio vaccination campaigns in several countries. Fears of links between multiple sclerosis (MS) and hepatitis B vaccination, chiefly in France, have reduced uptake in those countries. The long running claims by a handful of workers that MMR vaccination is associated with Crohn's disease and autism has seen a dramatic reduction in vaccination rates in some parts of the UK and a consequential rise in measles cases. This has occurred in the face of universal and global refutation of the link from nearly every respected body of scientific opinion both in the UK and overseas (Stephenson 2002). Furthermore, the high media profile given to the introduction of genetically modified foods has resulted in many genetically engineered products being given the 'Frankenstein' label (Rowland 2002). Cost, lengthy regulatory procedures and fear of crippling litigation for occasional adverse events have led to an increasing risk-averse culture when decisions have to be made about vaccination policies.

Summary and conclusions

The development of successful flavivirus vaccines, especially those against dengue, continues to present us with significant challenges. As with all RNA viruses, vaccines based on live attenuated agents are subject to rapid genetic mutation and recombination, giving rise to concerns about reversion to a virulent phenotype. However, these concerns can be allayed by careful genetic design, thorough preclinical research and extensive human trials which employ modern molecular techniques to assess the frequency of these events. Dengue virus occurs as four distinct serotypes, all of which occur in most endemic areas. Therefore any vaccine must contain all four components and for any live vaccine, whether made by traditional or new molecular technologies, the replication of each of these four components and the immune response to it must not suppress that of any other component in the vaccine. As significant cross-reaction between serotypes can be demonstrated for both B cell and T cell epitopes this could be difficult to achieve. For any vaccine dependent on stimulating responses to the major virion envelope protein, the spectre of generating enhancing antibodies will always be a potential threat unless a complete and balanced immune response to all four serotypes can be guaranteed. Even if this is achieved the kinetics of antibody decline must also be matched to prevent an excess of enhancing over neutralizing antibodies occurring. These concerns have lead to the development of vaccines based on non-structural proteins such as NS1 and NS3. Nevertheless, several vaccines, developed through conventional technologies or by genetic engineering, are undergoing commercial development and many of the initial assessments contain promising results.

Infections with dengue viruses continue to present a major and escalating global public health problem. Vector control programmes are expensive, largely unsuccessful or of only short-term benefit, and thus vaccine development continues to be the most effective strategy for tackling this problem. The complex pathogenesis of dengue infections and the apparent involvement of the immune response in severe disease has raised questions over the safety and efficacy of every vaccine design strategy developed so far. However, when industrialised nations have been faced with theoretical threats of the same magnitude as dengue much less risk-averse policies have been adopted. Following the terrorist attack on September 11th 2001 in New York, many countries became fearful of a terrorist attack using smallpox. Consequently large stocks of vaccine were ordered and administration policies developed, even though the vaccine had not gone through the normal regulatory pathway and the incidence of adverse events where above the accepted norm (Belongia & Nalway 2003). More recently several governments have purchase stocks of an H5N1 influenza vaccine, even though its efficacy against a pandemic influenza strain is unknown (Daems et al 2005). Furthermore regulatory agencies in Europe are actively considering proposals to accelerate their normal procedures to enable the rapid (less than 6 months) introduction of a human pandemic influenza vaccine. In both these cases the risk of not vaccinating the population was judged to outweigh all the other risks associated with vaccination. The question then arises—how bad does dengue have to get before less risk-averse policies are adopted to dengue vaccination?

In this rather pessimistic environment there has been one recent encouraging development. At a meeting in Vietnam in 2001 (Almond et al 2002) it was recognized that an important step towards accelerating the introduction of a dengue vaccine would be the establishment of an international network of phase III clinical trial centres. This meeting lead to the establishment of the Paediatric Dengue Vaccine Initiative. Has the time now come to stop worrying about theoretical objections to the use of experimental dengue vaccines and, after suitable safety testing, assess their efficacy in the populations most at risk?

References

Alcon S, Talarmin A, Debruyne M, Falconar A, Deubel V, Flamand M 2002 Enzyme-linked immunosorbent assay specific to Dengue virus type 1 non-structural protein NS1 reveals circulation of the antigen in the blood during the acute phase of disease in patients experiencing primary or secondary infections. J Clin Microbiol 40:376–381

Almond J, Clemens J, Engers H et al 2002 Accelerating the development and introduction of a dengue vaccine for poor children, 5–8 December 2001, Ho Chi Minh City. Vaccine 20:3043–3046

Amanna I, Slifka MK 2005 Public fear of vaccination: separating fact from fiction. Viral Immunol 18:307–315

Belongia EA, Naleway AL 2003 Smallpox vaccine: the good, the bad, and the ugly. Clin Med Res 1:87–92

Brotherston JG, Boyce JB 1969 Vet Rec 84:514–515

Daems R, Del Giudice G, Rappuoli R 2005 Anticipating crisis: Towards a pandemic flu vaccination strategy through alignment of public health and industrial policy. Vaccine 23:5732–5742

Freestone DS, Ferris RD, Weinberg AL, Kelly A 1977 Stabilised 17D strain yellow fever vaccine: dose response studies, clinical reactions and effect on hepatic function. J Biol Standard 5:181–186

Grosenbaugh DA, Backus CS, Karaca K, Minke JM, Nordgren RM 2003 Equine vaccine for West Nile virus. Dev Biol 114:221–227

Guirakhoo F, Pugachev K, Arroyo J et al 2002 Viremia and immunogenicity in nonhuman primates of a tetravalent yellow fever-dengue chimeric vaccine: genetic reconstructions, dose adjustment, and antibody responses against wild-type dengue virus isolates. Virology 298:146–159

Halstead SB 1982 Immune enhancement of viral infection. Prog Allergy 31:301–364

Halstead SB, Deen J 2002 The future of dengue vaccines. Lancet 360:1243–1245

Heinz FX, Kunz C, Fauma H 1980 Preparations of a highly purified vaccine against tick-borne encephalitis by continuous flow zonal ultracentrifugation. J Med Virol 6:103–221

Jacobs SC, Stephenson JR, Wilkinson GWG 1994 Protection elicited by a replication-defective adenovirus vector expressing the tick-borne encephalitis virus non-structural glycoprotein NS1. J Gen Virol 75:2399–2402

Jacobs M, Young P 2003 Dengue vaccines: preparing to roll back dengue. Curr Opin Investig Drugs 4:168–171

Klockmann U, Bock HL, Franke V, Hein B, Reiner G, Hilfenhaus J 1989 Preclinical investigations of the safety, immunogenicity and efficacy of a purified, inactivated tick-borne encephalitis vaccine. J Biol Standard 17:331–342

Lin CF, Lei HY, Shiau AL et al 2003 Antibodies from dengue patient sera cross-react with endothelial cells and induce damage. J Med Virol 69:82–90

Lobigs M, Arthur CE, Muellbacher A, Blanden RV 1994 The flavivirus non-structural protein NS3 is a dominant source of cytotoxic T cell peptide determinants. Virol 202:195–201

Monath TP 2002 Japanese encephalitis vaccines: current vaccines and future prospects. Curr Topics Microbiol Immunol 267:105–138

Mongkolsapaya J, Dejnirattisai W, Xu X et al 2003 Original antigenic sin and apoptosis in the pathogenesis of dengue haemorrhagic fever. Nature Med 9:921–927

Ng T, Hathaway D, Jennings N, Champ D, Chiang YW, Chu HJ 2004 The anamnestic serologic response to vaccination with a canarypox virus-vectored recombinant West Nile virus (WNV) vaccine in horses previously vaccinated with an inactivated WNV vaccine. Vet Ther 5:251–257

Popov OV, Sumarakov AA, Shkol'nik RYa, El'bert LB, Vorob'era MA, Krutyanskaya GL 1985 Study of the reactogenicity and antigenic activity of a chromatographic cultural purified concentrated inactivated dry vaccine against tick-borne encephalitis, Zh. Mikrobiologii, Epidemiologii, Immunobiologii 62:34–39

Rothman AL 2003 Immunology and immunopathogenesis of dengue disease. Adv Vir Res 60:397–419

Rowland IR 2002 Genetically modified foods, science, consumers and the media. Proc Nutr Soc 61:25–29

Seligman SJ, Gould EA 2004 Live flavivirus vaccines: reasons for caution. Lancet 363: 2073–2075

Stephenson JR 1988 Flavivirus vaccines. Vaccine 6:471–480

Stephenson JR 2001 Genetically modified viruses; vaccines by design. Curr Pharmaceut Biotech 2:47–76

Stephenson JR 2002 Will the current measles vaccines ever eradicate measles? Expert Rev Vaccines 11:355–362

Stephenson JR 2005 Understanding dengue pathogenesis: implications for -vaccine design. Bull WHO 83:308–314

Stephenson JR, Bailey N, Lee JM, Shepherd AG, Melling J 1991 Adjuvant effect of HGH with an inactivated flavivirus vaccine. J Inf Dis 164:188–191

Stephenson JR, Lee JM, Easterbrook LM, Timofeev AV, Elbert LB 1995 Rapid vaccination protocols for commercial vaccines against tick–borne encephalitis. Vaccine 13:743–746

Timofeev AV, Butenko AV, Stephenson JR 2004 Genetic vaccination of mice with plasmids encoding the NS1 non-structural protein from TBE virus and dengue 2 virus. Virus Genes 28:81–93

Venugopal K, Gould EA 1994 Towards a new generation of flavivirus vaccines. Vaccine 12:996–975

Volpina OM, Volkova TD, Koroev DO et al 2005 A synthetic peptide from the NS1 non-structural protein of tick-borne encephalitis virus induces protective immune response against fatal encephalitis in an experimental animal model. Virus Res 112:95–99

DISCUSSION

Canard: I have a comment on the dirty vaccine for smallpox. No one knows how to cope with the vaccination accidents, which are supposed to be quite high, at 1 out of 100 000 people. There is an antiviral drug for adverse effects, though, and this could also be applied to dengue if we accept an imperfect vaccine. The issue is different with dengue because of the secondary infection. If you monitor people, looking for an adverse effect, and you have an antiviral drug, it is a lot easier. Even if the viraemia is short, you can be there. With the secondary infection, the problem is again the short window for seeing the viraemia.

Stephenson: I'm not saying it is going to be easy. This isn't just a question of clinical judgement or science, it is a question of how society looks at these things in general. In the UK, litigation over pertussis vaccine has driven companies away from vaccine development. This is a bigger question for governments, who need to give clear guidance on no fault compensation for vaccine adverse events. Unless we get this pharmaceutical companies will be nervous about developing vaccines because all their profits could wiped out by one lawsuit. Your point about antiviral drugs is well made. It could be that we need both, and when there are problems with vaccine supply we can come in with the antiviral drug. This is the strategy the UK government will eventually take with pandemic flu. Everyone is doing both.

Hombach: One needs clear communication over risk. We don't want to be risk taking, but rather minimize risk. But it is clear that there is no such thing as a zero risk product. It would be beneficial to have a vaccine injury compensation system like that in the USA. One important issue that we must keep in mind is that current vaccines have mostly been introduced in industrialized countries where there are reasonably strong post-marketing surveillance schemes. Now we are talking about vaccines that might first be introduced in developing countries. We need to develop the infrastructure, regulatory mechanisms and post-marketing surveillance capacity in these countries for follow-up. For a vaccine against dengue, this will be extremely important.

Stephenson: Communication of risk has been difficult. We have been unsuccessful in the UK over MMR, despite our best efforts. The question I was posing is one of risk versus benefit. Just from this meeting, I wonder whether we have got this right. Everyone seems very risk averse in terms of putting the vaccines through the regulatory systems. How bad does dengue have to get before we become less risk averse?

Keller: What is the status of these two vaccines that GSK and Sanofi have?

Stephenson: They are going through the normal pathway. Scott Halstead may know where they are in terms of progress.

Halstead: There's bad news and good news. With the Mahidol tetravalent dengue vaccine, when it was combined as a mixture, the dengue 3 component consistently interfered with the immune response. The company (SanofiPasteur) never assigned a high enough priority to this vaccine that they could invest in fixing the problem. Finally they have given up, so it is not on the Sanofi development list. The GSK vaccine is very much alive, though. It is not perfect—the seroconversion rates of all four serotypes from a single dose may be 25%—so it is being administered and tested as a two-dose vaccine at a long interval. There isn't a lot of pull and push in these big companies. They have completed a trial in 15 month old infants. Imperfect as it may be, there seems to be a head of steam behind this vaccine, and we may see the first phase III trial in 2006. Meanwhile, the NIH vaccine is progressing: they have taken a page out of the US Army's book and aren't spending a lot of time in preclinical studies, but are testing candidates in people. They have already tested vaccines against all four viruses independently. Even though things aren't moving as fast as many would like, progress is respectable. I'd like to say a few words about the Pediatric Dengue Vaccine Initiative (PDVI). One of the first papers commissioned by the PDVI was an attitude survey of the major countries in this region.

Of all the possible vaccine options, dengue was top of the list. This disease may not kill a lot of people, but it upsets people seriously. You have already described some of the regulatory, immunological and political issues. As we prepared the proposal for the Gates Foundation, what haunted me the most was the prospect that a dengue vaccine might fail a regulatory test, or there might be negative events observed within the first 10 000 vaccinees. This would be enough, especially in countries where the legal profession is active or where there is political sensitivity, to end the project. We could have a great dengue vaccine that could fail because of adverse events, and there wouldn't be the data base for us to argue otherwise. So, in our effort, we are trying to build a database of answers to the key questions. We want to know what the dengue virus entry mechanisms are, and how the antibody interferes to protect against dengue infection. To do this we have to go back to square-one—to autopsies. We don't know definitively where dengue viruses replicate. We need good animal models for this. There are several PDVI colleagues here and we are part of an integrated effort to take the virus apart to look at mechanisms of attachment and entry, and we are moving in the direction of antibodies. We want to discriminate between good antibodies and bad antibodies. One of the things that would help a tetravalent dengue vaccine is if we could immunize people and have a test to discriminate between protective and non-protective antibody. If we could do this we might be ahead of the game. A good efficacy trial will do away with a lot of doubts. The worst news I have had in my several trips to Cuba was when I visited in 1997 and found that adults who had severe dengue haemorrhagic fever (DHF) had all been infected with dengue 1 in 1977.

Stephenson: I thought originally the main thrust of the PDVI was to stimulate phase III trials. Is that still on the cards?

Halstead: No, we are different from the other Gates' initiatives which were funded to bring vaccines through phase I, II and possibly III. We have two big pharmaceutical companies that have their own budget and domain, GSK and Sanofi. They have never asked for or showed any interest in Gates money to support phase III trials. We are trying to smooth the way so that we can negotiate some of the regulatory issues that are certain to come up in phase III.

Heinz: The concern with dengue vaccines has been the same for decades. This is that immunization would leave some of the vaccinees with an immunological status that not only fails to prevent infection but also increases disease by some mechanism. There is no clinical trial that will sort that out. It will only be long-term follow up that can resolve this question. What about the liability issue with companies producing such a vaccine?

Evans: I think the liability issues are huge. In the USA during development there is no protection from liability. It is a real problem for developing new vaccines in the USA. On the other hand, look at the Japanese encephalitis (JE) experience. China decided to produce an unbelievably toxic vaccine that would never be licensed by any other country and they essentially wiped JE out. They reduced JE to an extremely low level with a field trial where they used 40 000 people and had 20 cases. Doing dengue sounds easy to me! We should look at places where people are willing to take that risk ratio. The GSK rotavirus vaccine has been licensed in Mexico, where the field trials were done.

Hombach: You have to do the trials in the disease endemic countries if you do field studies, so we won't be doing this in the USA with dengue. It also depends on what you want to prevent: dengue fever, or severe dengue. There are many issues. You need to develop reasonable assumptions: what type of profile does the vaccine need to have? From the Mahidol vaccine there is a cohort of 120 children who have been immunized who were followed up for six years. There is no real alternative. I am not sure that the situation is all that different with dengue from other vaccines. In rotavirus the vaccine was withdrawn because of cases of intussusception. These things can happen, and these days vaccine development requires large-scale efficacy trials, which is expensive and time consuming. Finally, let me say the statement on the Chinese JE vaccine is very harsh and unjustified.

Halstead: There are a lot of worries. There are some interesting, innovative things we could do that wouldn't cost a lot of money. One would be to take a vaccine, no matter how good or bad, and vaccinate until you stop transmission. It may sound fanciful, but smallpox and polio have been eradicated this way.

Stephenson: The answer to the question of liability doesn't rest with companies, but with government departments. There are some countries where they would be prepared to grasp that nettle. What Scott Halstead has described is attractive, but someone has to cover the liability of the company that will manufacture the vaccine to do that.

Halstead: You might find that the pressure on the part of countries like China and Indonesia to vaccinate their population, so that the reverse would occur of what you are seeing in Africa, where there would be demand for product even though it is not licensable in the UK.

Gubler: Vietnam is one of those countries that is anxious to get a vaccine. They are currently thinking of doing this. A small biotech company has licensed the CDC chimeric vaccine and is looking at the possibility of doing clinical trials.

Farrar: Yes, in Vietnam it is high on the list of priorities. The Cuban data are worrying though. Follow-up following vaccine has to be for the long term (5–20 years) and I doubt that regulatory approval would happen without that follow-up. While we are waiting for this vaccine we can do quite a lot for patients, and in the modern world with better fluid management and nursing care the dengue mortality rate should be close to zero. You can make a big difference to patients by looking after them well and with experience.

We always write in every review that immunoprotection against all four serotypes is needed at the time of vaccination. In real life, though, this is not what happens. Most people have sequential infections and probably don't generate immune protection against all four serotypes and yet don't get disease. Are you sure you need cross-protection against all four at the start?

Halstead: No, that's what we need to find out. In real life these experiments are happening all the time, so we can observe this.

Zinkernagel: If we could know that to be immune against 1–4 prevents haemorrhagic fever, (which is possible) then you would go for all four serotypes. Then the question is at what titre do you need to keep these to have protection for the next 50 years? These are all details we don't know.

A genomics approach to understanding host response during dengue infection

Martin L. Hibberd*, Ling Ling*, Thomas Tolfvenstam*, Wayne Mitchell*, Chris Wong*, Vladimir A. Kuznetsov*, Joshy George*, Swee-Hoe Ong*, Yijun Ruan*, Chia L. Wei*, Feng Gu†, Joshua Fink†, Andy Yip†, Wei Liu†, Mark Schreiber† and Subhash G. Vasudevan†

**Genome Institute of Singapore, Biopolis, Singapore 138670 and †Novartis Institute for Tropical Diseases, 10 Biopolis Road #05-01 Chromos, Singapore 138670*

Abstract. Dengue infection results in a wide clinical spectrum, ranging from asymptomatic, through fever (DF), to the life threatening complications haemorrhagic fever (DHF) and shock syndrome (DSS). Although we now understand that factors such as repeat infections and the type or magnitude of the host response are important in determining severity, the mechanisms of these actions remain largely unknown. Understanding this host–pathogen interaction may enable outcome prediction and new therapy options. Developments in biology now allow a 'systems approach' to be applied to this problem, utilizing whole genomes of both human and virus, *in vitro* and *in vivo* to enable a more complete picture of their interplay to be built up. We have developed a chip-based approach to viral sequencing, to increase efficiency and enable large numbers of genomes to be completed, together with a web-based interpretation tool. We have also applied human whole genome expression arrays (24 000 genes) to characterize the types of host response made to infection and plan to investigate the role of host variation using human whole genome genetic association studies in the future. These technologies have identified novel host pathways involved in viral replication *in vitro*, and also host immune responses, such as the interferon signalling pathway, that are influenced by viral sequence. This data will be further refined through interlinking with similar data obtained from a large study of dengue patients, initiated in Singapore, that is able to look at the early host response to infection.

2006 New treatment strategies for dengue and other flaviviral diseases. Wiley, Chichester (Novartis Foundation Symposium 277) p 206–217

Infection with dengue virus can cause a spectrum of disease ranging from dengue fever (DF) to dengue haemorrhagic fever (DHF) and dengue shock syndrome (DSS). It has been estimated that dengue infection results in 50–100 million cases of DF annually and, at present there are no drugs to treat the disease or preventative vaccines (Kroeger et al 2004). The complexity of the dengue disease process, combining both viral and host factors as well as geographic considerations, has made its understanding very difficult. By comparison, genomic approaches were quickly applied to a more recent virus pathogen, the SARS Coronavirus, where they rapidly enabled a number of key developments to be made. For example at the Genome Institute of Singapore, we applied rapid re-sequencing approaches (Ruan et al 2003, Wong et al 2004, Liu et al 2005) to reveal a detailed map of the introduction of the virus into Singapore and its spread through the community. From this we developed a diagnostic that was later licensed to Roche and sold commercially (Ng et al 2004a). In further work, through the development of a microarray host response model system (Ng et al 2004b), we showed a virus directed down-regulation of the interferon pathway, that we also identified as a productive target for inhibiting viral replication using commercially available drugs (Tan et al 2004).

Similarly for dengue, we now have the opportunity to begin undertaking a more complete, systems approach, which should enable new treatment insights. Beginning from a genomics perspective—investigating every gene involved (human and virus) without a prior prediction of their involvement—allows a data-directed hypothesis to be built up. While still at an early stage, this work aims to put together some important components of the 'dengue virus–human host' interaction jigsaw.

To contribute to the interpretation of our *in vitro* work we have also partnered with the 'Early DENgue (EDEN) study', a prospective, longitudinal study of Singapore patients with fever of less than 3 days (and without signs of upper respiratory tract infection) to capture patients at the early stage of dengue infection and follow their disease development. This study was established specifically to link clinical data to research samples that include those that will allow us to undertake whole genome approaches to both the virus and the host; and will permit us to develop a new level of systems biology interrogation *in vivo.*

Identification of host pathways regulating viral replication using *in vitro* analysis

In order to identify host factors involved in dengue replication, we have undertaken an analysis of the changes in gene expression in a human hepatocytic cell line following infection with dengue virus. The HepG2 cell line was infected with TSV01, a serotype 2 dengue virus (GenBank AY037116; McBride & Vasudevan 1995), and cells harvested over a succession of time points up to three days post-infection. Comparisons between live virus infection and heat-inactivated virus were used to provide information about the host factors involved in virus replication. We analysed the changes in host response using an in-house microarray presenting 19800 Compugen oligos and compared the relative expression of the host compared to a universal reference sample (Stratagene). Genes that were

TABLE 1 In-house, Compugen, microarray-identified pathways, significantly differently expressed between live and heat inactivated dengue 2 strain TSV01 at 72 h following infection of HepG2 cells

Biological Process	*Genes in Pathway*	*Genes found to be significant*	*Pathway P value*
Interferon-mediated immunity	41	9	7.19E-11
Immunity and defence	786	26	3.74E-10
NF-kappaB cascade	43	3	5.43E-03

Genes differentially expressed were identified by SAM analysis and then used to identify pathways using the ABI Panther analysis software. The statistical analysis (pathway *P* value) column represents the probability that the genes in the SAM list occurred in this pathway by chance.

significantly differentially expressed between live and heat inactivated virus were determined using a 'statistical analysis of microarray' (SAM) approach (*http://www-stat.stanford.edu/~tibs/SAM/*, Tusher et al 2001). SAM-identified gene lists were initially analysed to identify specific host response pathway analysis using Applied Biosystems Panther microarrray analysis tools (*https://panther.appliedbiosystems.com*). This tool determines the probability that a specific pathway is involved by comparing the total number of genes in the pathway investigated, with the number of genes in that pathway identified by SAM. Comparing TSV01 with heat-inactivated TSV01 we identified three significant pathways (Table 1).

More detailed manual analysis identified several other interesting gene clusters, including an integrin-mediated cell adhesion pathway, the ubiquitin proteosome pathway, p53-mediated apoptosis pathways, and cell structure and motility pathways. Specific genes that were up-regulated more than threefold included 2′-5′oligoadenylate synthetase 2, β-tubulin, complement component 3 and cytokeratin 8, consistent with previous reports (Warke et al 2003, Liew & Chow 2004, Moreno-Altamirano et al 2004). We are now undertaking quantitative PCR confirmation of the involvement of these genes and will assess some of them as potential drug targets in appropriate virus infection assays.

Ex *vivo* investigation of the innate immune response

Monocytes/macrophages and dendritic cells are the proposed reservoir of dengue viral replication, while innate and adaptive immune responses all contribute to the immunopathogenesis of the infection. Since dengue is thus mainly a blood-borne disease (also we use peripheral blood to investigate *in vivo* response in our prospective study (EDEN)), we sought to establish an *ex vivo* primary human peripheral blood mononuclear cell (PBMC) infection model system to assess the host response to infection in the naturally infected cells.

TABLE 2 Illumina microarray-identified pathways, significantly differently expressed in PBMCs at 24 h following infection and mock infection with dengue 2 strain TSV01

Biological process	*Genes in pathway*	*Genes found to be significant*	*Pathway P value*
Immunity and defence	1318	45	9.21E-11
Interferon-mediated immunity	66	11	5.81E-10
Macrophage-mediated immunity	118	6	2.87E-03

Genes differentially expressed were identified by SAM analysis and then used to identify pathways using the ABI Panther analysis software. The statistical analysis (pathway *P* value) column represents the probability that the genes in the SAM list occurred in this pathway by chance.

As in other studies, we observed that the monocyte/macrophage/dendritic cells are the principle targets to take up dengue virus. We compared the host response of the PBMCs over several time points to infection and mock infection with TSV01. Using SAM analysis as before, we observed 254 genes that were differently expressed at 24 h and identified three key pathways with the Panther Pathways Analysis package (Table 2).

Interestingly, we observed a large overlap at the gene level with the *in vitro* host response seen in the HepG2 cells. These *in vitro* models will enable us to do detailed analysis of the genes and pathways involved in the host response as they permit the use of specific manipulations, including inhibiting them with selected siRNA, that may reveal key host components involved in virus replication.

The early *in vivo* host response to dengue infection

The pathogenesis of human dengue disease is not well characterized, particularly during the early stages of the infection. In these early stages, dengue is difficult to distinguish from other causes of fever, ensuring that it is rarely diagnosed at this stage. In many countries where dengue is endemic there are also resource issues, meaning that it is only the most sick, those that are admitted to hospital, that are recognized as having the disease. And yet these early stages can give us many clues as to how the human host deals with the infection as the majority of infections do not result in hospitalization.

Components of the early markers of dengue may be predictive of disease progression or outcome and become useful targets for therapeutic approaches.

Using the Illumina microarray platform, we investigated the expression pattern of 24 000 genes in eight fever subjects with, and eight fever subjects without dengue fever (patients from the EDEN prospective study). Blood samples from dengue

PCR-positive patients at their first time point were compared to blood samples from PCR-negative patients at the same time point and further controlled by selecting only those genes that were also differentially expressed compared to the sample 3 to 4 weeks later.

Using this approach, 262 genes were found to be significantly differentially expressed in dengue patients at their first time point, using a SAM analysis developed by Illumina, with a false discovery rate prediction of less than 1. Using the Panther system (described above), the largest grouping of these genes were characterized as immunity and defence and in particular interferon genes (Table 3). Interestingly, this list included more than 15% of the genes identified by the *in vitro* work in HepG2 cells, including eight of the interferon regulation genes.

This process was repeated for the second time point of the PCR positive patients (3 days later), finding 964 genes to be significantly differentially expressed, again with a very low false discovery rate. In this list, the most significantly represented genes were in the ubiquitin proteasome pathway and there was no significant grouping of immune genes. At this time point the list contained more than 20% of the genes identified by the *in vitro* work in HepG2 cells. This second patient sampling may overlap with previous studies of 'early' dengue, which tend to use entry to hospital as the first time point (Avila-Aguero et al 2004), and agrees with many of these studies finding low levels of interferon involvement.

Future work will assess the differences in host response between primary infection and secondary infection, between mild and severe disease, between different serogroups and between similar strains with different severity indicators.

TABLE 3 Illumina microarray identified pathways, significantly differently expressed between fever patients, with and without dengue

Biological process	*Genes found to be significant*	*Pathway P value*
Day 1		
Interferon-mediated immunity	11	6.35E-10
Immunity and defence	33	3.85E-05
Day 3		
Ubiquitin proteasome pathway	14	1.38E-04

Genes differentially expressed were identified by SAM analysis and then used to identify pathways using the ABI Panther analysis software. The statistical analysis (pathway *P* value) column represents the probability that the genes in the SAM list occurred in this pathway by chance.

A role of viral variation in determining the host response?

The role of dengue virus variation in determining host responses and clinical outcome to infection is largely unknown due to the paucity of data on the viral genome structure. Previous studies have concentrated on structural genes, often from an epidemiological perspective. With the development of high-throughput capillary sequencing for viral genomes, investigation of their functional aspects becomes possible, but only with large commitments of time and money.

We are developing a rapid and cost-effective mechanism for dengue resequencing, utilizing an array platform approach that has already been developed to track the genetic diversity of the SARS CoV (Wong et al 2004), using high-density custom microarrays manufactured by Nimblegen Systems. For the pilot array, we first created reference consensus sequences to the four serotypes of dengue, derived from all available dengue genomes in GenBank. Resequencing probes were synthesized *in situ* onto the array using our previous bioinformatics approaches. While the probes were able to distinguish between the four serotypes of dengue, the resultant sequence was not complete, giving sequencing errors primarily in regions where there were multiple polymorphisms within the same resequencing probe. To overcome this problem, we generated a second consensus sequence derived from viruses isolated in Southeast Asia and conducted allelic association analysis of single nucleotide polymorphisms which occur in close proximity using a novel bioinformatics approach. This new approach is now able to sequence over 95% of the genome (Fig. 1), with further improvement possible with a development of the bioinformatics interpretation of the hybridization pattern. With this new tool, rapid and cost effective whole genome sequencing is becoming a reality, allowing the investigation of viral variation in host–pathogen interaction to become routine.

To enable the community of dengue researchers to make use of this new data flow we have developed the DengueInfo website and database (*www.dengueinfo.org*) that is specifically focused on the genomic aspect of dengue biology. It will serve as a repository for full length and near full length dengue genome sequences. It also unifies the sometimes poorly standardised GenBank annotations. The website offers a powerful query interface that allows one to find sequences that match very flexible search criteria. To help researchers investigate any link between genome and severity we also store, where available, some clinical information such as disease outcome.

In a preliminary investigation, we compared the genomic sequence of the prototype strain New Guinea C (NGC) dengue 2 virus, with a clinical isolate TSV01 (GenBank AY037116) and identified nine variations leading to non-conserved amino acid substitutions (Fig. 2). We then investigated possible consequences of these genomic variations on the host response using our HepG2 *in vitro* model and microarray readout (described above). SAM analysis revealed 88 genes that were

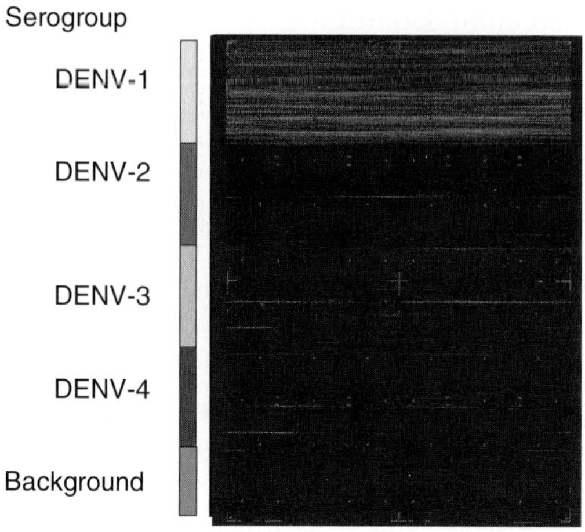

FIG. 1. Dengue resequencing by microarray hybridization. The figure shows the result of a hybridization to a dengue 1 PCR amplification to a chip containing probes to all four serogroups (partitioned into sections, as labelled to the left). In this hybridization, accurate calls were made to 98.3% of the whole genome.

TABLE 4 Expression profile differences between NGC and TSV01 in HepG2 cells

Biological process	*Genes in pathway*	*Genes found to be significant*	*Pathway P value*
Interferon-mediated immunity	41	9	8.36E-13
Immunity and defence	786	18	2.78E-08

differentially expressed between these strains, predominantly because NGC initiated a relatively small host response (Table 4). Interestingly, the most statistically important difference was in interferon mediated immunity (Table 4).

We predict that there are likely to be naturally occurring variations in the dengue genome, most likely in the non structural genes, that manipulate the host response. Future work will look at this variation and also investigate specific variations using a site directed approach.

Conclusions

The work presented here points towards the beginning of a deeper understanding of the dynamics of this human–virus interaction.

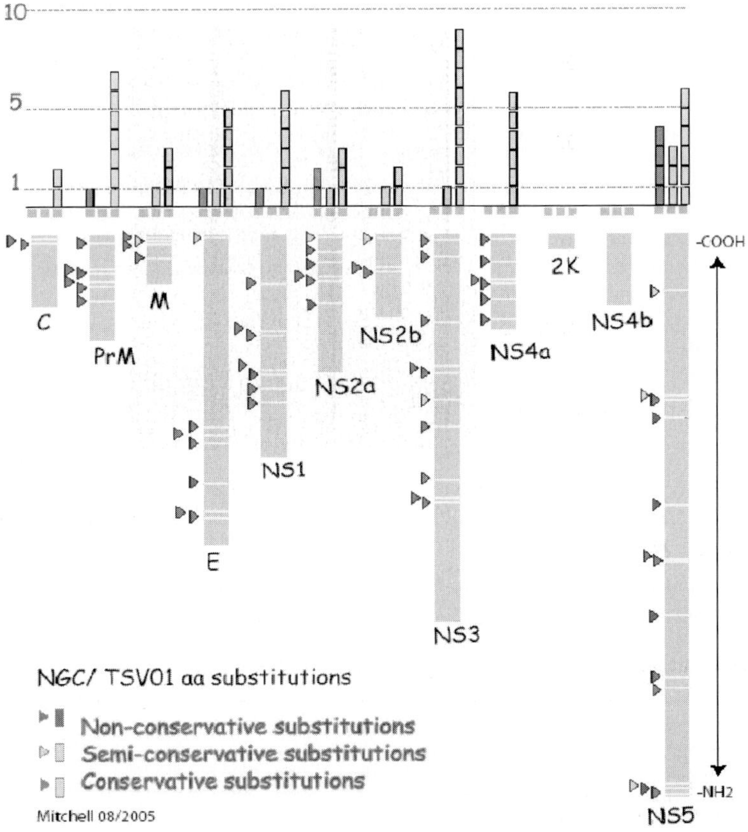

FIG. 2. Protein variation as a result of genomic sequence difference between dengue 2 strains NGC and TSV01.

Acknowledgements

We would like to fully acknowledge our partners in the Singapore EDEN study, in particular the study leader Adrian Ong from Tan Tock Seng Hospital, the study nurse Diana Tan Bee Har and Lee Ching Ng, Tim Hart and Bijon Kumarsil from the Environmental Health Institute and Ooi Eng Eong from the Defense Science Organization. In addition at GIS we would like to thank Frans Verhoef, for Bioinformatic support and LEONG Wan Yee, Aw Poh Kim Pauline and Khoo Chen Ai, for technical assistance. We would also like to thank the participants in this study for their willingness to be involved. This work is in part funded through a grant from the Singapore Tissue Network and partly from a grant from NITD. At GIS the work is funded through A*STAR.

References

Avila-Aguero ML, Avila-Aguero CR, Um SL, Soriano-Fallas A, Canas-Coto A, Yan SB 2004 Systemic host inflammatory and coagulation response in the Dengue virus primo-infection. Cytokine 27:173–179

Kroeger A, Nathan M, Hombach J 2004 World Health Organization TDR Reference group on Dengue disease update. Nat Rev Microbiol 2:360–361

Liew KJ, Chow VT 2004 Differential display RT-PCR analysis of ECV304 endothelial-like cells infected with dengue virus type 2 reveals messenger RNA expression profiles of multiple human genes involved in known and novel roles. J Med Virol 72:597–609

Liu JJ, Lim SL, Ruan Y et al 2005 SARS Transmission pattern in Singapore reassessed by viral sequence variation analysis. PLoS Med 2:162–168

McBride WJH, Vasudevan SG 1995 Relationship of a dengue 2 isolate from Townsville, 1993, to international isolates. Comm Dis Intel Aus 19:522–523

Moreno-Altamirano MM, Romano M, Legorreta-Herrera M, Sanchez-Garcia FJ, Colston MJ 2004 Gene expression in human macrophages infected with dengue virus serotype-2. Scand J Immunol 60:631–638

Ng FPL, Wong MS, Koh S et al 2004a Detection of SARS coronavirus in blood of infected patients. J Clin Micro 42:347–350

Ng FPL, Hibberd ML, Ooi EE et al 2004b A human in vitro model system for investigating genome-wide host responses to SARS coronavirus infection. BMC Infec Dis 4:34–45

Ruan Y, Wei CL, Ling AE et al 2003 Comparative full-length genome sequence analysis of 14 SARS coronavirus isolates and common mutations associated with putative origins of infection. Lancet 361:1779–1785

Tan ELC, Ooi EE, Lin CY et al 2004 Inhibition of SARS coronavirus in vitro with clinically approved anti-viral drugs. Emerg Infect Dis 10:581–586

Tusher VG, Tibshirani R, Chu G 2001 Significance analysis of microarrays applied to the ionizing radiation response. Proc Natl Acad Sci USA 98:5116–5121

Warke RV, Xhaja K, Martin KJ et al 2003 Dengue virus induces novel changes in gene expression of human umbilical vein endothelial cells. J Virol 77:11822–11832

Wong CWC, Albert TJ, Vinsensius B et al 2004 Tracking the evolution of the SARS coronavirus using rapid, high-throughput, high-density resequencing arrays. Genome Res 14:398–405

DISCUSSION

Rice: I have a question about the differences between the two different dengue isolates in terms of how they perform. Thinking about this from a virological standpoint, whether it is the HepG2 cells or the pBMCs, do you also monitor the percentage of cells that become infected? With the gene response patterns that you are seeing, how many of those reflect the changes that occur in an infected cell versus cytokine-mediated changes that are occurring in bystander cells?

Hibberd: In the HepG2 cells we are looking at a cell line, which consists of one cell type. We are looking in something very specifically involved in that process.

Rice: Are you doing this under conditions where it is a synchronous high multiplicity infection?

Gu: We have tried to achieve a synchronized wave of infection, but it is very difficult. We infected cells at a high multiplicity of infection (MOI) of 10, and under this condition, about 50% of the HepG2 cells were infected after 48 h.

Rice: In the strain-specific differences, if there are differences in the infectivity with different strains for different cell types in a mixed population experiment, could this be a multiplicity dependent difference, rather than something due to the constellation of genetic changes?

Hibberd: I didn't look at the comparison between them in the PBMC experiment. This was only in the HepG2 cells. We can get a bit of an angle on this by looking at the expression profile over time period in the different strains.

Rice: Or by repeating the experiment with different multiplicities.

Harris: Or sorting. There is some nice work on dendritic cells that show different responses in bystander versus infected cells. If you are doing *in vitro* work you could sort the population.

Jans: I'd like to ask about the arrays. By using arrays of only 24 000 annotated genes, aren't you limiting yourself by leaving out the unknowns?

Hibberd: 24 000 isn't a bad number to look at in an array experiment! There is an option of looking at 24 000 unknown genes; it depends on what you are trying to do. At this stage just getting a grip on anything in this process is important. I am not sure I want to find an unknown gene responding because there would be a lot of work then needed to find out what it is.

Farrar: In our array work with patients, the most striking thing is a difference in the interferon (IFN)γ pathway between non-severe and severe patients. The severe patients have a profound reduction in their IFNγ pathways which we are now trying to correlate with viral load.

Hibberd: This is a quick look-see in our patients, and I would predict that these are relatively mild compared with the patients you are seeing.

Harris: There are a number of sites doing this work simultaneously. It will be interesting to compare the results, and to tease out any differences caused by age or geography.

Hibberd: We do have an earlier time point in our study than most people will be able to look at. Hopefully, this will give us some useful information. In work that I did with SARS coronavirus, we looked directly in patient samples and used minisequencing to look at specific sites over the genome. This technology is designed for human genotyping, so we were able to monitor the amounts of different species within an individual. We found that when we cultured the virus inevitably we ended up with only one strain at the end of the process.

Vasudevan: That information of the various species (quasispecies) should be on the chip; it is just a matter of mining the data in that case.

Evans: Are you sequencing virus directly from serum?

Hibberd: At the moment we are sequencing grown virus, although the technology can be used directly on serum. Because of our early time point, there is a lot of virus in the serum.

Rice: Is the PCR sequence independent or are you introducing some bias by your choice of primers?

Hibberd: There are two sites where a consensus sequence is needed for sequencing, and you are not sequencing those two sites.

Holmes: You showed the severity cluster: was this a cluster of patients with a particular disease syndrome?

Hibberd: The problem was that there wasn't enough clinical information for us to say that, but from the information we got it didn't appear to be a clustering-based.

Holmes: What serotype is that?

Hibberd: Dengue 3. This is whole-genome sequencing.

Rice: Do you also have the capabilities for looking at serum profiles at a protein level, as well as at an mRNA level?

Hibberd: We are saving the samples. We have a fair amount of mass spectrometry technology at our institute so this is an option we'll be looking at in the future.

Rice: The questions that you can think about addressing with these kinds of samples are a moving target. The quality of your database and your sample preservation will be a fantastic resource in the future as some of these technologies get better.

Fairlie: What did you think about the Toll-like receptor showing up on day 1?

Hibberd: It was interesting. Clearly there is some innate immune response to the virus in the blood. DC-SIGN is probably not the only thing that is interacting with the virus.

Jans: Are your array data going to be on the website?

Hibberd: They are likely to head there eventually. The sequencing is already there.

Jans: Eva Harris, will your array data become public?

Harris: The sequencing is going public right away, and the array data will wait for publication of a paper.

Keller: Array data are of relatively limited use. We know many immunological diseases where drugs were made against up-regulated proteins and they do nothing.

Hibberd: What arrays allow us to do is to look without a prior hypothesis. If something comes out we then have to work hard to see whether it is really involved.

Keller: My experience from working 10 years in asthma is that many drugs were made against things that were strongly up-regulated, but most of them had no impact on disease progression.

Hibberd: One of the things I think is important is that when you look at expression profiles it is just an observation. But if you can link that with some genetic information, genetics is telling you what the consequences of these variations are. If you can show that in the host, variation in these genes genetically leads to a variation in some kind of favourable outcome, then you have a much harder angle on this real gene that is involved.

Evans: Do you plan sequencing across the immunological element of the human genome?

Hibberd: Our human genetics program is using a whole-genome approach. We are using 250 000 markers over the whole genome, with an average of about 5 or 6 SNPs per gene in every one of the 24 000 genes we are looking at. This is a prioritizing approach: we can look at the most associated genes that come from that initial screening process. We won't just be looking at the immune genes that come out.

Mouse and hamster models for the study of therapy against flavivirus infections

Nathalie Charlier, Erik De Clercq and Johan Neyts

Laboratory of Virology and Chemotherapy, Rega Institute for Medical Research, Minderbroedersstraat 10, B-3000 Leuven, Belgium

Abstract. Small animal models that are reminiscent of flaviviral disease in human will be instrumental in identifying therapeutic strategies against flavivirus infections. Here we review models in mice and hamsters for the most clinically important flaviviruses: dengue virus, yellow fever virus, West Nile virus, Japanese encephalitis virus and tick-borne encephalitis virus. In addition, models are discussed that employ no known vector viruses such as the Modoc virus. These viruses can be manipulated in BSL-2 laboratories and in infected mice and hamsters they mimic flaviviral disease in human.

2006 New treatment strategies for dengue and other flaviviral diseases. Wiley, Chichester (Novartis Foundation Symposium 277) p 218–232

The *Flavivirus* (family *Flaviviridae*) genus contains (i) viruses that are transmitted by mosquitoes or ticks (arthropod-borne) and (ii) viruses with no known vector (NKV) (Chambers et al 1990). Currently, more than 70 flaviviruses have been reported. All flaviviruses of human importance belong to the arthropod-borne group. The NKV group holds a few viruses which have been isolated from mice or bats and for which no arthropod-borne or natural route of transmission has (yet) been demonstrated (Kuno et al 1998).

One of the most important flaviviruses causing disease in humans is dengue virus (DENV). It is estimated that there are worldwide annually as many as 50–100 million cases of dengue fever and several hundred thousand cases of dengue haemorrhagic fever (DHF) or dengue shock syndrome (DSS), the latter with an overall case fatality rate of about 5% (Gubler 1997, Thomas et al 2003). Yellow fever virus (YFV) is, despite the availability of a highly efficacious vaccine, still a leading cause of haemorrhagic fever worldwide. The World Health Organization has estimated that there are annually 200 000 cases of YF, including 30 000 deaths, of which over 90% occur in Africa (*http://www.who.int/inf-fs/en/fact100.html*). Japanese encephalitis (JE), a mosquito-borne arboviral infection, is the leading cause of viral encephalitis in Asia (Hennessy et al 1996). Approximately 50 000 sporadic and epidemic cases of JE have been reported annually. The infection results in high mortality (30%), and about half of the survivors develop long-lasting neurological sequelae (Kalita & Misra 1998, Misra et al 1998). Tick-borne encephalitis virus (TBEV) is believed to cause annually at least 11 000 human cases of encephalitis in Russia and about 3000 cases in the rest of Europe (*http://www.who.int*) (Gritsun et al 2003). In 1996 an outbreak of West Nile (WN) encephalitis was reported in Romania (Han et al 1999, Tsai et al 1998). In 1999, the disease appeared for the first time in the northeastern USA and has continued to spread across the USA and Canada (*http://www.cdc.gov/ncidod/dvbid/westnile/*). Outbreaks of WN encephalitis occurred in recent years also in Southern Russia and Israel (Lvov et al 2000, Siegel-Itzkovich 2000). Several other flaviviruses, including St. Louis encephalitis virus (SLEV) and the Murray Valley encephalitis virus (MVEV) cause severe disease in human (Kramer et al 1997, McCormack & Allworth 2002).

Here, we review animal models for the most important flavivirus infections. Some models may be readily amenable for the evaluation of antiviral strategies.

Dengue virus

SCID (severe combined immune deficiency) mice reconstituted with human peripheral blood lymphocytes (hu-PBL-SCID mice) and infected subsequently by the intraperitoneal route with DENV-1, developed viraemia, although virus production was highly variable (Wu et al 1995). DENV-4 replicates in SCID mice that

TABLE 1 Mouse or models for dengue virus infection

Model	*Reference*
SCID mice transplanted with human peripheral blood lymphocytes and infected with DENV-1	Wu et al 1995
SCID mice transplanted with human K562 cells and infected with DENV-2	Lin et al 1998
SCID mice transplanted with a human hepatocarcinoma cell line and infected with DENV-2	An et al 1999
SCID mice transplanted with human liver cells and infected with DENV-4	Blaney et al 2002
AG129 mice (IFNα/β and IFNγ receptor genes knockout) infected with DENV-2	Johnson & Roehrig 1999
NOD/SCID mice xenografted with CD34$^+$ cells and infected with DENV-2	Bente et al 2005
A/J mice infected with DENV-2	Huang et al 2000
BALB/C mice infected with DENV-2	Atrasheuskaya et al 2003
BALB/C mice infected with a mouse-neuroadapted strain	Ben Nathan et al 2003

have been transplanted with Huh-7 cells (human liver cells) (Blaney et al 2002). SCID mice that had been engrafted with human K562 cells (an erythroleukemia cell line) and that were subsequently inoculated with DENV-2 developed paralysis and died. A high titre of DENV-2 was found in the brain, which correlated well with the progression of encephalopathy (Lin et al 1998). Mouse-adapted DENV-2 replicates in intraperitoneally infected AG129 mice (which lack interferon [IFN]α/β and IFNγ receptor genes). Infected animals developed neurological abnormalities, including hind leg paralysis and blindness at 7 days, and died within two weeks following infection (Johnson & Roehrig 1999). An et al (1999) transplanted SCID mice with a human hepatocarcinoma cell line (HepG2) and inoculated the animals intraperitoneally with DENV-2 about 7–8 weeks after transplantation. In the early stage postinfection, virus was detected in the liver and in serum of the HepG2-engrafted mice. At later time-points, virus was also detected in the brain, and this was corroborated by the fact that the animals developed paralysis. The mice also presented with thrombocytopenia and a prolonged partial thromboplastin time. Also an increase in hematocrit, blood urea nitrogen and tumour necrosis factor (TNF)α was observed in paralysed mice. TNF-α might be involved in the aetiology of shock in DHF/DSS (An et al 1999).

Huang and colleagues developed another infectious model that mimics to some extent DHF/DSS (Huang et al 2000). A/J strain mice infected intravenously with DENV-2 (PL046), developed manifest thrombocytopaenia and produced anti-platelet antibody. Anti-platelet antibodies lyse platelets in the presence of complement and interfere with thrombin-induced platelet aggregation (Huang et al 2000).

A lethal mouse model for DENV-2 infection was reported in which infected mice developed several clinical, histopathological, virological and haematological similarities to human DENV-2 infection. Young BALB/c mice (haplotype H-2d) infected with DENV type 2 (strain 23085), developed anaemia, thrombocytopaenia, pre-terminal paralysis and shock. Levels of TNFα abruptly and steeply increased 24 h before death (mean at day 6). Serum levels of interleukin (IL)1β, IL6, IL10, IL1 receptor antagonist and soluble TNF receptor I continuously increased with time after infection. A 100% mortality rate was noted in the infected animals; treatment with anti-TNFα serum reduced mortality from 100% to 40% (Atrasheuskaya et al 2003).

The main targets for dengue infection and replication seem to be dendritic cells (mainly monocyte derived) and macrophages (mainly Langerhans). A full repertoire of dendritic cells develops in non-obese diabetic/severely compromised immunodeficient (NOD/SCID) mice xenografted with human cord blood haematopoietic progenitor ($CD34^+$) cells. The NOD/SCID strain of mice lacks C5, resulting in a deficiency in haemolytic complement. Following subcutaneous inoculation, xenografted mice developed fever, rash and thrombocytopaenia (comparable to dengue fever in human). Viraemia peaked between day 2 and 6 post-inoculation. Spleen, liver and skin tested positive for dengue RNA (by quantitative PCR) in some of the infected xenografted animals, but in none of the infected non-xenografted animals. It will need to be assessed whether or not this model is amenable to antiviral studies (Bente et al 2005).

Yellow fever virus (Table 2)

Various strains of YFV killed young (4- to 5-week-old) BALB/c and TO mice within 2 weeks post intracerebral inoculation. Following intranasal inoculation, only the Asibi virus, the French neurotropic vaccine, two out of three 17D vaccine substrain viruses (Brazil & Colombia) and some wild-type isolates were able to cause mortality (Barrett & Gould 1986). Passage of the vaccine strain YFV17D in SCID and ICR mice resulted in a neuroinvasive YFV variant (SPYF) (Chambers & Nickells 2001). This virus caused uniform mortality in young adult CD-1 mice following intracerebral inoculation of as little as 1 pfu of virus, replicated in mouse brain tissue more rapidly than the non-neuroadapted virus, and resulted in higher peak titres of brain-associated virus and earlier death (Schlesinger et al 1996). Also adult SCID mice proved very sensitive to the neuroadapted virus, and developed a fatal encephalitis as early as 8 days post intraperitoneal inoculation (Chambers & Nickells 2001, Nickells & Chambers 2003).

Following intraperitoneal inoculation with the virulent YFV strain (Jimenez), golden hamsters developed a high-titred viraemia (lasting 5–6 days) and had abnormal liver function tests. The mortality rate in YFV-infected hamsters was variable and dependant on the virus strain and the age of the animals (Xiao et al 2001b). These clinical and pathological changes in hamsters were very similar to those described in experimentally infected macaques and in fatal human cases of YF (Tesh et al 2001, Xiao et al 2001b, Arya 2001).

TABLE 2 Models for yellow fever virus infection

Model	*Reference*
BALB/c, TO mice infected with various strains	Barrett & Gould 1986, Burke & Monath 2001
CD-1, SCID, ICR mice infected with a neuroadapted YFV 17D variant	Chambers & Nickells 2001, Nickells & Chambers 2003
Golden Syrian hamsters infected with the Jimenez strain	Tesh et al 2001, Xiao et al 2001b, Arya 2001
Golden Syrian hamsters infected with the Asibi strain	McArthur et al 2003, 2005

A hamster model for YFV infection was also developed using the well characterized YFV Asibi strain. An Asibi derived strain that is viscerotropic for hamsters was obtained after serial passage of the virus through hamsters. The parental Asibi/hamster p0 virus caused only a mild and transient viraemia in hamsters with no obvious clinical signs of illness. In contrast, the strain obtained after 7 passages was viscerotropic and caused a robust viraemia, severe illness and death in subadult hamsters. The majority of mutations in the viscerotropic strain, as compared to the wild-type strain, fall within the envelope protein (McArthur et al 2003). The same group also succeeded in generating an Asibi derived strain that was passaged in hamster (denominated P7b) and did not cause clinically detectable signs of YFV infection yet produced a high viraemia as well as histopathological lesions in the liver consistent with YFV infection (McArthur et al 2005).

West Nile virus

Mortality rates in BALB/c mice that had been infected intraperitoneally with mouse-neuroadapted WN virus was 100% (Ben Nathan et al 2003). The WNV strains of Middle Eastern or African origin proved highly lethal to adult ICR mice following intraperitoneal inoculation (Lustig et al 2000). Also a New York isolate and strain IS-98-ST1, isolated from a stork in Israel in 1998, and closely related to strain Isr98/NY99 (Deubel et al 2001, Malkinson et al 2002) were highly virulent and neuroinvasive in 3- to 4-week-old NIH Swiss outbred mice or e.g. BALB/c, ICR mice following intraperitoneal inoculation (Mashimo et al 2002). In the brain, perivascular cuffing (with macrophages and lymphocytes) was found around small vessels. Neuronal degeneration and necrosis, neuronophagia, spongy degeneration

TABLE 3 Models for West Nile virus infections

Model	*Reference*
ICR mice infected with strains of Middle Eastern or African origin	Lustig et al 2000
Swiss mice infected with the New York isolate	Beasley et al 2002
BALB/c, ICR mice infected with strain IS-98-ST1 or a mouse-neuroadapted strain	Mashimo et al 2002, Ben-Nathan et al 2003
Hamsters infected with various strains	Tesh et al 2002, 2005, Beasley et al 2002, Xiao et al 2001a, Morrey et al 2005, Tonrey et al 2005, Ding et al 2005, Sbrana et al 2005

and focal haemorrhages were observed as well. The virus is mainly detected in the *hippocampus* and *cortex* areas of the mouse brain (Shrestha et al 2003, Deubel et al 2001). WN virus causes encephalitis and death in most laboratory inbred mouse strains after peripheral inoculation, most strains derived from recently trapped wild mice are completely resistant. WN virus sensitivity of susceptible mice was correlated with the occurrence of a point mutation in the 2′-5′-oligoadenylate synthase (Mashimo et al 2002).

Following intraperitoneal inoculation with WN virus strain 385-99 (a 1999 New York isolate), hamsters developed a viraemia of 5 to 6 days in duration, followed by the development of humoral antibodies (Tesh et al 2002). The appearance of viral antigen in the brain and neuronal degeneration began 6 days after infection and about half of the animals died during the acute phase of the infection (Beasley et al 2002). WN virus was cultured from the brain of convalescent hamsters up to 53 days after initial infection, suggesting that persistent virus infection occurred (Xiao et al 2001a). According to Morrey et al (2004) different strains of WN virus cause different degrees of mortalities in hamsters.

Golden hamsters experimentally infected with WN virus developed chronic renal infection and persistent shedding of virus in urine for up to 8 months, despite initial rapid clearance of virus from blood and appearance of high levels of specific neutralizing antibodies. Infectious WN virus could be recovered by direct culture from the urine and by cocultivation of kidney tissue for up to 247 days after infection (Tesh et al 2005, Tonry et al 2005). WN virus, like most members of the JE virus complex is assumed to be mainly neurotropic and little is known about the genetic basis for its renal tropism. Nucleotide changes in virus isolated from the urine of the infected hamsters were mostly in the coding regions with amino acid substitutions in the E, NS1, NS2B and NS5 proteins (Ding et al 2005). The same team also reported that hamsters can be infected with WN virus via the oral route (Sbrana et al 2005).

Japanese encephalitis virus

In mice, disease outcome following extraneural inoculation with JEV is strongly host factor (for example age) dependent (Monath 1986). Mice up to 3 to 4 weeks of age are highly susceptible to a low peripheral inoculum of virulent (neuroinvasive) JEV strains. In these mice, the virus is detected in the brain, as well as in various extraneural tissues as well as in the blood (McMinn et al 1996). In older mice the peripheral injection of viruses that belong to the JEV complex is mostly associated with low or undetectable viraemia and often fails to result in encephalitis and death. However, when injected directly into the brain, the viruses grow to high titres and cause a fatal encephalomyelitis (Licon Luna et al 2002, Lee & Lobigs 2002).

Tick-borne encephalitis virus

Outbred white mice (of various ages) were inoculated subcutaneously with the Skalica strain of TBEV. The appearance and severity of encephalitis and levels of virus in brain tissue decreased with increasing age of the animals (Rajcani & Gresikova 1982). TBEV (strain Neudörfl) caused a lethal infection following either subcutaneous or intraperitoneal infection of young adult BALB/c mice. Viraemia became first detectable at 24 h post infection. The virus was first detected in the brain about 6 days after virus inoculation; titres then increased steadily and remained high until death of the animals. The mean survival time in this model of TBEV infection was at about day 10 post infection (Kreil & Eibl 1997). The Oshima strain, isolated from a sentinel dog in Hokkaido, Japan, proved lethal for 87% of young adult ICR mice that had been inoculated subcutaneously. The Sofjin strain caused lethal encephalitis in 100% of the animals (Chiba et al 1999).

Modoc virus and Montana *Myotis* leukoencephalitis virus

The Modoc virus (MODV) and Montana *Myotis* leukoencephalitis virus (MMLV), were first isolated from the white footed deer mouse (*Peromyscus maniculatus*) (MODV) and bats (*Myotis lucifugus*). These viruses do not cause disease in human (Zarnke & Yuill 1985, Bell & Thomas 1964). Both viruses are neuroinvasive in immunodeficient mice. Although no major histopathological changes were observed in the brain of SCID mice that had obvious neurological symptoms following MODV (or MMLV) infection, ultrastructural analysis of the neurons revealed a cellular pathology characteristic of flavivirus infection. Viral RNA and/or antigens were detected in the brain (and was confined to neurons) and serum of SCID mice infected with MODV or MMLV. The interferon-α/β inducer poly(I). poly(C) and Ampligen® efficiently protected against MODV- and MMLV-induced mortality in SCID mice (Leyssen et al 2001, 2003b, Charlier et al 2002). Infection with MODV of NMRI (Naval Medical Research Institute) mice that had received various degrees of immunosuppressive therapy revealed that immunopathological factors also contribute to disease progression but that direct virus-induced damage to the neurons is the principal cause of virus-induced mortality. As a consequence, antiviral therapy should have a beneficial effect on flavivirus encephalitis (Leyssen et al 2003a). The MODV and MMLV SCID models may be convenient for the study of antiviral strategies against flavivirus encephalitis, because MODV and MMLV are (i) highly pathogenic to (SCID) mice following peripheral inoculation, (ii) classified as biosafety level 2 pathogens (by the Subcommittee on Arbovirus Laboratory Safety [SALS]; *http://www.cdc.gov/od/ohs/biosfty/bmbl4/bmbl4s71.htm*), (iii) have the same overall organization of the genome as flaviviruses that cause infections in human, and (iv) contain, in those genes that are considered to be

TABLE 4 Models for tick-borne encephalitis virus, Japanese encephalitis virus and Modoc virus infection

Model		*Reference*
TBEV	Outbred white mice	Rajcani & Gresikova 1982
	BALB/c mice	Kreil & Eibl 1997
	ICR mice	Chiba et al 1999
JEV	Various strains including Swiss and C57BL/6	McMinn et al 1996, Licon Luna et al 2002, Lee & Lobigs 2002
MODV	SCID mice	Leyssen et al 2001, 2003b
	NMRI mice	Leyssen et al 2003a
	Golden Syrian hamsters	Leyssen et al 2001, 2003c

interesting antiviral targets (e.g. the NS3 which encodes an NTPase/helicase and serine protease and the NS5 encoding an RNA-dependent RNA polymerase), and the same conserved motifs as flaviviruses that are infectious to humans (Charlier et al 2002, Leyssen et al 2001, 2002, 2003a,b) (Table 4).

The Modoc virus is also highly neuroinvasive in hamsters. Following intraperitoneal or intranasal inoculation with MODV, hamsters develop acute encephalitis. About 50–60% of the animals succumb during the acute phase of the infection, about 40% survive without neurological sequelae and the remaining 10% recover with obvious long-lasting neurological sequelae. The situation in hamsters that survive the acute phase of MOD encephalitis is reminiscent of that of patients who survived severe JE, SLE, WN encephalitis or TBE. In hamsters, viral RNA was detected during the acute phase of the disease in the brain and various other organs. Perivascular cuffing, one of the characteristics of flavivirus encephalitis, was observed around small or medium-size blood vessels in the brain (Leyssen et al 2003c). A particular characteristic of the MODV/hamster model is that MODV-infected hamsters, akin to WNV-infected hamsters, shed virus in the urine. This allows the evaluation of antiviral strategies over a period of several days in one and the same animal without having to use invasive sampling methods (Leyssen et al 2001).

Conclusion

It is important to have convenient small animal models at hand that allow us to rapidly assess the potential of novel strategies for the treatment or prevention of flavivirus infections. Monkey models of flavivirus infections have been reported (for example for YFV, DENV or JEV), but because of the cost involved and the restricted availability of monkeys, the number of studies that can be carried out using these animals is of course limited.

It has proven particularly difficult to elaborate mouse models of dengue infection. Most DENV mouse models suffer from the fact that only low levels of viremia are reached and that the symptoms caused by the infection do not, or only partially, mimic the clinical situation in man. It remains to be studied whether these models will be suitable to monitor the *in vivo* antiviral efficacies of antiviral drugs against DENV. As compared to dengue, it has been relatively more easy to obtain animals models for YFV infections. In particular the hamsters models that make use of the viscerotropic Asibi strain may be particularly useful for antiviral studies. Overall, hamsters inoculated with YFV develop clinical and pathologic changes (such as abnormal liver function) that are very similar to those described in fatal human cases of YF and these models may therefore be an excellent alternative for the non-human primate models for viscerotropic flavivirus infections.

Infection of mice or hamsters with WN virus has been studied in great detail. The virus causes viremia and encephalitis both in mice and in hamsters. Hamsters may develop persistent infection, with release of the virus in the urine. Measuring viral load in the urine may provide a practical approach to monitor the efficacy of novel antiviral strategies. As for WN virus, other neurotropic viruses, such as JEV and TBEV readily infect (young) mice. These mouse models should allow us to study the efficacy of antiviral drugs *in vivo* against JEV and TBEV.

The Modoc virus causes a pathology in mice (and hamsters) that is reminiscent of flavivirus encephalitis in human. This virus is in contrast to other encephalitic flaviviruses a BSL-2 pathogen and has a genomic organization and conserved motifs (in genes that encode antiviral targets) similar to flaviviruses that cause encephalitis in human. Like WN virus, hamsters infected with MODV shed the virus in the urine which thus allows monitoring of antiviral therapy without the need for invasive methods. The MODV models may therefore be particularly attractive for antiviral studies against flavivirus encephalitis.

The intense search for selective inhibitors of the hepatitis C virus (HCV), a virus that belongs, like flaviviruses, to the family *Flaviviridae*, will hopefully also lead to the discovery of selective inhibitors of flavivirus replication. It will be important to assess the efficacy of such compounds in reliable and relevant small animal models before further preclinical studies in monkeys or other larger animals are planned.

Acknowledgements

N. Charlier is a post-doctoral research fellow from the 'Onderzoeksfonds' of the University of Leuven.

References

An J, Kimura-Kuroda J, Hirabayashi Y, Yasui K 1999 Development of a novel mouse model for dengue virus infection. Virology 263:70–77

Arya SC 2001 Experimental model of yellow fever in the golden hamster (Mesocricetus auratus). J Infect Dis 184:1496–1497

Atrasheuskaya A, Petzelbauer P, Fredeking TM, Ignatyev G 2003 Anti-TNF antibody treatment reduces mortality in experimental dengue virus infection. FEMS Immunol Med Microbiol 35:33–42

Barrett AD, Gould EA 1986 Comparison of neurovirulence of different strains of yellow fever virus in mice. J Gen Virol 67:631–637

Beasley DW, Li L, Suderman MT, Barrett AD 2002 Mouse neuroinvasive phenotype of West Nile virus strains varies depending upon virus genotype. Virology 296:17–23

Bell JF, Thomas LA 1964 A new virus, 'MML', enzootic in bats (Myotis lucifugus) of Montana. Am J Trop Med Hyg 607–612

Ben Nathan D, Lustig S, Tam G, Robinzon S, Segal S, Rager-Zisman B 2003 Prophylactic and therapeutic efficacy of human intravenous immunoglobulin in treating west nile virus infection in mice. J Infect Dis 188:5–12

Bente DA, Melkus MW, Garcia JV, Rico-Hesse R 2005 Dengue fever in humanized NOD/SCID mice. J Virol 79:13797–13799

Blaney JE Jr, Johnson DH, Manipon GG et al 2002 Genetic basis of attenuation of dengue virus type 4 small plaque mutants with restricted replication in suckling mice and in SCID mice transplanted with human liver cells. Virology 300:125–139

Chambers TJ, Nickells M 2001 Neuroadapted yellow fever virus 17D: genetic and biological characterization of a highly mouse-neurovirulent virus and its infectious molecular clone. J Virol 75:10912–10922

Chambers TJ, Hahn CS, Galler R, Rice CM 1990 Flavivirus genome organization, expression, and replication. Annu Rev Microbiol 44:649–688

Charlier N, Leyssen P, Paeshuyse J et al 2002 Infection of SCID mice with Montana Myotis leukoencephalitis virus as a model for flavivirus encephalitis. J Gen Virol 83:1887–1896

Chiba N, Iwasaki T, Mizutani T, Kariwa H, Kurata T, Takashima I 1999 Pathogenicity of tick-borne encephalitis virus isolated in Hokkaido, Japan in mouse model. Vaccine 17:779–787

Deubel V, Fiette L, Gounon P et al 2001 Variations in biological features of West Nile viruses. Ann NY Acad Sci 951:195–206

Ding X, Wu X, Duan T et al 2005 Nucleotide and amino acid changes in West Nile virus strains exhibiting renal tropism in hamsters. Am J Trop Med Hyg 73:803–807

Gritsun TS, Lashkevich VA, Gould EA 2003 Tick-borne encephalitis. Antiviral Res 57:129–146

Gubler DJ 1997 Dengue and dengue hemorrhagic fever: its history and resurgence as a global public health problem. D.J. Gubler and G. Kuno (ed) Dengue and dengue hemorrhagic fever. CAB International, Wallingford, United Kingdom, 1–23

Han LL, Popovici F, Alexander JJ et al 1999 Risk factors for West Nile virus infection and meningoencephalitis, Romania, 1996. J Infect Dis 179:230–233

Hennessy S, Liu Z, Tsai TF et al 1996 Effectiveness of live-attenuated Japanese encephalitis vaccine (SA14-14-2): a case-control study. Lancet 347:1583–1586

Huang KJ, Li SY, Chen SC et al 2000 Manifestation of thrombocytopenia in dengue-2-virus-infected mice. J Gen Virol 81:2177–2182

Johnson AJ, Roehrig JT 1999 New mouse model for dengue virus vaccine testing. J Virol 73:783–786

Kalita J, Misra UK 1998 EEG in Japanese encephalitis: a clinico-radiological correlation. Electroencephalogr Clin Neurophysiol 106:238–243

Kramer LD, Presser SB, Hardy JL, Jackson AO 1997 Genotypic and phenotypic variation of selected Saint Louis encephalitis viral strains isolated in California. Am J Trop Med Hyg 57:222–229

Kreil TR, Eibl MM 1997 Pre- and postexposure protection by passive immunoglobulin but no enhancement of infection with a flavivirus in a mouse model. J Virol 71: 2921–2927

Kuno G, Chang GJ, Tsuchiya KR, Karabatsos N, Cropp CB 1998 Phylogeny of the genus Flavivirus. J Virol 72:73–83

Lee E, Lobigs M 2002 Mechanism of virulence attenuation of glycosaminoglycan-binding variants of Japanese encephalitis virus and Murray Valley encephalitis virus. J Virol 76:4901–4911

Leyssen P, Van Lommel A, Drosten C, Schmitz H, De Clercq E, Neyts J 2001 A novel model for the study of the therapy of Flavivirus infections using the Modoc virus. Virology 279:27–37

Leyssen P, Charlier N, Lemey P et al 2002 Complete genome sequence, taxonomic assignment, and comparative analysis of the untranslated regions of the Modoc virus, a flavivirus with no known vector. Virology 239:125–140

Leyssen P, Drosten C, Paning M et al 2003a Interferons, interferon inducers, and interferon-ribavirin in treatment of flavivirus-induced encephalitis in mice. Antimicrob Agents Chemother 47:777–782

Leyssen P, Paeshuyse J, Charlier N et al 2003b Impact of direct virus-induced neuronal dysfunction and immunological damage on the progression of flavivirus (Modoc) encephalitis in a murine model. J Neurovirol 9:69–78

Leyssen P, Croes R, Rau P et al 2003c Acute encephalitis, a poliomyelitis-like syndrome and neurological sequelae in a hamster model for flavivirus infections. Brain Pathol 13:279–290

Licon Luna RM, Lee E, Mullbacher A, Blanden RV, Langman R, Lobigs M 2002 Lack of both Fas ligand and perforin protects from flavivirus-mediated encephalitis in mice. J Virol 76:3202–3211

Lin YL, Liao CL, Chen LK et al 1998 Study of Dengue virus infection in SCID mice engrafted with human K562 cells. J Virol 72:9729–9737

Lustig S, Olshevsky U, Ben Nathan D et al 2000 A live attenuated West Nile virus strain as a potential veterinary vaccine. Viral Immunol 13:401–410

Lvov DK, Butenko AM, Gromashevsky VL et al 2000 Isolation of two strains of West Nile virus during an outbreak in southern Russia, 1999. Emerg Infect Dis 6:373–376

Malkinson M, Banet C, Weisman Y et al 2002 Introduction of West Nile virus in the Middle East by migrating white storks. Emerg Infect Dis 8:392–397

Mashimo T, Lucas M, Simon-Chazottes D et al 2002 A nonsense mutation in the gene encoding 2′-5′-oligoadenylate synthetase/L1 isoform is associated with West Nile virus susceptibility in laboratory mice. Proc Natl Acad Sci USA 99:11311–11316

McArthur MA, Suderman MT, Mutebi JP, Xiao SY, Barrett AD 2003 Molecular characterization of a hamster viscerotropic strain of yellow fever virus. J Virol 77:1462–1468

McArthur MA, Xiao SY, Barrett AD 2005 Phenotypic and molecular characterization of a non-lethal, hamster-viscerotropic strain of yellow fever virus. Virus Res 110:65–71

McCormack JG, Allworth AM 2002 Emerging viral infections in Australia. Med J Aust 177:45–49

McMinn PC, Dalgarno L, Weir RC 1996 A comparison of the spread of Murray Valley encephalitis viruses of high or low neuroinvasiveness in the tissues of Swiss mice after peripheral inoculation. Virology 220:414–423

Misra UK, Kalita J, Srivastava M 1998 Prognosis of Japanese encephalitis: a multivariate analysis. J Neurol Sci 161:143–147

Monath TP 1986 Pathobiology of the flaviviruses. In: Schlesinger S, Schlesinger MJ (eds) The Togaviridae and Flaviviridae. Plenum Press, New York, 375–440

Morrey JD, Day CW, Julander JG et al 2004 Modeling hamsters for evaluating West Nile virus therapies. Antiviral Res 63:41–50

Nickells M, Chambers TJ 2003 Neuroadapted yellow fever virus 17D: determinants in the envelope protein govern neuroinvasiveness for SCID mice. J Virol 77:12232–12242

Rajcani J, Gresikova M 1982 Pathogenicity of the skalica strain (from the tick-borne encephalitis complex) for white mice. Acta Virol 26:264–269

Sbrana E, Tonry JH, Xiao SY, da Rosa AP, Higgs S, Tesh RB 2005 Oral transmission of West Nile virus in a hamster model. Am J Trop Med Hyg 72:325–329

Schlesinger JJ, Chapman S, Nestorowicz A, Rice CM, Ginocchio TE, Chambers TJ 1996 Replication of yellow fever virus in the mouse central nervous system: comparison of neuroadapted and non-neuroadapted virus and partial sequence analysis of the neuroadapted strain. J Gen Virol 77:1277–1285

Shrestha B, Gottlieb D, Diamond MS 2003 Infection and injury of neurons by West Nile encephalitis virus. J Virol 77:13203–13213

Siegel-Itzkovich J 2000 Twelve die of West Nile virus in Israel. BMJ 321:724

Tesh RB, Guzman H, da Rosa AP et al 2001 Experimental yellow fever virus infection in the Golden Hamster (Mesocricetus auratus). I. Virologic, biochemical, and immunologic studies. J Infect Dis 183:1431–1436

Tesh RB, Travassos Da Rosa AP, Guzman H, Araujo TP, Xiao SY 2002 Immunization with heterologous flaviviruses protective against fatal West Nile encephalitis. Emerg Infect Dis 8:245–251

Tesh RB, Siirin M, Guzman H et al 2005 Persistent West Nile virus infection in the golden hamster: studies on its mechanism and possible implications for other flavivirus infections. J Infect Dis 192:287–295

Thomas SJ, Strickman D, Vaughn DW 2003 Dengue epidemiology: virus epidemiology, ecology, and emergence. Adv Virus Res 61:235–289

Tonry JH, Xiao SY, Siirin M, Chen H, da Rosa AP, Tesh RB 2005 Persistent shedding of West Nile virus in urine of experimentally infected hamsters. Am J Trop Med Hyg 72:320–324

Tsai TF, Popovici F, Cernescu C, Campbell GL, Nedelcu NI 1998 West Nile encephalitis epidemic in southeastern Romania. Lancet 352:767–771

Wu SJ, Hayes CG, Dubois DR et al 1995 Evaluation of the severe combined immunodeficient (SCID) mouse as an animal model for dengue viral infection. Am J Trop Med Hyg 52:468–476

Xiao SY, Guzman H, Zhang H, Travassos-da-Rosa AP, Tesh RB 2001a West Nile virus infection in the golden hamster (Mesocricetus auratus): a model for West Nile encephalitis. Emerg Infect Dis 7:714–721

Xiao SY, Zhang H, Guzman H, Tesh RB 2001b Experimental yellow fever virus infection in the golden hamster (Mesocricetus auratus). II. Pathology. J Infect Dis 183:1437–1444

Zarnke RL, Yuill TM 1985 Modoc-like virus isolated from wild deer mice (Peromyscus maniculatus) in Alberta. J Wildl Dis 21:94–99

DISCUSSION

Holmes: What is the reason to believe that ribavirin induces error catastrophe in hepatitis C virus (HCV)?

Neyts: Most of the information that suggests that ribavirin induces error catastrophe was collected with picornaviruses. There have also been some studies with

HCV, with the replicon system and with the purified polymerase. Some studies have also been carried out with the GBV B virus.

Holmes: But not genus *Flavivirus*. Are the results definitive in hepatitis C that it really does induce high error frequency? And have they used the same assay that you did?

Neyts: With HCV, up to now the replicon system that does not produce infectious particles had to be used. Now that the HCV replication models have become available this will of course allow us to use other experimental approaches for studying the antiviral effect of ribavirin on HCV. With YF we have the advantage that we can use the replicating virus and take infectious virus from the culture supernatant. I am not saying that what other people see with other virus is incorrect. I'm saying that for YFV and also for RSV there is a very tight correlation between the antiviral effect of ribavirin and depletion of intracellular GTP pools. For picornaviruses we don't see this type of correlation. Coxsackie virus, for example, in our hands is much less susceptible to mycophenolic acid than respiratory syncytial virus (RSV) or yellow fever virus (YFV).

Evans: Have you taken the NM283 that you have shown is active *in vitro*, and put it *in vivo* into your animal model?

Neyts: We did not, but Wouter Schul did using the 7-deaza-adenosine analogue. Wouter noted some antiviral activity of this compound in the dengue mouse model that he is using.

Canard: Do you have an idea about the targets for imidazopyridine compounds?

Neyts: For pestiviruses it is the polymerase. The funny thing is that Viropharma published years ago on a compound which was active against pestiviruses; a compound with a structure very different to the anti-pestivirus compound that we report. This compound was inducing the F224S mutation in the polymerase. With our class of anti-pestivirus compounds we do see the same mutation, although the structures of our compounds are very different from that of the Viropharma compound. But the mutation is exactly the same and this is corroborated by the fact that the two compounds are cross-resistant.

Schul: In your model of dengue/Modoc chimeras, similar to the yellow fever/ Modoc chimeras, it is probable that it will target to the nervous system and the brain. How do you think this will influence it as a model for testing compounds?

Neyts: I'm not saying that this will be a good model for the study of the pathogenesis of dengue. The aim is to have a model in which a chimeric virus that contains the entire replication machinery of dengue replicates efficiently in mice. Whether you hit the virus in the liver or the brain may in first instance not be that important in testing antiviral drug efficacy. The first aim is to see whether or not a compound that targets the dengue replication machinery is able to block the replication of this virus in an animal.

Rice: You are worried more about access to normally infected tissues.

Neyts: Imagine that we have a drug that was able to penetrate the brain and that would be effective against a CNS infection with flaviviruses, then this would be excellent. Drugs to treat JE, TBE and WN encephalitis would need to penetrate the brain.

Schul: If the target was a host target rather than a viral one, this would be different.

Neyts: Most antiviral drugs today target viral factors, but it is of course possible to target a host factor (just think of the CsA being active against HCV). If a drug targets a human host factor, it may indeed be a problem to test such drugs in mice. Animal models are of course not perfect.

Evans: Are you planning to do any experiments with siRNA?

Neyts: No, this is not really our field of expertise.

Rice: Which animal models are the best for looking at dengue?

Neyts: I think the AG129 model that Wouter Schul is using is a good one.

Schul: For viraemia, there are several specific strains replicating in mice; Ag129 or A/J can do this. You can use these mouse infection systems to test antivirals, but not for studying pathology.

Vasudevan: We found that DENV-2 strains NGC and TSV1 behave differently in the AG129 animal model.

Schul: We are trying to find clinical strains for all four serotypes that work in the Ag129. We would like a model for all four.

Rice: What is the current state of the art for models for the more severe pathology?

Harris: We have peripherally passaged virus in the Ag129 from the serum, and have ended up with a virus that essentially induces a vascular permeability like plasma leakage. This is much more present in the white blood cells and the bone marrow than its parent virus. There is some interesting genetics about it, and it looks as if we have at least part of a mechanism. It is much more virulent and has a paralytic phenotype.

Rice: Is that seen in the immunodeficient mice?

Harris: Yes. We are trying to learn from this what the significant changes are.

Halstead: Anna Durbin at Johns Hopkins School of Public Health has one of these miraculous events with dengue. She infected a rhesus monkeys with dengue 1 and challenged with dengue 2, 12 months later. One of the two animals was sufficiently sick that it had to be sacrificed. I had previously done many such experiments but at shorter intervals (the longest was 6 months). In humans the interval between dengue infections is usually at least a year.

Harris: We are using this model to explore sequential infection. We have been doing a few experiments with different viruses in the same mouse, and we are also setting up a passive of adoptive transfer experiment. In the meantime we are using two strains of dengue 1 and two of dengue 2 to look at the antibody response over a half year period.

Schul: Since we don't really know what DHF is, it is hard to model it.

Harris: We know that in theory it is because of cross-reactive B cells and/or T cells, and waning immunity. The idea is to characterize what the antibodies look like, and then look at the T cell responses. We hope to then be able to manipulate these at different times.

Zinkernagel: Is anyone looking in monkeys at simple early parameters of infection, such as localization, in a time-dependent fashion with various serotypes? These are data we would love to have in humans. Although the monkeys don't develop the haemorrhagic part of the disease, it would still be helpful if we could know where the virus is. We need to know this in order to understand the pathogenesis. For this the mouse is probably too different from humans.

Halstead: The problem is that it costs \$2000–\$5000 an animal.

Zinkernagel: I agree with you, this is very expensive. But it is essential, so it should be done.

Secretion of flaviviral non-structural protein NS1: from diagnosis to pathogenesis

S. Alcon-LePoder*[1], P. Sivard*, M.-T. Drouet*, A. Talarmin†, C. Rice‡ and M. Flamand*‡[2]

* *Institut Pasteur, Virology Department, 25 Rue du Dr Roux, 75015 Paris, France,* † *Institut Pasteur Madagascar, BP 1274, Antananarivo 101, Madagascar, and* ‡ *The Rockefeller University, Laboratory of Virology and Infectious Disease, 1230 York Avenue, New York, NY 10021, USA*

Abstract. Flaviviruses are major arthropod-borne human pathogens responsible for life-threatening encephalitis, hepatitis and haemorrhagic fevers. These enveloped, single-stranded, positive-sense RNA viruses encode a polyprotein precursor of about 3400 amino acids, processed into three structural and seven non-structural proteins. The non-structural glycoprotein NS1 is essential for flavivirus viability. During host-cell infection *in vitro*, NS1 is found associated with intracellular organelles as a requisite for its role in viral replication, or is transported to the cell surface where it may trigger specific signalling pathways. In addition, a secreted form of the protein is released from flavivirus-infected mammalian cells. We have previously shown that the NS1 protein circulates during the acute phase of the disease in the plasma of patients infected with dengue virus type 1 and have extended our retrospective studies to dengue type 2 and type 3 cohorts, confirming the value of the NS1 antigen as an alternative diagnostic marker. Interestingly, detection of the NS1 protein in yellow fever virus and West Nile virus infections suggests that NS1 secretion is a hallmark of human flavivirus infections. The objectives of our current studies are to define the biological properties of the secreted form of the NS1 protein, to evaluate its possible contribution to viral pathogenesis, and to validate this protein as a candidate target for passive immunoprophylaxis against flaviviruses.

2006 New treatment strategies for dengue and other flaviviral diseases. Wiley, Chichester (Novartis Foundation Symposium 277) p 233–250

Flaviviruses, which belong to the *Flaviviridae* family, are small enveloped positive-strand RNA viruses that have long been recognized as important human

[1]Present address: ENVA, 7 avenue du Général de Gaulle, 94704 Maisons-Alfort Cedex, France.

[2]This paper was presented at the symposium by M. Flamand, to whom correspondence should be addressed.

pathogens worldwide (Lindenbach & Rice 2003). Over recent years, however, members of the genus have been associated with an unprecedented rise in the incidence of epidemics (Mackenzie et al 2004, Thomas et al 2003, Weaver & Barrett 2004).

Yellow fever virus (YFV), the prototype flavivirus, is responsible for fulminant hepatitis in human populations of South America and Africa, with a high case fatality rate of 20–50% (Barrett & Monath 2003, Gubler 2004). It was recognized as a blood-borne filterable agent in 1900 and found to be transmissible to a rhesus monkey in 1927. The virus is maintained through an enzootic cycle involving primates and arboreal mosquitoes and an urban cycle involving human populations and mosquitoes that breed in close vicinity. Despite the availability of an effective live vaccine, the number of fatal cases has steadily increased over the past 20 years, reaching an annual total of 15000 deaths, although a tentative eradication of the mosquito vector in South America in the 1970s has significantly improved the situation on this continent (Monath 2001, Mutebi & Barrett 2002, Tomori 2004).

Dengue (DEN) is an acute febrile illness of the tropics, transmitted to human by mosquitoes of the *Aedes* species (Gubler 2004, Thomas et al 2003). Dengue virus was isolated for the first time in 1951 and has since then been classified into four distinct serotypes (DENV-1, -2, -3, -4). More than two billion inhabitants live in endemic areas and the number of individuals infected by the virus is thought to exceed 100 million per year. Most clinical infections consist of a flu-like syndrome (dengue fever, DF) that naturally resolve in a matter of days. The severe form of the disease (dengue haemorrhagic fever, DHF) is characterized by haemorrhage, acute inflammation, thrombocytopenia, coagulopathy and plasma leakage to which a risk of fatal hypovolemic shock is associated. DENV infections are the major cause of haemorrhagic fevers in humans, leading annually to an estimated 500000 hospitalizations and 50000 deaths that usually occur in less than a week after the onset of fever. There is no specific treatment but appropriate medical care can significantly improve the outcome of disease (Halstead 2002). In the absence of early pathognomonic signs of infection or prognostic markers of DHF, a rapid and accurate diagnosis of DENV infections within the clinical phase of the disease is critical.

Most available methods provide a diagnosis *a posteriori* (Guzman & Kouri 2004, Kao et al 2005, Shu & Huang 2004, Teles et al 2005). These are based on virus detection from plasma or evidence of seroconversion. Detection of viral antigens can be achieved by indirect immunofluorescence after virus inoculation from plasma onto cell cultures or into mosquitoes. It is still the most reliable method to confirm flavivirus infections, though time-consuming and difficult to perform. The presence of specific antibodies can be assessed by haemagglutination inhibition (HI) or neutralization tests on convalescent sera, but the most commonly used assay is an ELISA measuring specific immunoglobulins M (IgM) in patient sera, the presence of which denotes a recent infection. These antibodies become detectable 4–5 days after the onset of fever and a second blood-sampling during the convalescent phase is sometimes required to show an increase in the antibody response. Finally, the use of RT-PCR to detect viral RNA during infection has been successfully implemented over the last two decades and although it is a valuable diagnostic methodology, it still faces limitations due to cost, technical constraints and reliability.

West Nile virus (WNV) is responsible for epizootics among birds, the major natural reservoir for the virus, as well as horses, amphibians and reptiles (Mackenzie et al 2004, Petersen et al 2003). Viral transmission to humans is incidental and less than 1% of infected individuals will develop a meningo-encephalitic syndrome (Hayes et al 2005). WNV is transmitted to its vertebrate hosts by mosquito vectors of the *Culex* species. WNV was first isolated in 1937 in Uganda, and its circulation has since been reported in Africa, the Middle East and Europe. Recent outbreaks involving several hundreds to thousands of humans have occurred in Israel (2000), Romania (1996), and more recently in the United States (1999–2004) (Campbell et al 2002, Gould 2003, Komar 2003, Zeller & Schuffenecker 2004). WNV present in donors has also been suspected to be responsible for deadly encephalitis in transfused or transplanted patients, emphasizing the need for systematic screening using sensitive RNA amplification methods in high-prevalence regions (Busch et al 2005, Macedo de Oliveira et al 2004).

The flavivirus genome contains one long open-reading frame that encodes a polyprotein of approximately 3400 amino acids. Co- and post-translational cleavages give rise to three structural proteins C, prM and E, and seven non-structural proteins NS1, NS2A, NS2B, NS3, NS4A, NS4B and NS5 (Lindenbach & Rice 2003). Non-structural protein NS1 of flaviviruses is a highly conserved glycoprotein essential for virus viability. The NS1 protein is expressed as three discrete species in infected mammalian cells: an intracellular, membrane-associated form essential for viral replication (Lancaster et al 1998, Lindenbach & Rice 1997, Mackenzie et al 1996, Muylaert et al 1996, 1997, Westaway et al 1997), a cell surface-associated form that may be involved in signal transduction (Jacobs et al 2000), and a secreted form which role may be related to its ability to interact with target cells such as hepatocytes (Alcon-LePoder et al 2005). Secretion of the NS1 protein has been reported in cell cultures infected with various flaviviruses *in vitro*, including the DENV complex (Flamand et al 1999, Pryor & Wright 1993, Winkler et al 1988), the Japanese encephalitis virus complex (Fan & Mason 1990, Hall et al 1990, Lee et al 1991, Winkler et al 1988), YFV (Post et al 1991) and tick-borne encephalitis virus (Crooks et al 1990).

NS1 antigen is detected in the plasma of dengue virus-infected patients

In order to assess the biological relevance of NS1 secretion in the course of natural DENV infections, we developed a cross-reactive antigen-capture ELISA that displays similar sensitivity to all four serotypes of NS1 (Fig. 1A). Cohorts

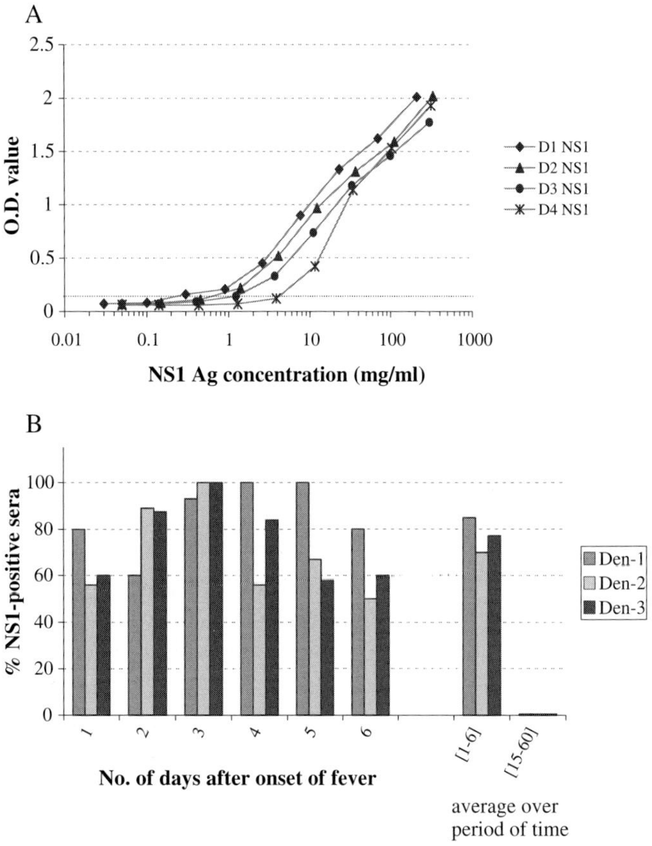

FIG. 1. NS1 antigen circulation in plasma during acute DENV infections. (A) Standard curves of the DENV NS1 antigen-capture ELISA. The sensitivity of the assay was assessed for each of the four serotypes using the corresponding purified secreted form of the NS1 protein. The positive threshold, indicated by a dotted line, represents twice the mean value of the negative controls. The detection sensitivity of the ELISA is about 0.5 ng/ml for type 1 and 2, 1 ng/ml for type 3, and 4 ng/ml for type 4. (B) Percentage of NS1-positive plasma samples in panels of DENV-infected patients. Infections occurred during the epidemics of DENV-1 and -2 in French Guiana in 1996–99, and of DENV-3 in Guadeloupe in 2000–01. Specimens were recovered within the acute phase of the disease (between the onset of fever, day 0, till day 6) or during the convalescent phase (day 10 onwards).

comprising 100–200 individuals were constituted within the Pasteur Institute Network during epidemics involving DENV-1 and -2 in French Guiana in 1996–99, and DENV-3 in Guadeloupe in 2000–2001 (collaboration with J.-L. Cartel, Y. Sanchis and F. Saintpère). Sera were recovered during the acute phase (usually less than a week after the onset of fever, set at day 0) and the convalescent phase (most frequently day 10 onwards). Acute sera were tested for viraemia by RT-PCR or virus isolation, and by MAC-ELISA for DENV-specific IgM detection. Some of the convalescent sera were tested by ELISA for the quantification of both IgM and IgG, or by HI in order to identify primary from secondary infections. The vast majority of the sera recovered between day 0 and day 9 were found to be positive in NS1 antigen, irrespective of the viral serotype involved in the infection (Fig. 1B) (Alcon et al 2002). Over the period of day 0 to day 6, the percentage of positive sera varied from 50% to 100%, with a mean value as of 87% for type 1, 70% for type 2, and 77% for type 3 (Fig. 1B). For type 4, only a limited number of sera were available due to a lower incidence of this particular serotype of DENV. Out of 14 blood samples collected in Vietnam within a week after the onset of fever (kindly provided by VTQ Huong), 11 were positive in NS1 antigen, representing 79% of the samples tested. As expected, none of the convalescent sera recovered after day 15 had detectable levels of NS1, presumably once infected cells have been cleared by the immune system, and the protein degraded or opsonized by specific antibodies. Not surprisingly, the NS1 antigen-capture assay showed a very high correlation with detection of virus by RT-PCR, up to 90% for type 1, 87% for type 2 and 95% for type 3 (data not shown). In addition, the NS1 protein could still be detected several days after viraemia had become undetectable, suggesting that NS1 secretion may occur at a higher magnitude than virus particle release or that the half-life of the NS1 protein may be substantially longer than virions as the immune system acts to eliminate the infection.

Production of the NS1 antigen in blood is virus and host dependent

In order to quantify the levels of the NS1 viral antigen in DENV-1, -2 and -3 infections, we made use of purified NS1 protein standards for each of the corresponding viral serotypes (Fig. 1A). Even in a comparable epidemiological setting, such as for DENV-1 and DENV-2 that co-circulated in French Guiana during the same period, NS1 concentrations varied depending on the individual, the viral serotype and the course of infection (Fig. 2). Concentrations of the NS1 antigen in plasma frequently reached values above 10–100 ng/ml, although for DENV-3-infected patients, the amounts of protein could be substantially lower. Accordingly, the mean value of NS1 concentration calculated over a 6 day period from the onset of fever was three orders of magnitude higher for type 1 compared to type 3, and 7.5-fold higher for type 2 than for type 3 (mean values of 285, 700 and 93 ng/ml

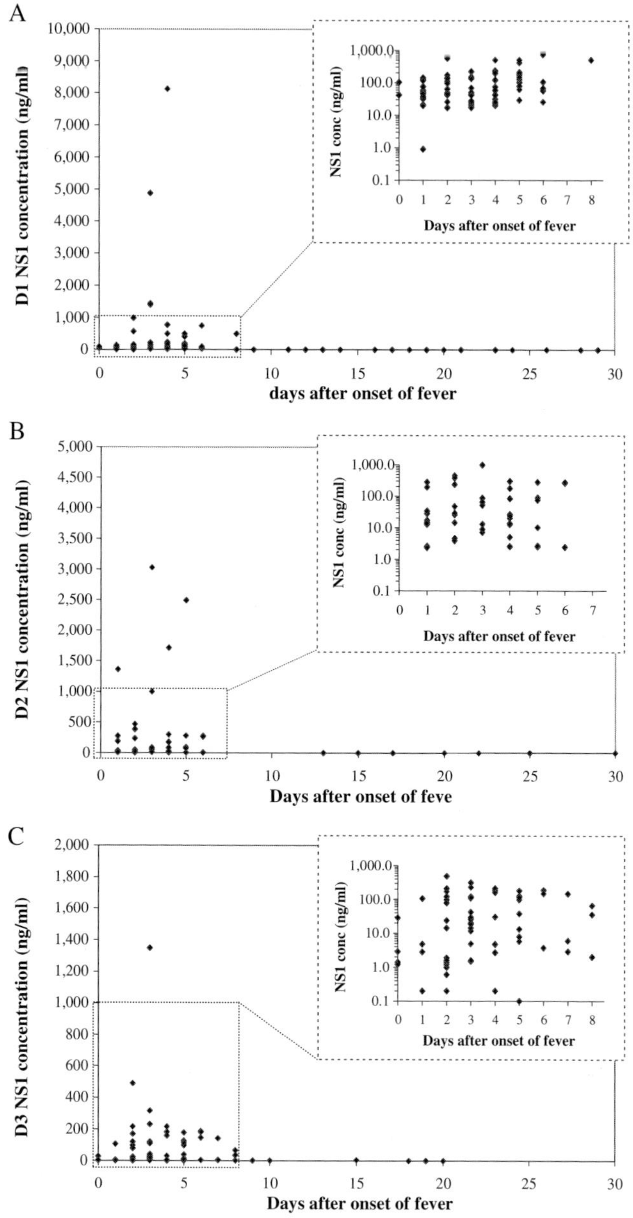

FIG. 2. NS1 concentrations in plasma from patients infected with (A) DENV-1, (B) DENV-2 and (C) DENV-3. Specimens were tested by the NS1 antigen-capture ELISA at different dilutions (1:2, 1:10 and 1:100) and the levels of NS1 estimated using values falling within the linear range of the assay.

for type 1, 2 and 3, respectively). It is not clear why the range of NS1 concentrations in DENV-3 infections is substantially lower than that of the other serotypes while the breadth of NS1 secretion is at least as wide for type 3 as for the two other viral serotypes (positive type 3 specimens found over a period of 8 days, Fig. 2). One possibility is that DENV-3 does not replicate as efficiently, either due to the genetic background of the population living in Guadeloupe compared to that of French Guiana, or to the particular strain involved in the epidemics. Another possibility is that DEN-3 NS1 might have a shorter half-life *in vivo*, or bind more efficiently to target cells or soluble host factors, thereby lowering the detectable pool of NS1 proteins in plasma. Quite notably, several individuals from the different serogroups peaked in the range of 1–10 μg NS1/ml, significantly above the average values. However, high NS1 concentrations did not seem to correlate with specific clinical symptoms or haematological markers, suggesting that NS1 level may be of poor prognostic value.

DENV-1 NS1 can form immune complexes with specific IgG produced during the acute phase of secondary infections

In the panel of DENV-1 infections, specific IgM could not be detected before day 3, seldom on day 4, and was prominent from day 5 onwards (Table 1), whereas more than half of the patients displayed significant levels of DEN-specific IgG by day 1, reaching a plateau of 100% by day 5 (Table 2). The presence of a higher

TABLE 1 Detection of NS1 antigen is not prevented by the rise of dengue-specific IgM during the acute phase of the disease in dengue virus type 1-infected patients[a]

	Capture-ELISA NS1⁺		*Capture-ELISA NS1⁻*	
No days after onset of fever	*MAC IgM⁻*	*MAC IgM⁺*	*MAC IgM⁻*	*MAC IgM⁺*
0	2			
1	12		3	
2	8	1	6	
3	10	4	1	
4	4	6		
5	1	6		
6		4		1
7		1		1
8				1
9		1		2
11–66				33

[a]The number of positive (+) or negative (−) sera in each of the assays, the IgM MAC-ELISA (MAC IgM) or the NS1 antigen-capture ELISA (Capture-ELISA NS1), is reported day by day, from the onset of fever day 0 onwards.

TABLE 2 Detection of NS1 antigen is not prevented by the presence of dengue-specific IgG during the acute phase of the disease in dengue virus type 1-infected patients[a]

No. days after onset of fever	*Capture-ELISA NS1⁺*		*Capture-ELISA NS1⁻*	
	IgG⁻	*IgG⁺*	*IgG⁻*	*IgG⁺*
0	1			
1	2	2	1	2
2	1	3	3	2
3	4	4		
4	2	4		
5		2		
6		1		1
7		1		1
8–24				22

[a]The number of positive (+) or negative (−) sera in each of the assays, the IgM MAC-ELISA (MAC IgM) or the NS1 antigen-capture ELISA (Capture-ELISA NS1), is reported day by day, from the onset of fever day 0 onwards.

amount of IgG compared to IgM indicates a recall of the antibody response primed during former episodes of dengue virus infections. A high proportion of secondary infections could indeed be confirmed by elevated HI titres on a subset of paired convalescent sera. Not surprisingly, DENV IgG-positive specimens of the acute phase also contained NS1-specific IgG (Fig. 3), with an inverse correlation between the amount of detectable NS1 antigen and NS1-specific IgG (Fig. 4). Although this suggested the formation of immune complexes, the presence of NS1-specific IgG did not appear to preclude NS1 detection for a week after the onset of fever since 79% of the acute sera positive in NS1-specific IgG were positive in NS1 antigen. Taken together, these results show that NS1 circulation in blood usually precedes the rise in specific antibodies. During secondary infections however, NS1-specific IgG is produced in a significant number of specimens in the acute phase, concomitantly with secretion of NS1 and before the appearance of NS1-specific IgM.

Secretion of the NS1 protein in plasma is a hallmark of flavivirus infections

We investigated NS1 secretion in clinical flavivirus infections other than DEN, such as YF, and more recently WN. With the use of YF NS1-specific antibodies (generous gift from JJ Schlesinger), it was possible to capture the protein by ELISA in 7 out of 18 serum samples recovered from patients infected with YFV in Senegal (kindly provided by C. Mathiot) (Table 3). The NS1-positive sera clustered in the

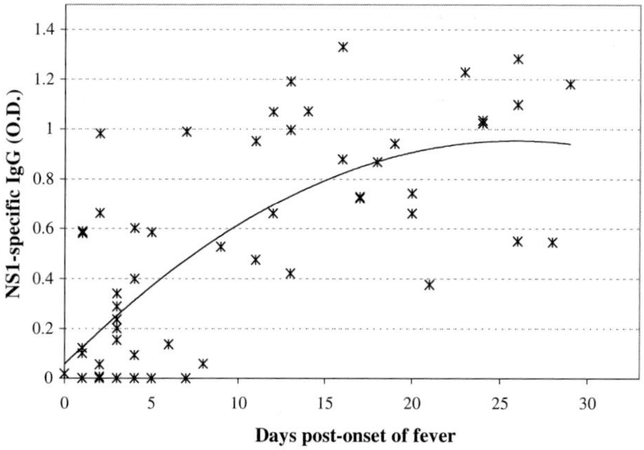

FIG. 3. Status of the NS1-specific IgG response during the course of DENV-1 infections. The NS1-specific response was analysed by ELISA for paired acute and convalescent sera using purified DEN-1 NS1 antigen in coating. Relative values are reported on the graph (with a mean value of the negative control of 0.1). The dotted line represents the average intensity of the NS1-specific IgG response over time.

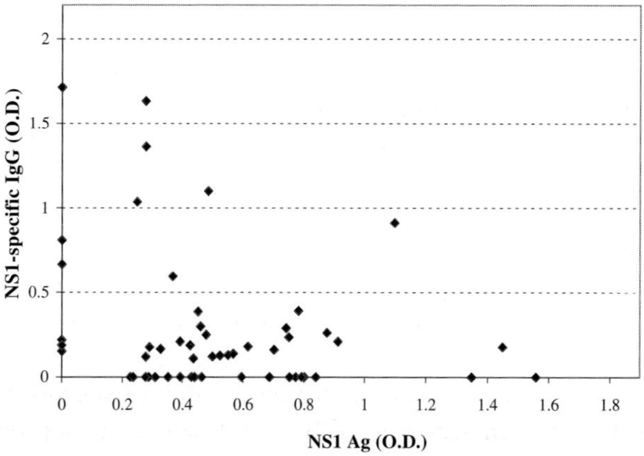

FIG. 4. Inverse correlation between the levels of NS1 antigen and NS1-specific IgG in DENV-1 infected patients. The NS1 antigen and the NS1-specific IgG were quantified by ELISA among the group of acute specimens recovered between day 0 and day 6. Relative values are reported on the graph (with a mean value of the negative controls of 0.05 for 'NS1 Ag' and of 0.1 for 'NS1-specific IgG').

lowest—albeit positive—concentrations of IgM. For the detection of the NS1 protein of WNV in natural infections, we made use of monoclonal antibodies raised against DENV NS1 that cross-reacted with WNV NS1 or against NS1 of a closely related flavivirus, Murray Valley encephalitis virus (collaboration with R.

TABLE 3 Detection of NS1 antigen in sera from patients infected with yellow fever virus[a]

Patient reference #	*IgM MAC-ELISA (OD)*	*NS1 Ag-Capture (OD)*	*Virus isolation*
1	0.31	1.13	+
2	0.53	1.20	+
3	0.36	0.60	
4	0.58	0.65	
5	0.74	0.75	
6	0.97		
7	1.01		
8	1.10		
9	1.16		
10	1.28	0.61	
11	1.34		
12	1.45		
13	1.47		
14	1.50		
15	1.52	1.19	
16	1.59		
17	1.94		
18	2.11		

[a] For each patient, serum was tested by the IgM MAC-ELISA or the NS1 antigen-capture ELISA assays. Positive values of the corresponding optical densities (OD) are reported, as well as positive virus isolation (+). NS1-positive sera are highlighted in bold.

Hall), to generate optimized conditions of NS1 capture. We processed biological samples (serum and cerebrospinal fluid, CSF) from 10 patients who developed a meningo-encephalitic syndrome during a recent WN outbreak in Romania (1996, collaboration with C. Ceianu and A. Ungureanu-Alexse) (Table 4). All samples, apart from one that remains unknown, have been recovered 2–5 days after the first clinical symptoms appeared and WNV infections were diagnosed by a positive signal for specific IgM. Interestingly, WNV NS1 could be detected in half of the serum samples but in none of the corresponding CSF samples.

In conclusion, secreted NS1 can be readily detected in acute flavivirus-associated hepatic or encephalitic infections in humans.

Discussion

Secretion of the NS1 protein is a hallmark of human flavivirus infections. We have shown that NS1 circulates in humans infected by each of the four DENV serotypes, as well as YFV and WNV. NS1 release in plasma is concurrent with viraemia, and coincides with the clinical phase of the disease. Viral antigen could be

TABLE 4 Detection of NS1 antigen in sera of patients presenting a meningo-encephalitic syndrome during the West Nile virus 1996 outbreak (Romania)[a]

Patient Reference No.	*Age*	*Biological sample*	*No. of days with clinical symptoms*	*IgM MAC-ELISA*	*NS1 Ag-capture (OD)*
1	76	Serum	2	+	0.20
		CSF	1	+	0.08
2	64	**Serum**	2	+	**1.50**
		CSF	2	+	0.07
3	25	**Serum**	3	+	**0.29**
		CSF	2	+	0.09
4	61	Serum	3	+	0.22
		CSF	2	+	0.08
5	56	Serum	3	+	0.20
		CSF	3	+	0.08
6	28	**Serum**	3	+	**0.43**
		CSF	5	+	0.09
7	73	Serum	3	+	0.16
		CSF	5	+	0.08
8	67	**Serum**	4	+	**0.33**
		CSF	4	+	0.08
9	18	Serum	5	+	0.15
		CSF	5	+	0.08
10	71	**Serum**	?	+	**0.83**
		CSF	?	+	0.08

[a] For each patient, serum and CSF were tested by the IgM MAC-ELISA or the NSl antigen-capture ELISA assays. Optical densities (OD) are reported when available and NSl-positive sera are highlighted in bold.

detected in a majority of DENV infections from the onset of fever until day 6. Interestingly, a significant number of specimens were NS1-positive between day 7 and 11. Due to its early and sustained release in the blood stream, NS1 represents a promising diagnostic marker of acute infection. A commercial dengue diagnostic kit based on an antigen-capture ELISA is expected to be available in 2006 from Bio-Rad. NS1 capture from plasma specimens can be achieved within hours with specific antibodies, enabling a sensitive and reliable diagnosis that is expected to compete favourably with the most widely utilized serological assays, including RT-PCR. By broadening the specificity of the assay, one can envision a similar approach to identify any ongoing flavivirus infections, in particular subclinical infections, which would be of benefit to epidemiological surveys.

Unexpectedly, we found that DEN NS1 concentrations were highly variable, with values ranging from less than 0.1 ng protein/ml to more than 1 µg protein/ml, depending on the infected individual, the nature of the causative virus and the conditions of sample recovery (time of infection, conditions of preservation). Although the breadth of NS1 secretion was comparable for all three serotypes, the mean NS1 concentration over a period of 6 days after the onset of fever was significantly lower in DENV-3 infections compared to DENV-1 or DENV-2 (2.5- and 7.5-fold respectively). The reason for this discrepancy is still unclear. A previous study conducted by Libraty et al (2002) on a cohort of DENV-2-infected patients showed that NS1 plasma levels were significantly higher in DHF than in DF. However, most patients in our study only developed DF and we did not find any consistent correlation between high levels of NS1 in plasma (in the range of 1–10 μg/ml) and the most severe clinical symptoms. Considering that any of the four serotypes of DENV can produce DHF in humans, a more extensive comprehension of NS1 secretion during DENV infections will be required to assess whether NS1 levels in plasma may constitute a reliable predicator of disease severity.

Nonetheless, the question of the potential contribution of the secreted NS1 protein to the pathophysiology of flavivirus disease is of great importance. According to the present view, DF/DHF is considered as an acute immunopathological disorder but no key element has been consistently identified as the main trigger of severe infections. Antibodies and antibody/antigen complexes have long been suspected to play a role in disease pathology. In particular, antibodies directed against the envelope protein may allow opsonized virus particles to infect Fc-receptor-bearing cells such as macrophages and exacerbate the infectious process (hypothesis of antibody-dependent enhancement) (Halstead 2003). In addition, some anti-E and anti-NS1 antibodies have been reported to cross-react with self-antigens (Falconar 1997, Lin et al 2003, 2001, Markoff et al 1991), possibly promoting adverse effects on the host and enhancing inflammatory processes. However, the correlation between the specificity or intensity of the antibody response and the clinical outcome remains uncertain and one may argue that most flaviviral infections resolve naturally with the onset of virus-specific adaptive immune responses.

DENV is undoubtedly a complex and multifactorial disease and we suggest that secreted NS1 plays a pivotal role in viral pathogenesis. We have recently found that the purified secreted form of DENV NS1 shows a marked tropism for the mouse liver upon intravenous injection (Alcon-LePoder et al 2005). The liver is known to play an important role in coagulopathy and alteration of haemostasis, two major features of DHF (Lisman et al 2002). Moreover, the accumulation of exogenous NS1 in naïve human hepatocytes *in vitro* alters their physiology and enhances subsequent DENV infection, suggesting that, in addition to the previously recognized involvement of the intracellular form of NS1 in viral replication, the secreted form of the protein may facilitate viral propagation in the infected organism (Alcon-LePoder et al 2005). The determination of secreted NS1 function has, up to now, been hampered by the absence of sequence homology with proteins of known activity and by the difficulty of studying the effects of point mutations or deletions without affecting viral viability. We are currently pursuing studies on the biological activity of NS1 protein from several complementary angles: the resolution of its three-dimensional structure, the characterization of its primary receptor and the putative downstream effectors, and the analysis of the pathological implications of NS1 secretion in animal or relevant cell models.

Altogether, our observations highlight the importance of the secretion process of the non-structural NS1 protein during flavivirus infections in humans, emphasizing the value of NS1 antigen as a surrogate diagnostic marker. They also point to alternative roles for NS1 during viral infection, such as promoting viral propagation and triggering long-range signalling pathways in uninfected cells. By gaining insight into the biological function of NS1 and its role in viral pathogenesis, we may uncover new therapeutic approaches against flaviviruses, which are the causative agents of life-threatening diseases, such as dengue, yellow fever, Japanese encephalitis, tick-borne encephalitis and West Nile encephalitis.

Acknowledgements

We thank J.-L. Cartel, Y. Sanchis, F. Saintpère, V.T.Q. Huong, P. Dussart, C. Mathiot, C. Ceinau, A. Ungureanu-Alexse, J.J. Schlesinger, R. Hall, D. Laune, M. Puynière, A. Pouzet, M. Martin and J.-F. Delagneau and V. Deubel. This work was supported financially by the Pasteur Institute and the Greenberg Medical Research Institute and thank you to BioRad for their contribution to this study.

References

Alcon S, Talarmin A, Debruyne M, Falconar A, Deubel V, Flamand M 2002 Enzyme-linked immunosorbent assay specific to dengue virus type 1 nonstructural protein reveals circulation of the antigen in blood during the acute phase of disease in patients experiencing primary or secondary infections. J Clin Microbiol 40:376–381

Alcon-LePoder S, Drouet MT, Roux P et al 2005 The secreted form of dengue virus nonstructural protein NS1 is endocytosed by hepatocytes and accumulates in late endosomes: implications for viral infectivity. J Virol 79:11403–11411

Barrett AD, Monath TP 2003 Epidemiology and ecology of yellow fever virus. Adv Virus Res 61:291–315

Busch MP, Caglioti S, Robertson EF et al 2005 Screening the blood supply for West Nile virus RNA by nucleic acid amplification testing. N Engl J Med 353:460–467

Campbell GL, Marfin AA, Lanciotti RS, Gubler DJ 2002 West Nile virus. Lancet Infect Dis 2:519–529

Crooks AJ, Lee JM, Dowsett AB, Stephenson JR 1990 Purification and analysis of infectious virions and native non-structural antigens from cells infected with tick-borne encephalitis virus. J Chromatography 502:59–68

Falconar AKI 1997 The dengue virus nonstructural-1 protein (NS1) generates antibodies to common epitopes on human blood clotting, integrin/adhesin proteins and binds to human endothelial cells: potential implications in haemorrhagic fever pathogenesis. Arch Virol 142:897–916

Fan W, Mason PW 1990 Membrane association and secretion of the Japanese encephalitis virus NS1 protein from cells expressing NS1 cDNA. Virology 177:470–476

Flamand M, Megret F, Mathieu M, Lepault J, Rey FA, Deubel V 1999 Dengue virus type 1 nonstructural glycoprotein NS1 is secreted from mammalian cells as a soluble hexamer in a glycosylation-dependent fashion. J Virol 73:6104–6110

Gould EA 2003 Implications for Northern Europe of the emergence of West Nile virus in the USA. Epidemiol Infect 131:583–589

Gubler DJ 2004 The changing epidemiology of yellow fever and dengue, 1900 to 2003: full circle? Comp Immunol Microbiol Infect Dis 27:319–330

Guzman MG, Kouri G 2004 Dengue diagnosis, advances and challenges. Int J Infect Dis 8:69–80

Hall RA, Kay BH, Burgess GW, Clancy P, Fanning ID 1990 Epitope analysis of the envelope and non-structural glycoproteins of Murray Valley encephalitis virus. J Gen Virol 71: 2923–2930

Halstead SB 2002 Dengue. Curr Opin Infect Dis 15:471–476

Halstead SB 2003 Neutralization and antibody-dependent enhancement of dengue viruses. Adv Virus Res 60:421–467

Hayes EB, Sejvar JJ, Zaki SR, Lanciotti RS, Bode AV, Campbell GL 2005 Virology, pathology, and clinical manifestations of West Nile virus disease. Emerg Infect Dis 11:1174–1179

Jacobs MG, Robinson PJ, Bletchly C, Mackenzie JM, Young PR 2000 Dengue virus nonstructural protein 1 is expressed in a glycosyl-phosphatidylinositol-linked form that is capable of signal transduction. Faseb J 14:1603–1610

Kao CL, King CC, Chao DY, Wu HL, Chang GJ 2005 Laboratory diagnosis of dengue virus infection: current and future perspectives in clinical diagnosis and public health. J Microbiol Immunol Infect 38:5–16

Komar N 2003 West Nile virus: epidemiology and ecology in North America. Adv Virus Res 61:185–234

Lancaster MU, Hodgetts SI, Mackenzie JS, Urosevic N 1998 Characterization of defective viral RNA produced during persistent infection of Vero cells with Murray Valley encephalitis virus. J Virol 72:2474–2482

Lee T, Watanabe K, Aizawa C, Nomoto A, Hashimoto H 1991 Preparation of Japanese encephalitis virus nonstructural protein NS1 obtained from culture fluid of JEV-infected Vero cells. Arch Virol 116:253–260

Lin CF, Lei HY, Liu CC et al 2001 Generation of IgM anti-platelet autoantibody in dengue patients. J Med Virol 63:143–149

Lin CF, Lei HY, Liu CC et al 2003 Antibodies from dengue patient cross-react with endothelial cells and induce damage. J Med Virol 69:82–90

Lindenbach BD, Rice CM 1997 *trans*-Complementation of yellow fever virus NS1 reveals a role in early RNA replication. J Virol 71:9608–9617

Lindenbach BD, Rice CM 2003 Molecular biology of flaviviruses. Adv Virus Res 59:23–61

Lisman T, Leebeek FW, de Groot PG 2002 Haemostatic abnormalities in patients with liver disease. J Hepatol 37:280–287

Macedo de Oliveira A, Beecham BD, Montgomery SP et al 2004 West Nile virus blood transfusion-related infection despite nucleic acid testing. Transfusion 44:1695–1699

Mackenzie JM, Jones MK, Young PR 1996 Immunolocalization of the dengue virus nonstructural glycoprotein NS1 suggests a role in viral RNA replication. Virology 220:232–240

Mackenzie JS, Gubler DJ, Petersen LR 2004 Emerging flaviviruses: the spread and resurgence of Japanese encephalitis, West Nile and dengue viruses. Nat Med 10:S98–109

Markoff L, Innis B, Houghten R Henchal L 1991 Development of cross-reactive antibodies to plasminogen during the immune response to dengue virus infection. J Infect Dis 164:294–301

Monath TP 2001 Yellow fever: an update. Lancet Infect Dis 1:11–20

Mutebi JP, Barrett AD 2002 The epidemiology of yellow fever in Africa. Microbes Infect 4:1459–1468
Muylaert IR, Chambers TJ, Galler R, Rice CM 1996 Mutagenesis of the N-linked glycosylation sites of the yellow fever virus NS1 protein: Effects on virus replication and mouse neurovirulence. Virology 222:159–168
Muylaert IR, Galler R, Rice CM 1997 Genetic analysis of the yellow fever virus NS1 protein: identification of a temperature-sensitive mutation which blocks RNA accumulation. J Virol 71:291–298
Petersen LR, Marfin AA, Gubler DJ 2003 West Nile virus. Jama 290:524–528
Post PR, Carvalho R, Galler R 1991 Glycosylation and secretion of yellow fever virus nonstructural protein NS1. Virus Res 18:291–302
Pryor MJ, Wright PJ 1993 The effects of site-directed mutagenesis on the dimerization and secretion of the NS1 protein specified by dengue virus. Virology 194:769–780
Shu PY, Huang JH 2004 Current advances in dengue diagnosis. Clin Diagn Lab Immunol 11:642–650
Teles FR, Prazeres DM, Lima-Filho JL 2005 Trends in dengue diagnosis. Rev Med Virol 15:287–302
Thomas SJ, Strickman D, Vaughn DW 2003 Dengue epidemiology: virus epidemiology, ecology, and emergence. Adv Virus Res 61:235–289
Tomori O 2004 Yellow fever: the recurring plague. Crit Rev Clin Lab Sci 41:391–427
Weaver SC, Barrett AD 2004 Transmission cycles, host range, evolution and emergence of arboviral disease. Nat Rev Microbiol 2:789–801
Westaway EG, Mackenzie JM, Kenney MT, Jones MK, Khromykh AA 1997 Ultrastructure of Kunjin virus-infected cells: Colocalization of NS1 and NS3 with double-stranded RNA, and of NS2B with NS3, in virus-induced membrane structures. J Virol 71:6650–6661
Winkler G, Randolph VB, Cleaves GR, Ryan TE, Stollar V 1988 Evidence that the mature form of the flavivirus nonstructural protein NS1 is a dimer. Virology 162:187–196
Zeller HG, Schuffenecker I 2004 West Nile virus: an overview of its spread in Europe and the Mediterranean basin in contrast to its spread in the Americas. Eur J Clin Microbiol Infect Dis 23:147–156

DISCUSSION

Harris: In your injection experiments did you use a control protein such as E or another dengue viral protein?

Flamand: The difficult thing in having a pertinent control is to find a protein of similar size to the protein of interest but without any effect on the organism. This is difficult to find since NS1 is a hexamer of about 300 kDa. We used purified GFP as a negative control, injected it at a similar level to NS1 in mice and never found any signs of GFP anywhere.

Canard: How did you purify your NS1?

Flamand: The protein was recovered from the infected cell supernatants, the virus was eliminated by precipitation, and the protein further purified by a series of steps involving immunoaffinity. We monitored the absence of endotoxins in the purified preparations.

Canard: Do you have any tests to see whether it is formed correctly?

Flamand: Most of the monoclonal antibodies we generated have been used in ELISA and all recognize the purified protein in this assay. Some of the antibodies are directed against conformational epitopes.

Stephenson: In the experiments where you injected NS1 into the mice, did you see any pathology? Did you find NS1 anywhere else apart from the liver?

Flamand: We did not observe any obvious pathological effect associated to a single injection of NS1 *in vivo*, although we have some preliminary indications that it may trigger the expression of certain cytokines. We have occasionally observed isolated positively-labelled cells in other tissues than the liver, but it is difficult to tell whether the signals only represent background. Moreover, some of these cells have such a small cytoplasm that it may be difficult to detect NS1-positive punctate structures. At this stage, we cannot exclude that other cell types than hepatocytes may be positive in NS1 antigen.

Stephenson: You mentioned in passing that there was an increase in virus replication. Was this just in the liver or was this in general?

Flamand: These experiments have been done in human hepatocytes in culture. Cells were pretreated with purified NS1 for 24 h, and subsequently infected for 2 h with dengue virus. Virus production was then determined at 24 h and we found up to a sevenfold increase in virus titres when cells were pre-treated with NS1 compared to control cells.

Farrar: When you correlate NS1 levels with severity, do you correct for the degree of leak? When you detect raises levels, you may be measuring the loss of water. This applies to all small molecules. We have looked at cytokine levels in dengue. If you don't correct for the leak you can misinterpret raised levels, especially for these bigger molecules that are inevitably retained.

Flamand: The different cohorts that have been analysed so far include classical DF patients only, even though some of the individuals did show a certain level of haemorrhage. Plasma leakage may be negligible in this set of patients.

Farrar: You need to be careful with that. People leak in DF. This isn't a different disease. Even if you can't see it on a chest X-ray, these patients are leaking potentially almost as much as the severe patients. It is crucial to correlate with some other marker of enterovascular capacity.

Flamand: You are raising an important point. NS1 would be a valuable marker if one could prove its involvement in DF/DHF, which is currently uncertain. Libraty et al (2002) have shown that the plasma concentrations of NS1 are higher in DHF than in DF, but as you suggest, it might be related to a higher leakage in DHF patients. Concentration of NS1 may also be dependent on other factors, such as the number of infected cells, their ability to secrete NS1, or the presence of antibodies that might opsonize the protein.

Farrar: I should add that you can't use shock as your marker of leak, which goes on for many days before shock develops.

Zinkernagel: Does NS1 stick to cells other than infected cells in the infected host?

Flamand: We have tried many different cell types *in vitro*. Most internalize NS1 very efficiently, suggesting a ubiquitous mode of interaction of the viral protein with the cell surface. We have not investigated the endocytosis of NS1 by infected cells. NS1 has been detected on their plasma membrane but it very likely represents an endogenous population that has been transported along the secretory pathway.

Satchidanandam: Are the antibodies to NS1 capable of cytolysis?

Flamand: It is not clear whether NS1–antibody complexes have a role to play during dengue virus infection. It has been reported that anti-NS1 antibodies alone might be deleterious to the infected host by cross-reacting with self-antigens. There have also been several studies describing the fixation of complement by NS1-specific antibodies in the 1970s, which is one of the reasons why this antigen was originally called the soluble complement-fixing antigen.

Malasit: One of our recent publications (Avirutnan et al 2006) has demonstrated that soluble NS1 activates complement to completion and activation is enhanced by anti-NS1 antibody. Complement is also activated by cell associated NS1 in the presence of specific antibodies. We believe that complement activation mediated by NS1 leads to local and systemic generation of complement anaphylatoxins which may contribute to the pathogenesis of vascular leakage occurring in DHF/DSS patients.

Vasudevan: What about NS1 as a diagnostic? Will this come soon?

Flamand: BioRad is involved in the development of an NS1-based dengue diagnostic kit. It should be available in 2006.

Young: Panbio is also progressing with an NS1 diagnostic.

Rice: From the clinical data it looks like this will be useful. In many of these areas one isn't going to be able to use a PCR-based assay for RNA detection.

Young: Two recent WHO meetings have specialized on this topic.

Flamand: This should also have important implications in terms of epidemiology and clinical care. Detection of dengue virus infections at an early stage of the disease should facilitate the follow-up of patients and the study of relevant cohorts in a more appropriate manner.

Padmanabhan: Wong et al (2003) published a paper describing a microsphere immunoassay for detecting West Nile virus (WNV) NS5 antibodies which discriminates WNV infection from infections by Dengue and St. Louis encephalitis viruses and this assay is proposed to be of diagnostic value.

Flamand: To my knowledge, the NS1 protein is the major viral protein released in the bloodstream during acute phase of the disease, prior to the rise of specific antibodies. The NS1 protein is probably the most potent diagnostic marker to date.

Stephenson: We have seen NS5 antibodies in experimentally infected animals, but not in patients.

Young: We looked at this about 15 years ago in a small cohort of patients by Western blot. The predominant response in almost all of them is against prM and E and then there is a developing response against NS1. Coming up the rear are NS3 and NS5.

Keller: Hepatocytes are needed for metabolism, so is NS1 doing something useful in these hepatocytes?

Flamand: We have shown that the protein stimulates the endocytic activity of hepatocytes *in vitro*. Furthermore, hepatocytes are very receptive to cytokines and chemokines, and are important producers of complement and coagulation factors. It would certainly be interesting to look for an effect of NS1 on one of these components or their related signalling pathways.

Keller: You said that the NS1 levels in plasma could be used for diagnostics, but you also said it goes easily into cells. What are the pharmacokinetics of NS1 protein? How quickly does it go up, and how quickly is it metabolized?

Flamand: In mice, after a single intravenous injection, the protein is rapidly sequestered by the liver, perhaps in less than an hour. The protein can then be detected up to 24 h on liver sections probed with anti-NS1 antibodies. In patients, it is difficult to know because by the time fever occurs the production of NS1 is already high and sustained over a period of days.

Young: When you took your sections, how did you know that it was actually hepatocytes taking NS1 up?

Flamand: We treated sections adjacent to those labelled for NS1 with a haematoxylin-eosin stain, revealing a massive number of NS1-positive hepatocytes. Hepatocytes have typical rounded nuclei that can also be observed by immunofluorescence after propidium iodide staining. However, this does not exclude the possibility that other cell types have internalized NS1. This is why we are interested in analysing the interaction of NS1 with endothelial cells as an example.

Young: Another likely cell population is the Kupffer cells.

Flamand: Absolutely.

References

Avirutnan P, Punyadee N, Noisakran S et al 2006 Vascular leakage in severe Dengue virus infections: a potential role for the non-structural viral protein NS1 and complement. J Infect Dis 193:1078–1088

Libraty DH, Young PR, Pickering D et al 2002 High circulating levels of the dengue virus nonstructural protein NS1 early in dengue illness correlate with the development of dengue hemorrhagic fever. J Infect Dis 186:1165–1168

Wong SJ, Boyle RH, Demarest VL et al 2003 Immunoassay targeting nonstructural protein 5 to differentiate West Nile virus infection from dengue and St. Louis encephalitis virus infections and from flavivirus vaccination. J Clin Microbiol 41:4217–4223

FINAL DISCUSSION

Ng: I have a question about the feasibility of immunotherapy: is this a promising approach?

Rice: It has been a long road but there are some therapeutic antibodies being used in the clinic today. This is something people are considering as a viable option.

Fairlie: The cost is an issue, and in this case it might not be the best way to proceed. There has been a paradigm shift of interest by pharmaceutical companies in the last few years in antibodies and immunotherapy. Even the medicinal chemistry community is starting to appreciate the value of antibodies because of their exquisite selectivities and because, in certain cases, of their longevity *in vivo*. There is interest in this, but in this case because of the target populations I think we are looking for something cheaper than antibodies.

Zinkernagel: The strategy we are proposing here is not going to be cheap. The other issue is that you had better choose your monoclonal so as not to have enhancement.

Young: You wouldn't have enhancement with NS1. NS1 is a strong immunogen, and many of the monoclonal antibodies generated to it afford passive protection.

Fairlie: What do these antibodies protect against?

Young: We don't really know what the mechanism of protection is. The most likely possibility is antibody-dependent, cell-mediated cytotoxicity (ADCC).

Fairlie: We know that they interfere with shock. Do they prevent it altogether?

Young: It has never been tried in humans.

Rice: There are fewer dead mice!

Zinkernagel: The prediction would be that in a serotypically defined group of viruses these common antigenic determinants would not be protective. Otherwise coevolution wouldn't have selected for serotypes. The same argument has been made in influenza.

Young: This is a slightly unique situation in that the antibodies are reacting with a non-structural protein. In fact, we have some monoclonals that cross-react with all four serotypes that do passively protect, but it is via some other mechanism than neutralization.

Neyts: I have a question about the ChimeriVax vaccine. Should we be relaxed about using that in patients?

Rice: The chances of recombination occurring in a vaccine setting giving rise to virulent yellow fever, for example, are small.

Young: There is another potential issue here, which is the non-scientific one of public perception. With ChimeriVax, the issue is whether the community will accept something they have been informed is a chimera between two viruses.

Hibberd: You have already shown that you can get non-pathogenic viruses piggybacking on virulent ones.

Holmes: Never say never is the watchword in emerging viruses. I think they do recombine, but the key question is what is the rate of recombination related to other evolutionary processes? Successful recombination where the virus is viable is very rare. Mutation pressure is a much more powerful evolutionary force in dengue and other RNA viruses.

Heinz: I agree that in the context of using this as a vaccine, recombination is not really the issue. The point is that you have a newly generated virus which you are releasing into nature. By vaccination you put it into humans and the virus wants to change, and will. The issue is what will happen to this live, replicating virus once it is released into humans.

Holmes: I agree.

Rice: There seems to be a fear of live attenuated vaccines. In terms of vaccination in general, even if you take something that is a subunit vaccine, if you put it into a large enough population at different ages, this will influence their immune responses against a cavalcade of immunogens that they are then exposed to over their lifetime. The chances are it could have adverse effects in some people.

Heinz: In this case it is a live replicating virus. This is different.

Stephenson: I don't think we should be relaxed about ChimeriVax, but also I do not think we should concentrate on the potential problems. It is a shame that Tom Monath isn't here because he would have shown some good data proving just how good ChimeriVax is. But I fear that if we just discuss the potential problems nobody will go out and test if they are important issues in the real world. Alongside the vaccination programme you make sure that you do as much as you can to survey for major problems. There will be concerns in the population about reversion and also concerns about the fact that it is a genetically engineered vaccine. The best way of dealing with suspicion and fear is truth. We should not ignore potential problems, but should be robust in assessing whether they actually occur.

Heinz: Still, it should be discussed, at least.

Stephenson: We are past discussion.

Heinz: Here we are at a scientific discussion. One point is that such a vaccine also induces a chimeric immune response one against the non-structural proteins of yellow fever backbone and another one against the virus targeted by the vaccine.

Zinkernagel: That is irrelevant, i.e. only neutralizing antibodies protect as shown so far.

Holmes: I think Franz Heinz has a point. In the 1940s with the development of antibiotics, if there had been more discussion about resistance before they were developed, perhaps the problem would have been less serious.

Rice: I wouldn't say that we should be relaxed about any kind of clinical trial. But this is where we find out whether things work.

As this meeting draws to a close I am not going to give a long summary, but probably the only thing that we have agreed on is that dengue and flavivirus diseases are important! The usual control measures will all be important in terms of global control, whether they be mosquito control, vaccines or therapies. One of the things that struck me at this meeting was a new enthusiasm for tackling both the vaccine side and developing therapeutics. Despite the fact that these viruses are pretty good targets, there hasn't been a lot of action on the therapeutics front. It is refreshing to see resources now being devoted to this. It will be interesting to see how this all pans out. Antivirals that emerge for these flaviviruses may not only be clinically useful, they will also be very useful tools for understanding the biology of these viruses. I don't have any great words of wisdom, except to wish you all good luck in your various enterprises. I thank you all for your participation.

Contributor index

Non-participating co-authors are indicated by asterisks. Entries in bold indicate papers; other entries refer to discussion contributions.

Subject Index

E

K

L

M

N

O

P

Q

R

S

T

X

Y

Z